Proceedings in Life Sciences

650th anniversary

Bacterial Chromatin

Edited by
Claudio O. Gualerzi and Cynthia L. Pon

With 129 Figures

Springer-Verlag
Berlin Heidelberg New York Tokyo

Dr. Claudio O. Gualerzi
University of Camerino
Department of Cell Biology
Laboratory of Genetics
I-62032 Camerino (MC)

and

Max-Planck-Institut
für Molekulare Genetik
Ihnestraße 73
D-1000 Berlin 33

Dr. Cynthia L. Pon
Max-Planck-Institut
für Molekulare Genetik
Ihnestraße 73
D-1000 Berlin 33

Cover illustration:
Drawing of the *T. acidophilum* nucleoprotein particle. From D. G. Searcy, this Volume, page 178, Figure 2.

ISBN-13: 978-3-642-71268-5 e-ISBN-13: 978-3-642-71266-1
DOI: 10.1007/978-3-642-71266-1

Library of Congress Cataloging in Publication Data.Bacterial chromatin. (Proceedings in life sciences) Proceedings of the Symposium "Selected Topics on Chromatin Structure and Function" held at the University of Camerino, Italy, in May 1985. 1. Bacterial chromatin – Congresses. 2. Nucleoproteins – Congresses. I. Gualerzi, Claudio O., 1942–. II. Pon, Cynthia L., 1942–. III. Symposium "Selected Topics on Chromatin Structure and Function" (1985: University of Camerino) IV. Series. [DNLM: 1. Bacterial Proteins – congresses. 2. Chromatin – congresses. 3. DNA, Bacterial – congresses. QW 52 B1306 1985]

2131/3130-543210

Preface

This book brings together various contributions aimed at the elucidation of the structural and functional organization of the bacterial nucleoid. Most of these papers, spanning the fields of physical chemistry through biochemistry to genetics, were presented at the session on bacterial chromatin during the Symposium "Selected topics on chromatin structure and function" held at the University of Camerino, Italy, at the end of May 1985.

Times when the bacterial DNA was regarded as "naked" or, at most, complexed with polyamines, and when the absence of histones and organized chromatin was considered to be a distinct feature of the prokaryotic cell, now appear remote.

Our concepts of how DNA is packaged in bacteria are changing rapidly.

Studies on the structure of the bacterial nucleoid are not new. Recently, however, investigations in this field have flourished again, leading to some important contributions such as the elucidation of the three-dimensional structure of what appears to be the major protein constituent of the bacterial nucleoid or the development of methods to titrate the extent of DNA supercoiling within the bacterial cell.

In the meantime, our view of the organization of eukaryotic chromatin has also become less dogmatic with the realization that a "dynamic" type of chromatin, probably similar to the bacterial one, co-exists in the cell with a "static" chromatin frozen in its canonical nucleosomal organization. Thus, the conceptual gap between bacterial and eukaryotic chromatin is being bridged, at least in part, by the realization that the differences in the physical organization of the DNA in the two systems may after all be more apparent than real.

Rapid progress is also being made in the development of techniques, which will eventually lead to the identification of the components of the nucleoid in vivo, as well as in the biochemical characterization of the proteins associated with the bacterial DNA.

Given the present state of the art, the title "Bacterial Chromatin" chosen for the session of the symposium and, consequently, for the book may sound either provocative or presumptuous. This title reflects our perhaps over-optimistic anticipation that such a structure, no matter how elusive it may appear today, will in spite of its complexity and dynamic nature soon be understood as a well-defined physical entity. The first necessary step will be the establishment of a satisfactory common nomenclature for the proteins associated with the nucleoid, which in turn depends on the resolution of some controversies which, as this book will show, exist today concerning the properties, structure, function and cellular localization of these proteins. We hope that this collection of papers will represent a step forward in this direction and provide a stimulus for further study.

We are sincerely grateful to the many people on the staff of the University of Camerino who contributed to the organization of this symposium. We are particularly indebted to Professors D. Amici and G. Gianfranceschi for their invaluable support and counsel.

CLAUDIO O. GUALERZI
CYNTHIA L. PON

Contents

List of Contributors

You will find the addresses at the beginning of the respective contribution

What is the Logic of DNA Packing in Bacteria?

P. WOOLLEY[1]

Before taking a physical chemist's naive look at DNA in the bacterium, I should like to explain briefly what I mean by the word "logic" in the title of this contribution. The sciences of physics, chemistry and molecular biology are shot through by an interplay between the two types of logic exemplified by our two kinds of computing machine: analogue and digital. These kinds of logic are respectively those of the continuous and the discrete, the measurable and the countable, the deterministic and the probabilistic, the wave and the particle, the potentiometer and the switch. The interplay between them is the result, on the one hand, of the adding up of large numbers of small, discrete events to produce apparently continuous manifestations of them on a larger scale, and, on the other hand, of non-linearities in apparently (better: approximately) continuous systems that result in instabilities or singularities and thus in some phenomenon which appears in a sudden or discontinuous way.

It may often be thought that discontinuity is associated primarily with events on the level of atoms or quanta, such as the energy levels arising from a continuous electrostatic force field in the Schrödinger atom, or the laser effect, in which excited atoms or molecules above a certain population density fall into step and emit light in synchrony. However, this is not so: any sufficiently complex, non-linear system can show such discontinuity (when a very small electric current is passed through an electric door-bell, Ohm's law is obeyed, but, when the current is raised beyond a critical value, the continuous behaviour pattern is replaced by an oscillatory one).

An excellent example of the interplay between the continuous and the discontinuous is the well-known Zhabotinsky reaction, the metal-ion-catalysed oxidation of malonic acid by bromate. The participant atoms, which can be counted and which occupy discrete energy levels, collide by diffusion in such large numbers per reaction vessel per second that the energies and orientations of collision can be averaged so as to yield a rate constant of reaction – a continuous quantity, which can be measured, but not counted. However, the malonic acid and the bromate pass through a series of intermediates whose concentrations build up until their reactions with each other present a pattern so complex that the linear system breaks down and the well-known discrete, red and blue waves in time and space are produced.

Examples of this complex interplay in biological systems are not hard to come by – in fact, they are hard to avoid (unless one chooses to evade the issue and to work in vitro). Purely continuous, deterministic systems are exemplified by isolated enzymes,

1 Kemisk Institut, Aarhus Universitet, 8000 Århus C, Denmark

Bacterial Chromatin
(ed. by C.O. Gualerzi and C.L. Pon)

whose turnover obeys quantitative laws based on (continuous) concentrations of added substrates, co-factors, etc. However, even a small assembly of enzymes, passing intermediates one to another, can lead to an oscillatory rather than a continuous behaviour pattern, reminiscent of the Zhabotinsky reaction or the door-bell.

The regulation of gene expression may present a somewhat different case: statements such as "the concentration of repressor rises and thus the gene is switched off" hint at the assumption of a transition from the quantitative (concentration) to the qualitative (gene on – gene off), even though such a physical transition has not actually taken place. This interpretation of a gradual change in terms of switching demonstrates the preference which scientists have for seeking digital rather than analogue logic. The specificity of an enzyme for one rather than another substrate is one of the many cases in biology that we call "recognition" and which we think of as being a qualitative yes/no decision on the part of the recogniser, when in fact it is the result of a particular constellation of binding constants and concentrations (or: of energy levels and probabilities of encounter). The psychological advantages of preferring qualitative information and "relevant" to "irrelevant" data are clear, but it is also important to recognise the importance of the shift from quantitative to qualitative logic: if the concentration c of a ligand bound with an association constant K to a particular protein rises from $c \ll K^{-1}$ to $c \gg K^{-1}$, then, as far as both the organism and the biologist are concerned, the state of the protein really has been switched, even though concentration is a continuous quantity. This kind of switching is based upon extreme values rather than upon instability. It remains to be seen to what extent the discontinuities in processes and systems yet more complex, such as those involving cellular shape, the co-ordinated construction and positioning of organelles, the regulation of growth and the differentiation of cells, are due to "instability switching" or to "extreme-value switching".

A case that is particularly interesting, not only because of its central rôle in molecular biology, is that of the genetic code. At first sight, the replication or transcription of a nucleic acid or the translation of a nucleic acid sequence into a protein sequence would appear to fall clearly into the category of discrete events, because each of these involves a logical decision: one or three bases of the parent sequence is/are read and a new unit is added to the daughter sequence according to fixed rules that can be found in tables in biochemistry books. However, cells do not have access to text-books; instead, they undertake a large amount of trial and error, and finally make their decisions on the basis of energies of complementary-base-pair and other sorts of interaction. It would be more complicated, but more nearly correct, to say that they decide on the basis of differences between sums of different kinds of interaction energy (considering the many points of interaction between, for example, a tRNA molecule and a ribosome). Furthermore, the amount of trial and error involved should not be under-estimated. In general, every active or binding site is approached by every wrong substrate present in the same cellular compartment. Our present-day understanding of the enzyme-kinetic parameter K_m lies well behind our understanding of V_{max}. Thus, what looks like a sharply-defined law turns out to be a complex and rather blurred set of differences between energies and of ratios between rate constants; the resulting inaccuracies of replication, transcription and translation, charging, processing, splicing, repair and so forth are well known. But the interesting feature of the genetic code is that, in spite of the error-ridden nature of every step of its operation, its function is constrained by evolutionary pressure to be self-sustaining. In the short term, deleterious mutants are eliminated; thus, in the middle term, a stationary state is reached, in which the accuracy of the processes of the genetic code suffice for the amount of material to be handled without imposing too great a burden on the energetic resources of the organism. (In the long term, the population may exhaust its nutrient and thereby exterminate itself, but I shall not develop this point now; in any case, as far as the human race is concerned, that situation will, by definition, not be reached during our lifetime.) Once a genetic apparatus has arisen that is good enough to be self-sustaining – and the Göttingen school has argued convincingly that this was *the* critical step which preceded the nucleation of what we might recognise as a living entity – then it, in conjunction with the errors that it makes and the selection pressure to which it is exposed, may be regarded as a digital rather than an analogue machine.

This machine is albeit a metastable one, because it is always vulnerable to the possibility of evolutionary improvement by the incidence of a favourable mutation. But even in the favourable mutation we see the interplay of the qualitative and the quantitative, occurring in three not quite distinctly separable phases. First, the appearance of the mutant is clearly a stochastic event, as is its initial propagation; secondly, when the mutant reaches a reasonable number, its incursion into the population as a whole is described equally well by laws of continuous change; thirdly, after the mutant has taken over the entire population, the change may be regarded in retrospect as a qualitative one based upon a critical inquality between two slightly different rate constants of the propagation "reaction".

Let us now take a bacterial cell and consider the possible dispositions which its chromosomal DNA might adopt. One of the most well-publicised wonders of Nature is the encapsulation of 4×10^6 base pairs of genomic DNA, whose total length is 1.4 mm, in a procaryotic cell of typical length 2 μm. It is instructive to try and pack the DNA into the cell, as it were, ourselves. First, let us imagine distributing it uniformly. The ratio of the lengths is 700, and 700 double strands of DNA passing lengthways up and down the cell would, if they were spaced equally one from another, be some 200 Å apart – more for a hexagonal lattice of threads, less for a square one. Since exponentially-growing cells can possess four complete chromosomal molecules, we should really revise this figure to a spacing of 100 Å, which means a gap between adjacent double helices of only 80 Å. This is enough to allow most macromolecules of the cell to pass between the DNA helices, but it would restrict them to a very low rate of movement. Larger entities, such as mRNA and ribosomes, would be virtually immobilised. Everyday laboratory experience makes this clear enough: the concentration of DNA in the cell with four chromosomes is around 17 mg ml^{-1}, and the very high viscosity of freely-dispersed, double-stranded DNA at this concentration is well-known to those who have to pipette it.

Laboratory usage tells us something more about the likelihood of being able to find a bacterial chromosome in the form of free, dissolved DNA. If a solution of DNA is made 10% (w/v) in polyethylene glycol, in the presence of 0.5 M salt, then the DNA is precipitated efficiently. For comparison, the concentration of macromolecules in the bacterium is some 10%, the salt concentration is in the range of tenths of one molar and polyamines – effective co-precipitants for DNA – are present. The insoluble pellet of a genome thus expected would avail a bacterium nothing.

Another situation can be envisaged, based on the observed immiscibility of aqueous solutions of certain soluble polymers. Could the cytosol perhaps be composed in such a way that spontaneous separation into DNA-rich and DNA-deficient regions takes place, irrespective of DNA packing, on the basis of physical interactions in solution? This may be the case, but I find it very hard to imagine a spontaneous phase separation of this kind being of any help in solving the problem of the disposition of DNA, even for a hypothetical, primitive cell without the ability to pack its genome. The reason is quite simple: in the DNA-rich phase, the distance between the filaments would be even smaller than it would be with the DNA distributed uniformly throughout the cell. Consequently, the amount of free space available to proteins, etc. that have business in the DNA compartment would become even smaller, and the viscosity even greater.

Thus the procaryotic cells, which for reasons of scale and economy cannot regiment its DNA as the eucaryote does, and generally allows considerable freedom of movement within its cytosol, is still faced with a law-and-order problem with regard to its DNA. The dilemma to be resolved is: how can the cell ensure that the entire length of its genome is accessible when needed, while at the same time preventing the DNA filaments from hampering the diffusion of other molecules?

It is at this point, I hope, that the need for a principle or a logic of DNA packing, intimately connected with the mechanism(s) of gene expression, becomes clear. It could, of course, be that the packing/accessibility problem is solved in a different way for each of the several thousand genes of the bacterial chromosome. We are used to imposing upon Nature our anticipations of economy and elegance, even in regard of systems which evolved *ad hoc*, and we therefore decry the latter possibility as "unlikely"; I shall do so too, and shall try to set up some simple kinetic calculations with a view to defining the general packing/accessibility problem more narrowly.

Let us imagine a protein molecule of a certain kind seeking a particular nucleotide sequence that is unique on the chromosome, in order to turn on or off – i.e., apply switching logic to – a particular gene. A protein that does just that is the well-known *lac* repressor, and I shall later borrow some numbers from it in order to obtain a representative result. If the number of copies of this protein per cell is N_p, then the concentration C_p of the protein is $N_p/500$ micromolar in a bacterium of typical size. There are said to be $\sim$ 10 copies of the *lac* repressor, making a concentration of 2×10^{-8} M.

An association reaction is fastest when it is not impeded by barriers of activation energy – that is, when it is diffusion-controlled. For such reactions, the second-order rate constant is given by Smoluchowski's equation

$$k_2 = 24\,\pi\,(D_A + D_B)(r_A + r_B) \times 10^{12}\ M^{-1}\ s^{-1}$$

where D_A and D_B are the translational diffusion coefficients for the reaction partners in $cm^2\ s^{-1}$, while r_A and r_B are their radii in Ångstrøm units. For reactions between biological macromolecules, this equation is hard to apply, because of the irregular shape and large size of the molecules. The sum $(r_A + r_B)$ can be approximated to by taking a typical collision distance – say 5 Å – for two groups that match each other on the two complementary sites. However, this still fails to express the slow rotation of the macromolecules, which impedes recognition by making many of the collisions ineffective: before the correct orientation has been reached, the molecules may well have lost contact by diffusing apart again. If both reaction partners are small molecules, k_2 reaches its calculated value of around $10^9\ M^{-1}\ s^{-1}$. For two small molecules $D_A \approx D_B$, whereas if one, say A, is a protein, then $D_A \ll D_B$. We would therefore expect for reactions between a small molecule and a protein (e.g. the first enzyme-substrate binding step) a value of k_2 around one-half of $10^9\ M^{-1}\ s^{-1}$, whereas in fact about one order of magnitude is lost. The additional retardation is due to the slow rotation mentioned above. For a protein binding to a specific site on a molecule of DNA, we may regard the DNA as virtually stationary ($D_A \approx 0$), so that the diffusion observed is that of the protein. It may reasonably be assumed that another order of magnitude is lost, especially in view of the viscosity of the cytosol, which is higher than that of water. We thus have a probable value for k_2 of $\sim 10^7\ M^{-1}\ s^{-1}$.

The rate of occupation of the protein-binding sites in the cell, in the presence of an intracellular concentration C_p of binding protein, is given by

$$k_1 = k_2 C_p = k_2 N_p/(5 \times 10^8)$$

We obtain, for the N_p and k_2 values discussed above, $k_1 \approx 0.2\ s^{-1}$. The time taken for the recognition reaction to proceed to completion is thus of the order of magnitude of one second, and this, being shorter than times associated with the induction and repression of the *lac* operon, may be regarded as a realistic result.

An experimental value for k_2 three orders of magnitude greater than that estimated here is sometimes quoted. I believe that this should be regarded with caution until it is reliably confirmed, for two reasons. First, it is unusual for one step in a biochemical reaction path to be very much more rapid than the others, since this as a rule confers no evolutionary advantage. Second, the extremely large value demands a special physical explanation, which is usually formulated in terms of non-specific binding to the DNA at a random point followed by diffusion along it. In view of the vast number of other proteins attached in various ways to the bacterial genome, it is not clear how such diffusion could proceed unimpeded. Both of these issues will be addressed again at a later point.

A more pressing problem associated with the binding of proteins to specific sequences in the genome appears to be that of accessibility. As discussed above, the bacterium's dilemma lies in finding a compromise between keeping its DNA open to inspection by interested ligands, on the one hand, and preventing it from hindering the diffusion of macromolecules on the other. It is thus forced to invest some ingenuity, and possibly a little ATP, in finding a suitable logic of packing.

Some elements of this logic will, I believe, include the following.

1. The proteins responsible for packing DNA will bind by simple adhesion (i.e., not assisted by disphosphate hydrolysis or accompanied by important changes of their conformation) to DNA that is biased by torsion to make it amenable to negative supercoiling. Possibly the torsional force is not great enough to impose supercoiling in the absence of packing protein. Co-operativity between adjacent units of the packing protein may be expected to make the middle of a packed stretch of DNA stable, while allowing rapid association and dissociation of the packed protein at the ends.

2. The steady-state action of gyrase activities that relax supercoils and of those that produce them will keep a constant torque in all the non-packed regions of the DNA. In the packed regions the DNA is clamped by the stiff, multimolecular complex, but, since the complex is free to rotate, the torque will be transmitted through it. In order to keep the torque constant there will be a steady, balanced turnover of winding and relaxing, which consumes ATP and the rate of which has been optimised by achieving the best balance between efficacy and ATP cost.
3. An open question is the sequence specificity of the packing protein. It would be attractive to imagine a degree of sequence selectivity so high that promoters and other necessary regulatory elements were exposed specifically. However, it is hard at present to see how such a specificity could exist, since promoter sequences do not contain universally identical elements (cf. the variation in the composition of the Pribnow box); the degree of local perturbation in secondary structure due to sequence microheterogeneity has not yet been determined. Furthermore, whilst it is easy to imagine proteins which prefer to bind to a certain small sub-class of all possible oligonucleotide regions, it is harder to picture proteins that prefer to avoid such a sub-class.
4. A key question will be that of whether a genetic control element is recognisable in the packed form. Possible answers are (i) never, because the packed conformation is inappropriate for recognition and/or because the part of the surface area to be recognised is covered up; (ii) always, because there is always a sufficiently large part of the area to be recognised lying on the outside of the packed structure; (iii) sometimes, because DNA is packed in bulk, without account being taken of individual features of the sequence, so that the area to be recognised has a certain probability of being accessible.
5. The problem of the simultaneous replication, transcription, packing and unpacking, coiling and uncoiling of the genome, along with specific interactions and separation of daughter DNA after reduplication, may well appear far simpler if it can be established that the cell has certain right-of-way rules. Some of these can already be formulated with reasonable plausibility. For example, a replication fork should be able to move unimpeded through packed DNA, displacing the packed molecules, sequentially or in groups, on its way. Conversely, the torque-balancing gyrases can operate on any available non-packed region of DNA and, therefore, do not need to be able to displace packing protein. Consequently, we can postulate the rule of the road "DNA polymerase before packing protein, packing protein before gyrase". There may be many others.
6. Elements (1) and (2) above predict that after the passage of a replication fork or a transcription enzyme through a region of packed DNA, the combined presence of gyrases and free packing protein will cause the DNA to re-enter the packed state. However, this last process must take a finite time. The interval that elapses before the DNA is re-packed may provide an opportunity for recognition of a particular region which is otherwise inaccessible. If the remarkably high rate constant of association between the *lac* repressor and its corresponding promotor, discussed above, should turn out to be correct, and perhaps typical, then such momentary exposure of the DNA to be bound may at least explain the need for the high rate constant. However, such an exposure takes place only once per chromosome

per generation and can, therefore, only be of relevance – if at all – for the expression of genes whose products need not be called up at short notice.

7. Another open question remains that of how packed DNA is searched for recognition sequences, if they are not readable in packed form. It is conceivable that the equilibrium of packing allows random, thermal disruption of packed units, followed by short periods of exposure before they are re-packed. This is, however, highly improbable in view of the likelihood of a high degree of co-operativity in the packed structure and of a high torque (cf. below). The possibility that the genetic control elements are selectively omitted from packing has been considered above – it would be very elegant, but at present neither biological nor physical data can be found to support it. A possible alternative might be the existence of an unpacking mechanism, such as an energy-consuming enzyme that moves along the DNA with no purpose other than to open up packed structures for examination by other proteins. There is as yet no evidence for the existence of such an enzyme, but it might not be easy to detect – it would have no reaction product, its copy number would be very low and its consumption of chemical energy would be low in comparison with that of other cellular processes. The principal reason at present for disbelief in such an enzyme is a certain disdain for the sheer inefficiency of its action, which might be likened to paying an employee to spend all his time reading all the contents of all the cards in a card-file in order to find out where the titles were.

In this contribution I have examined some possible rules that may one day be found to govern the interplay of packing and transcription of the bacterial genome. The logical nature of these possibilities will now be summarised.

The first two possibilities concern forces at work determining the nature and strength of the packed structure: the binding constant between DNA and a molecule of packing protein, the degree of co-operativity in packing and the activities and concentrations of gyrases are clearly quantities which can be defined and measured, but which do not in themselves lead to any kind of switching function. If, however, the two elements are combined, then the combination of high torque and a stiff skeleton may enable the quantitative question "how tight?" to be reduced to the qualitative question "packed or not?" (see below).

The next two aspects of specificity and recognition may perhaps be regarded as "extreme-value digital" in nature, but up to now no quantitative data are available to show, for example, that certain types of binding proceed to virtual saturation while other, similar, competing types do not take place at all.

A possible set of "right-of-way" rules, which would reflect events on the bacterial DNA while simplifying our understanding of them, would clearly be qualitative or digital in nature. The same applies to "overshot" rate constants of binding and to possible mechanisms of searching.

One can try to assemble these logical elements, in an abstract way, into some kind of a pattern of what may be going on in the packing of the bacterial genome. I shall refrain from using the word "model"; perhaps "model for possible models" or "metamodel" would be better. It is based upon an energetic consideration which will first be developed in the following five steps.

1. It is generally accepted that the persistence length of double-helical DNA, a measure of the stiffness of the polymer, is around 450 Å. This means that the radius of bending inducible by thermal energy – i.e., Brownian motion – is of the same order of magnitude. At 37°C the energy of thermal motion is 620 cal mol^{-1}.
2. The energy required to bend a spring – or otherwise deform an object mechanically – is, for reasonably small distortions, proportional to the square of the extent of distortion; in this case, to the square of the angle of bending. It is easily shown that this is equivalent to saying that the energy is proportional to the inverse square of the radius of curvature.
3. A gyrase molecule introduces negative supercoiling into DNA by breaking and twisting the double helix and consuming thereby one or two molecules of ATP. This means that it can supercoil DNA until the torque of the DNA's natural tendency to untwist balances the winding torque produced by the enzyme. This winding torque, based on the hydrolysis of ATP is, $7000/2\pi$ cal/mole/radian if one molecule is hydrolysed, double this if two are used. If relaxed DNA is wound into supercoils, the torque opposing the enzyme is initially low, so the winding proceeds rapidly (better; irreversibly, in the thermodynamic sense) and is wasteful of ATP energy. As the winding proceeds, the opposing torque rises until the point of balance is reached. At this point the winding becomes thermodynamically reversible, so the energy put into one turn by the enzyme equals the energy taken up by the supercoiled helix. (The rather unconventional unit of torque, cal/mole/radian, is equal to 7×10^{-17} dyne-cm. This means that if the DNA double helix has a radius of 20 Å, then the force used by the enzyme to wind it is about 0.4 ng.)
4. The geometry of supercoiling requires that each turn of negatively supercoiled structure corresponds to one turn of unwinding of the double helix (this presupposes that the DNA cannot untwist by variation in relative orientation of adjacent bases, which is roughly true). Furthermore, the energy of untwisting equals the energy put into creating supercoils. The energy in a turn of supercoil is, therefore, also around 7000 cal mol^{-1}, perhaps twice this [cf. (3) above].
5. If all the energy of the ATP is put into twisting the DNA double helix, then the degree of twisting available can be estimated on the basis of the inverse square law [(2) above]. Taking persistence length as a measure of radius of curvature, we obtain

$$\frac{r_m^{\,2}}{450} = \frac{620}{7000} \text{ or } \frac{620}{14000}$$

where r_m is the minimum radius of any segment of DNA in the supercoil and the two values depend on the numbers of ATP molecules hydrolysed in creating a single turn of supercoil.

The respective r_m values calculated are 135 Å and 95 Å: the more ATP's that are hydrolysed, the more tightly the DNA can be coiled. Realistic estimates of the outside diameter of DNA coils in eucaryotic and procaryotic chromatin lie around 140 Å, giving a radius of curvature of some 60 Å, and this would correspond to several times the amount of energy provided by one or two molecules of ATP. Our consideration of the energy of supercoiling thus leads to the conclusion that *the energy coming from the ATP-driven gyrases is not sufficient to pack DNA into the tightly-coiled form found in the cell.*

Our metamodel of DNA packing in the eucaryotic cell, therefore, comprises the following facts, features or assertions.

1. Gyrases on their own are unable to induce a sufficient degree of supercoiling, as shown above.
2. DNA-packing proteins are also unable to induce a sufficient degree of supercoiling. This guess is based on the presumption that, if they were, the ATP-driven gyrases would be redundant and the packing of DNA in the cell would be thermodynamically spontaneous in the presence of nicking-closing enzymes. This in turn would make the DNA (thermodynamically) reluctant to come out and express itself.
3. The action of ATP-driven gyrases combined with the presence of DNA-packing proteins *is* enough to supercoil DNA to the extent found in the cell. The sufficiency of these two factors, but not of either one, has important consequences for the geometry, for the energy and for the logic of DNA packing.
   First of all, it is clear that the involvement of binding proteins simplifies the geometry of the packaged DNA, quite straightforwardly by defining the form and the size parameters of the packaged structure.
   Secondly, the packing proteins effect an economy of energy. If attractive forces assist the super- or other coiling imposed upon the DNA by the packing, then the gyrases will be relieved of some of the requirement to produce a high torque, so they will be able to operate at a lower mechanical advantage and will consume less energy – fewer ATP molecules – per twist of packaged DNA. This economy is important, less because the packing proteins help the gyrases to wind up the DNA than because it represents a *reversible* contribution to the energy of winding: if a stretch of DNA is unwound transiently, then the packing proteins can diffuse off and on again as required, whereas the ATP energy consumed in winding up the stretch is lost irrecoverably. The exact size of the economy made is less important as long as the three conditions set out above – gyrases alone cannot, packing protein alone cannot, both together can wind up DNA to the extent required for packing – are fulfilled.
   Thirdly, the combination of an excess of torque and a rigid framework of the packed protein-DNA complex lead to a change in the logic of packing, from analogue (how tightly coiled?) to digital (packed or not?). The only hint of a "degree of packing" might emerge in connection with the rate of transition from the packed to the unpacked state. This would depend *inter alia* upon the concentration of packing protein and the concentrations and activities of the enzymes involved in winding and unwinding.
4. The exposure of DNA is sufficient to allow recognition by most sequence-recognising proteins. If the sequence to be recognised is a long one, well over one double-helical turn, then some part of the groove binding to the protein in question must be exposed, and if the protein can bind provisionally to a part of its binding site, then it can make use of thermal, structural fluctuations in the DNA-protein complex to prise open the packed region and, if the recognition was correct, to extend its hold upon the specific sequence. In doing so, it must work against the torque

generated by the packing proteins and by the steady-state action of ATP-dependent gyrases and nicking/closing enzymes. This may explain the seemingly unnecessarily high binding constants associated with repressor molecules (e.g. 10^{13} M^{-1}, which is $\sim 10^3$ times higher than would be needed to saturate binding sites to 90% at a free-ligand concentration of only one molecule per cell!). High association rates might be explained by the possibility of two-dimensional diffusion across the surface of packed DNA, and not just along its length.

5. Finally, the co-operativity of packing is such as to allow the dismantling of the packed structure at *any* point. Experimental studies have not yet led to general agreement on whether DNA in a bacterium is packed by independent, small protein aggregates or in a quasi-continuous manner reminiscent of the RNA in tobacco mosaic virus, and I do not wish publicly to take sides in the debate. But from the theoretical point of view, one thing is clear. *If* packed DNA – for example, on the surface of a solenoid – is recognised by proteins that can prise it open, then there may be a certain disadvantage associated with the quasi-continuous type of structure. This is because the definition of the size parameters requires a high energy of adhesion between individual DNA-packing proteins. If this high degree of adhesion passes uninterrupted along the packed structure, then a large amount of energy will be needed by a sequence-recognising protein in order to make the initial break in the middle of the co-operative structure. If, however, the DNA-packing proteins are arranged in small groups, for example as discs or other multimers stacked along the axis of a DNA supercoil, then the disadvantage is no longer present. The relatively strong interaction required to define the shape of the packed complex can be confined to within the multimers, while the interaction between individual multimers can be made much weaker, allowing the chain of multimers to be opened up without the useless expenditure of energy. By the same token, weak interactions between individually, highly stable multimers would allow the packed structure to bend readily and thus to be accommodated more easily in the cell.

We have seen that both macroscopic and microscopic systems may in different circumstances show behaviour that obeys either "digital" or "analogue" logic and that there is an intimate interplay between these two. We have also seen that even where analogue logic is fundamentally applicable, digital logic is often preferred as a means of expressing observed behaviour in a simplified way, and that deductive biology contains many examples of all these kinds of logic. Consideration of some basic properties of DNA has led to a description of the logical nature of various possible rules that may be associated with the packing of the prokaryotic genome.

Several colleagues who were kind enough to read the manuscript of this article have complained that its title raises false expectations, since at no point do I state what the logic of the packing of DNA actually is. I cannot go along with this criticism. As it should, the title expresses in the shortest possible way the contents of the material that it heads: a defined, but at present unanswered question.

Considerations on the Condensation and the Degree of Compactness in Non-Eukaryotic DNA-Containing Plasmas

E. KELLENBERGER[1], E. CARLEMALM[1], J. SECHAUD[2], A. RYTER[3], and G. DE HALLER[4]

Summary and Introduction

It is well-known that in eukaryotic cells the chromtin is relatively extended during interphase, while it is condensed into chromosomes during the mitotic cell division.

Nothing similar has been observed with bacteria. In light microscopy the nucleoids of growing bacteria appear lighter than the cytoplasm (Valkenburg et al. 1985), indicating a lower concentration of matter, as is also indicated by electron microscopy [Kellenberger (1953) and Fig. 1 of Bjornsti et al. in this volume].

Bacteriophages, however, show a condensation during packaging, as we will discuss below.

Dinoflagellates and *Euglena* exhibit chromosomes in the optical microscope at all times of the divisional cycle. In the electron microscope they are, however, fundamentally different (Thomas and Cox 1973; de Haller and Kellenberger, unpublished). While the fine structure of the *Euglena* nucleus is similar to that of eukaryotic nuclei, the dinoflagellate nucleus shows chromosomes of a structure and behavior very comparable to that of bacterial nucleoids (Grell and Wohlfarth-Bottermann 1957; Grassé and Dragesco 1957; Soyer 1981; Herzog and Soyer 1981; Giesbrecht and Drews 1982). In *Amphidinium elegans* a condensation occurs that might be related to division (de Haller et al. 1964).

The subject of the present paper is to summarize and discuss these phenomena associated with non-eukaryotic DNA-containing plasmas.

1 The Local Concentration of DNA in Bacteria and Bacteriophages

The DNA of a bacterium is in the order of 1 mm long (Cairns 1963) and is in one piece per nucleoid or genome, corresponding to 4.2×10^{-15} g per genome for *E. coli.* In

1 Department of Microbiology, Biozentrum of the University of Basel, Klingelbergstraße 70, 4056 Basel, Switzerland
2 Department of Molecular Biology, University of Geneva, 30 Quai Ernest Ansermet, 1211 Geneva 4, Switzerland
3 Institut Pasteur, Department of Molecular Biology, 25 Rue du Docteur Roux, 75 Paris, France
4 Institute of Animal Biology, University of Geneva, 1211 Geneva 4, Switzerland

Bacterial Chromatin
(ed. by C.O. Gualerzi and C.L. Pon)

general, several nucleoids, from two up to four, are found per cell (Ingraham et al. 1983). In rapidly growing cells (doubling time 20–30 min), the DNA accounts for a crudely estimated 4% of the dry mass, i.e. about 1.5×10^{-14} g. In a poorer medium, the doubling time increases as does the relative amount of DNA. The volume of an exponentially growing *E. coli* is 1.4×10^{-12} ml (Moncany et al., in preparation). From electron micrographs, one estimates that about one-half to three-quarters of the cell volume is occupied by ribosomes, leaving $0.35–0.7 \times 10^{-12}$ ml for the DNA. From this, a local concentration of $13–26 \times 10^{-3}$ g ml^{-1} can be calculated for DNA.

With viruses, one has to distinguish between the replicating and transcribing viral DNA, which – for bacteriophages – is also called the "vegetative phage DNA-pool" and the "packed" or "packaged" DNA within a virion. For the head of bacteriophage T4, one calculates the concentration from the mass of DNA ($50\ \mu = 1.9 \times 10^{-16}$ g) and the estimated volume (2.3×10^{-16} ml) to be 0.8 g ml^{-1}. It is, therefore, 30 to 60 times higher than in a nucleoid.

In particular cases (Kellenberger et al. 1958), where the formation of phage particles is inhibited, the vegetative phage pool, containing an estimated 100 to 200 phage equivalents, occupies an estimated 50–75% of the bacterial volume (Fig. 8 Pl, and Fig. 1). We reach a local concentration of $20–50 \times 10^{-3}$ g ml^{-1} which is in the same range as the DNA concentration in a nucleoid.

Bacteriophage DNA packaging is thus an obvious example of a DNA condensation by a factor of 15–30. This value corresponds to an estimate made for bacteriophage λ (Hohn 1976); it is higher than the factor of 5 estimated very early (Kellenberger et al. 1958).

The term "condensation" could sensibly be replaced also by "compaction". If we do so, we have to be careful because of another definition of the compaction factor that is frequently used when comparing linear dimensions, e.g. between a stretched

Fig. 1a–c. The intracellular pool of vegetative DNA of bacteriophage T2 and its condensation during packaging. After infection with phages T2 (T4, T6) the nucleoids of the host disintegrate and disappear. **a** The newly synthesised, replicating and transcribing phage-DNA becomes visible 12 min after infection as a lighter area of fibrillar structure, comparable to a nucleoid. In this thin section also a few finished, packaged phage heads are visible as "black particles" (*bp*). At later times, after full maturation, these particles will become complete, infective phages.

b The DNA pool has been artificially enlarged by inhibiting late protein synthesis by chloramphenicol added 8 min after infection. During the next 30 min, phage-DNA synthesis continues at full rate, but particles cannot be formed because of the inhibition of protein synthesis. After this period, chloramphenicol was removed and the cells observed 15 min later. Packaged phage heads (*bp*) are now visible within a large vegetative DNA pool.

Figures **a** and **b** are conventionally fixed with OsO_4 under R.K. conditions and embedded in Vestopal. In both cases the "pool" is clearly defined by its "lightness" and by the absence of ribosomes.

In **c** the modern method of rapid freezing and cryosubstitution has been employed (Hobot et al. 1985) on a mutant of bacteriophage T4 affecting head maturation. Again, the pool is enlarged and visible – with some effort – as ribosome-free space. Besides head precursor particles (*hp*) also packaged heads (*bp*) are recognised.

These three examples illustrate the factor of compaction (or condensation) of 15–30 incurred by the phage DNA when it is withdrawn from the pool and packaged into the head precursor. Figures **a** and **b** are from our archives, Nos. 21383 and 21455 respectively. These micrographs were taken 1958 by A. Ryter and J. Sechaud in the frame of a research project (Kellenberger et al. 1958). Figure **c** is due to the courtesy of W. Villiger and J. Hobot of the Basel laboratory and taken 1985

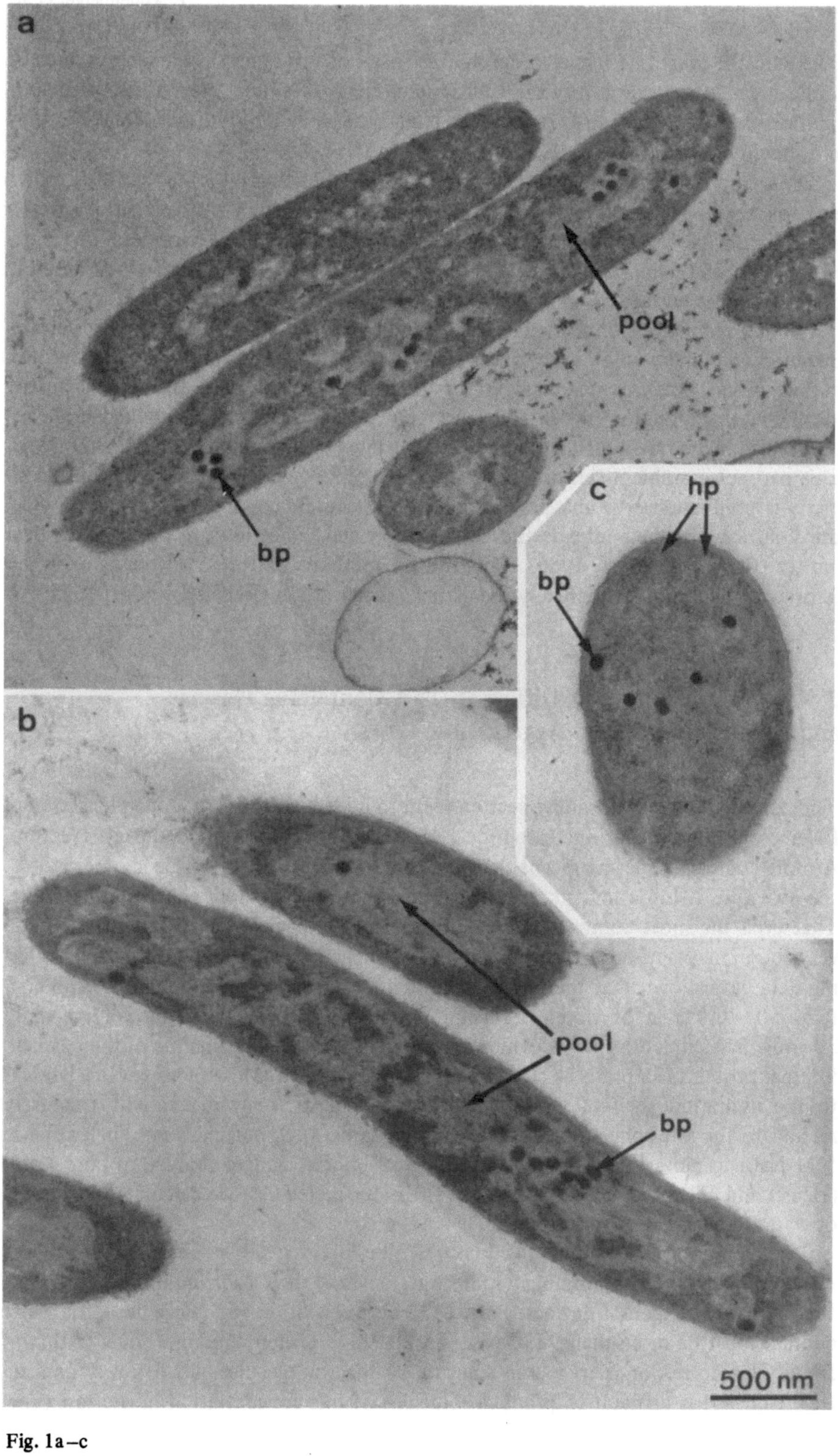

Fig. 1a–c

DNA and the length of higher order helices made of this same DNA, e.g. by a nucleosomic organisation and a possible additional coiling. It is quite clear from these considerations that the *degree of compactness* is best described by the local concentration and should not lead to any confusion.

DNA compaction in the case of bacteriophage is illustrated by Fig. 1, which shows electron micrographs of sections of infected cells (prepared by conventional techniques) with the vegetative pool and also containing a few phage particles. The different grey tones of the respective structures reflect the condensation of DNA, when it passes from the pool to become packaged into a virion.

In this example, conventional electron-microscopic staining of the sections is involved and one cannot state with absolute certainty that the "blackness" of the mature virion, when compared with the "lightness" of the pool, is due to different local concentrations; many people have argued that the virion is stained more by virtue of a higher stain affinity due to the proteins inside the phage head (about 8%). This particular problem can now be solved as soon as a new imaging mode (ratio-contrast) is used to observe unstained thin sections (Carlemalm and Kellenberger 1982); the contrast obtained reflects the composition and concentration of the biological matter (Carlemalm et al. 1985). Quantitative methods are presently being explored in order to determine numerically the actual concentration (Reichelt et al. 1985).

2 The Water Content of Different DNA-Containing Plasmas and Artificially-Induced Aggregation

The considerations in the above section seem to be straightforward: A condensation of DNA corresponds to an increased local concentration, to a more compact structure. A more compact structure means that it is penetrated less by water. Or in other words, the structure is loose and its different domains are not in close contact everywhere. That even the most compact DNA known, namely that contained in virions, is still very loose, may appear as a surprise to many. Indeed, calculations from X-ray scattering data have shown that the DNA inside of the capsid of bacteriophage T4 still contains 50–70% w/w of water (Earnshaw et al. 1978). Calculated by volume, only 30–40% is occupied by DNA, while the rest is water. The 8% volume of the residual, internal proteins (Onorato et al. 1978) is negligibly small. Most of this water belongs to the hydration shell of DNA, which, according to the water activity (inversely related to the salt concentration), varies from some 5–8 molecules per nucleotide in 5 M NaCl to more than 10 in a "physiological" buffer and to over 20 in pure water (Hearst and Vinograd 1961). If we take 20 water molecules, we calculate the volume of water to be 1.7 relative to that of DNA.

Less compact DNA-containing plasmas, like the vegetative phage pool and the nucleoid, contain correspondingly more water, even if they also contain proteins. With the estimated condensation factor of 15–30 between virion and phage pool, the water content must be more than 95%, even if we assume protein amounts equal to that of DNA. This corresponds to a true solution of macromolecules, albeit a very concentrated one. It is difficult to produce solutions of this concentration in the test tube.

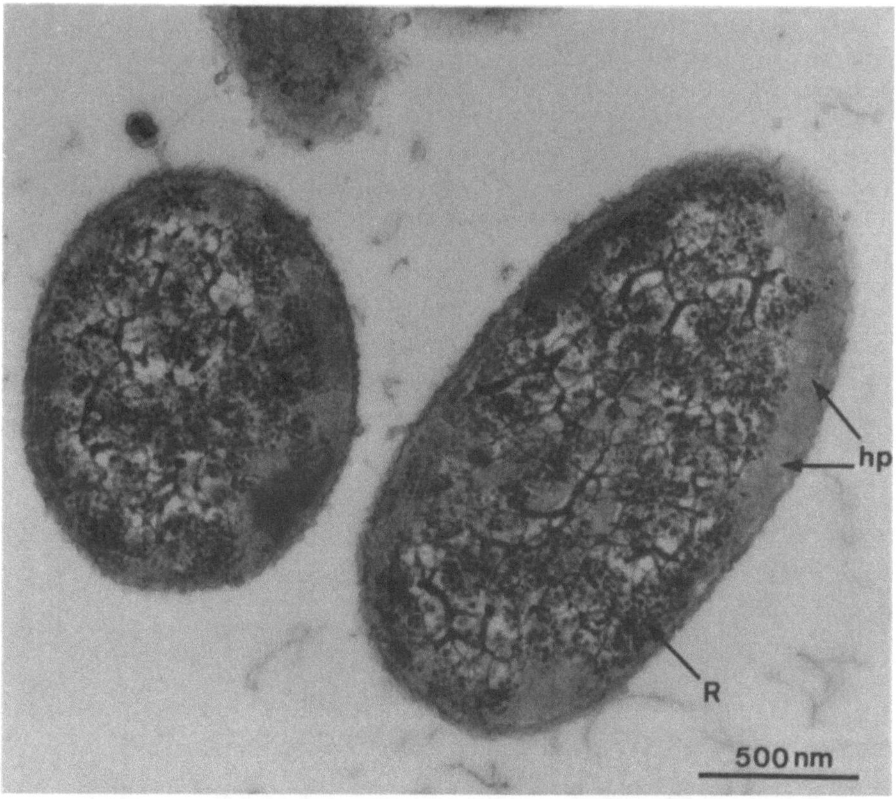

Fig. 2. Artificial aggregation of the DNA of a vegetative T4 phage pool. Cells infected with a mutant of T4 that produces a large DNA pool and precursor proheads without packaged DNA (Onorato et al. 1978), were fixed in formaldehyde, dehydrated in ethanol and embedded in Epon. By extensive experiments – reviewed in the text – it was demonstrated that the aggregation is a consequence of dehydration in organic solvents. In the figure these aggregates of DNA form the clearly visible strands. Ribosomes (*R*) are also seen, as well as head precursors (*hp*)

A solution of pure DNA of 3–6 mg ml^{-1} is already highly viscous, i.e. the consistency of honey. For such solutions, evidence was given that the distances between the molecules are too great to produce a gel by cross-linking with the usual fixatives like aldehydes and OsO_4 (Kellenberger et al. 1981), unless one performs it under the particular R.K. conditions (Kellenberger and Ryter 1964). In this study and others (Schreil 1964; Kellenberger 1962), it was shown that only gelled DNA plasms resist aggregation as a consequence of replacing water with organic solvents. Relatively dilute DNA plasms, as those of nucleoids and phage pools, thus easily undergo aggregation into compact bodies such as those seen in Fig. 2. These aggregates are artificial and should be carefully distinguished from physiological condensations (Kellenberger 1962).

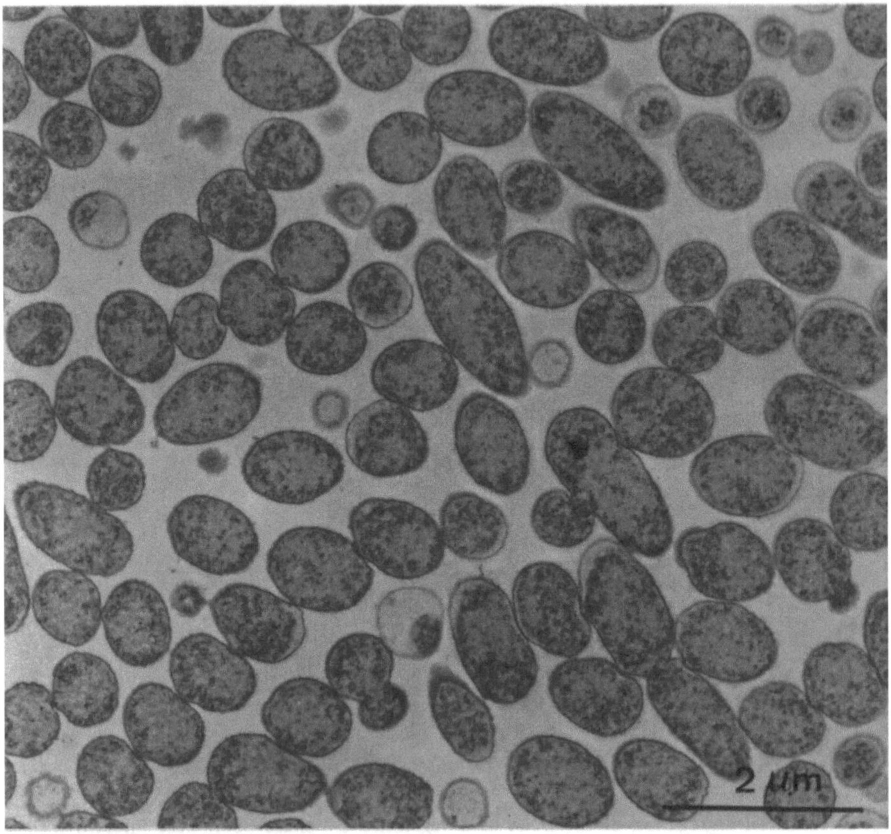

Fig. 3. Exponentially growing cells of *E. coli.* The cells were harvested on a membrane filter, transferred to the freezing apparatus and rapidly frozen on a copper block cooled with liquid helium; the frozen bacteria were then cryosubstituted and embedded in Epon (Hobot et al. 1985). The ribosome-free areas, identified as nucleoids, are visible. In no case do we observe a condensation. If it were to occur as part of division, condensed nucleoids should be visible on some 10% of a random culture. Preparation and microscopy by W. Villiger and J. Hobot of the Basel laboratory

3 Physiological Condensations and the Confusion Created by the Misleading Use of a Word

With the example of the bacteriophage above (Figs. 1 and 8), we have demonstrated a real, physiological condensation in a non-eukaryotic system. In the metabolising (replicating, transcribing, recombining) DNA plasma of the vegetative phage pool, the water content is so high that diffusion of metabolites and enzymes is possible. In the virion – as a sort of a storage form – the DNA becomes much more compacted and now occupies a 15–30 times smaller space.

Bacteriophage condensation is thus comparable to the condensation occurring with chromosomes of eukaryotic cells during mitosis. In interphase, the DNA is metabolis-

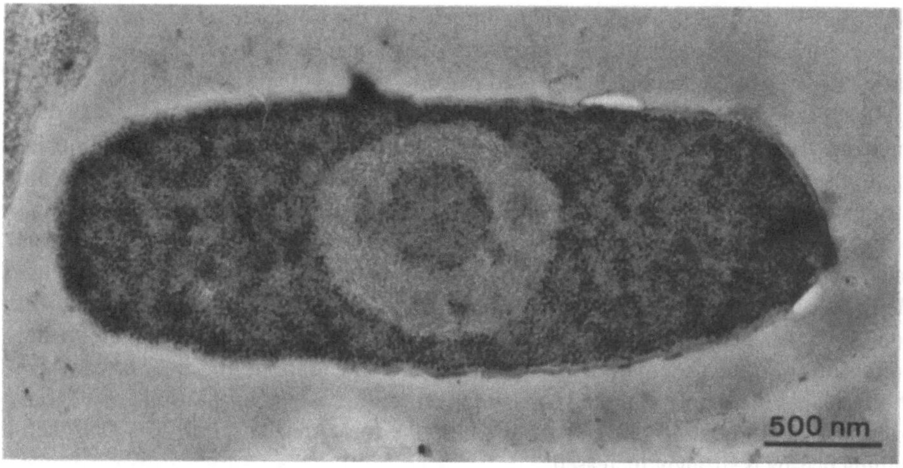

Fig. 4. The typical "annular" nucleoid obtained after inhibition of protein synthesis. This very typical shape of the bacterial nucleoid as a hollow sphere is obtained some 15–30 min after protein synthesis had been stopped. In the present case, we see a micrograph of exponentially growing *E. coli B* after 25 μg ml^{-1} of chloramphenicol had been added. Frozen as described in Fig. 3, cryosubstituted into acetone with 3% glutaraldehyde and embedded at low temperature into Lowicryl K4M; the same procedure was used for the immunocytochemistry described in Bjornsti et al. (this volume). Preparation and micrograph by J. Hobot, M.-A. Bjornsti and W. Villiger

ing, but in metaphase it has to be condensed (or compacted) so as to become disentangled and enabled to form individual chromosomes, which now can be separated with the help of the spindle in later mitotic phases. In this condensed state, DNA is no longer metabolising until the cell is again in interphase and the DNA decondenses. Unfortunately, local concentrations are difficult to estimate for chromosomes during the mitotic cycle, particularly because clear locations of the chromatinic material in interphase has not yet been achieved. This should become easier with the possibility of determining concentrations by ratio-contrast in scanning transmission electron microscopy.

We now ask the question about other cases of condensation in non-eukaryotic chromatins (Fig. 8). For a bacterium, with its single linkage group that represents a bacterial nucleoid or chromosome, condensation is not a necessity for understanding the separation of the two daughter chromosomes after division. Indeed, a non-synchronised bacterial culture of exponentially growing cells does not show any cells with an increased local concentration of the DNA (see Bjornsti et al., this volume and Fig. 3). These results are obtained by cryotechniques; they fully confirm the observations available from conventional techniques (reviewed in van Iterson 1984). There is thus no longer any doubt that bacteria do not undergo a condensation phase when they distribute their genetic material to the daughter cells during a preceding cell division.

Confusion was created, when, what we call a more confined arrangement of DNA, was misleadingly described as a "condensed form". This situation is illustrated in Figs. 4 and 8, which show the typical form of the nucleoid which is obtained after inhibition of protein synthesis. When compared to the ribosome-free space, which is rather dispersed throughout the cytoplasm of exponentially growing cells, we now have all the DNA-containing plasma confined to a hollow sphere. But the DNA plasma itself appears not to be more significantly concentrated than in the normal, dividing nucleoid The DNA-containing plasma is thus not really condensed, but the normally more dispersed, cleft ribosome-free area is now confined to a more restricted location. Strictly speaking, it is true that the nucleoid itself, as a whole, might be considered as being more condensed; but how are we then to describe metaphases of eukaryotic cells, where the (condensed!) chromosomes are distributed throughout the nucleus? Here, everybody agrees that the DNA-containing plasma representing a chromosome is condensed (or compacted) relative to that in interphase! In order to avoid further confusion, we thus propose not to use the term "condensation" for those cases where the DNA plasma is only confined to a more restricted area; "condensation" and "compaction" should be used only for cases where the local concentration of a DNA-containing plasma is strongly increased.

4 Characteristics of Non-Eukaryotic DNA

Are there other cases of condensation in non-eukaryotic, DNA-containing plasmas besides that of bacteriophages, where a condensation occurs when DNA is withdrawn from the vegetative pool and is packaged into preformed proheads? As we have seen above, no real condensation has been observed with bacteria. Apparently, the equipartition of the genome into the daughter cells is made without such a compaction.

A physiological condensation, clearly distinct from an artificial aggregation (Michel-Salamin et al. 1984), had been observed in a dinoflagellate (de Haller et al. 1964; Figs. 5 and 6), although it was not possible to prove unambiguously that this condensation was related to the divisional cycle rather than to another physiological process. Cells containing variably condensed chromosomes coexist in the same culture. All these cells show identically well-preserved cytoplasms. The degree of condensation is thus not related to a sort of necrosis of dead cells.

From other publications (Ris 1962; Kubai and Ris 1969; Spector et al. 1981; Blank and Trench 1985), we tend to understand that others made similar observations on the existence of a condensation cycle, while some species are reported to show only and always the same state of compactness (Herzog et al. 1984).

"Dinokaryotes" are of particular interest for bridging the gap between bacteria and eukaryotic cells (Giesbrecht and Drews 1982; Herzog et al. 1984). Dinoflagellates have several distinct chromosomes which are genetically not identical (Roberts et al. 1974). They have – in contrast to bacteria – a nuclear membrane and their divisional pattern is an atypical mitosis. With conventional methods, the fine structure of the chromosomes appears to be very similar to that of the bacterial nucleoid (Fig. 7). The results obtained with cryomethods again confirm their similarity with nucleoids (Michel-Salamin et al. 1984). So-called histonelike proteins have also been described for dinokaryotes (reviewed in Herzog et al. 1984; Rizzo 1981; Rizzo and Burghart 1982). As in bacteria, they are quite different in quality and quantity from the eukary-

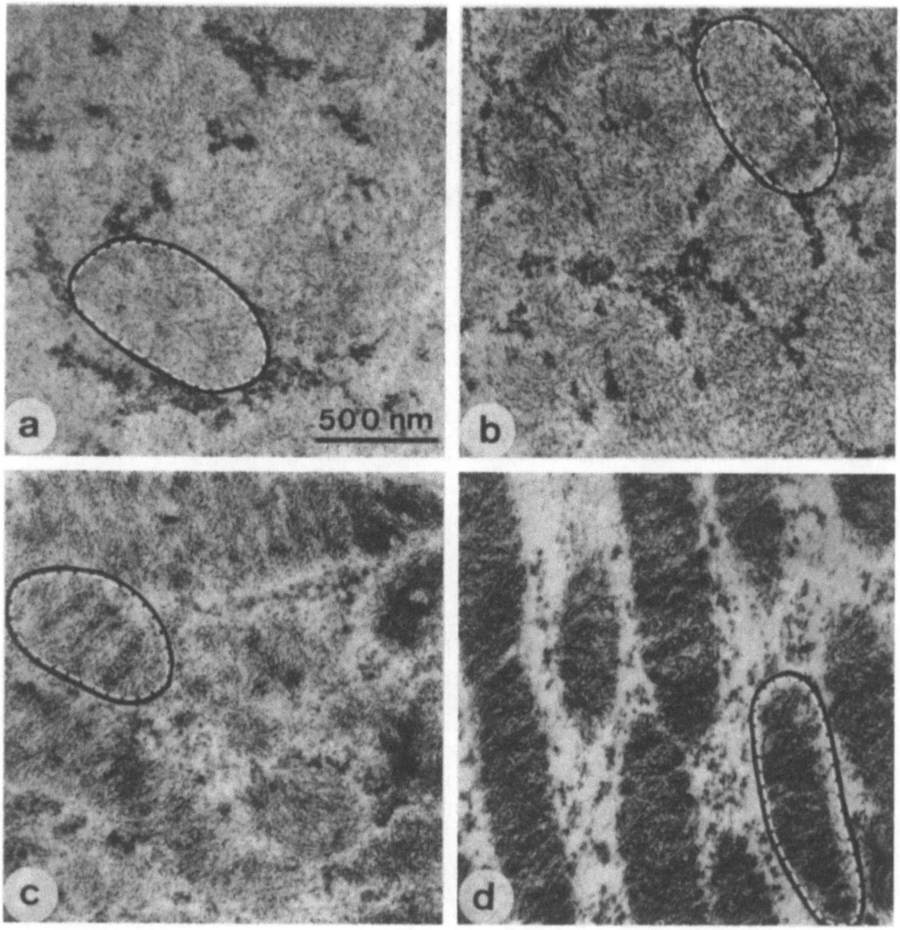

Fig. 5a–d. Different degrees of condensation of the chromosomes of the dinoflagellate *Amphidinium elegans* (de Haller et al. 1964). This dinoflagellate, grown in liquid culture in the laboratory, shows the chromosomes in the expanded form **a**, but also different intermediates of compaction are observed, **b** and **c**. The most compact form is seen in **d** and more enlarged in Fig. 6. Sections of single chromosomes are outlined for a better understanding of the changes of concentration. The sizes of the outlined chromosomes are obviously not significant because they do not represent longitudinal sections in every case. The material was fixed with OsO_4 under R.K. conditions and post-fixed in uranyl acetate and embedded in Vestopal

otic histones and it has, therefore, been postulated that their DNA is not organised into conventional nucleosomes. A study on the location of these DNA-binding proteins, in the way it had been described by Bjornsti et al. (this volume), would thus be of highest interest, as also in case of the nucleoids of archaebacteria.

Comparing the behavior of the DNA-containing plasmas of dinokaryotes, mitochondria, bacterial nucleoids and vegetative phase DNA pools, one finds that they behave similarly and very differently from the chromosomes of eukaryotic cells when judged by the following criteria: (1) Aggregation during dehydration occurs much

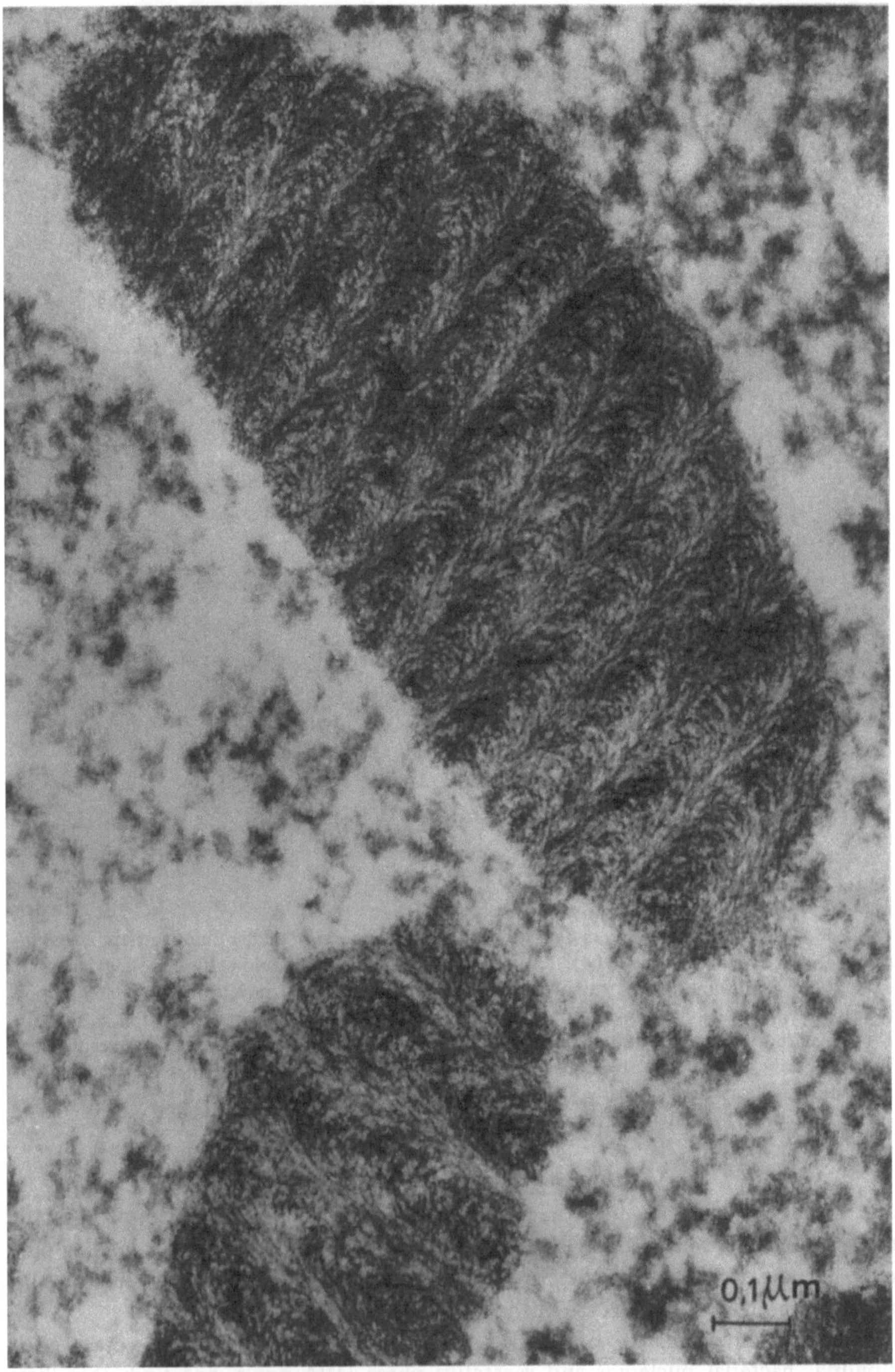

Fig. 6. A chromosome of *Amphidinium elegans* in its relatively compact form (de Haller et al. 1964). From a cell with condensed chromosomes, as shown in Fig. 5d, two chromosomes are shown at higher magnification. Note the arched arrangement of the fibrilla. The exact spatial arrangement is not yet clearly understood and still under investigation (Livolant et al. 1978)

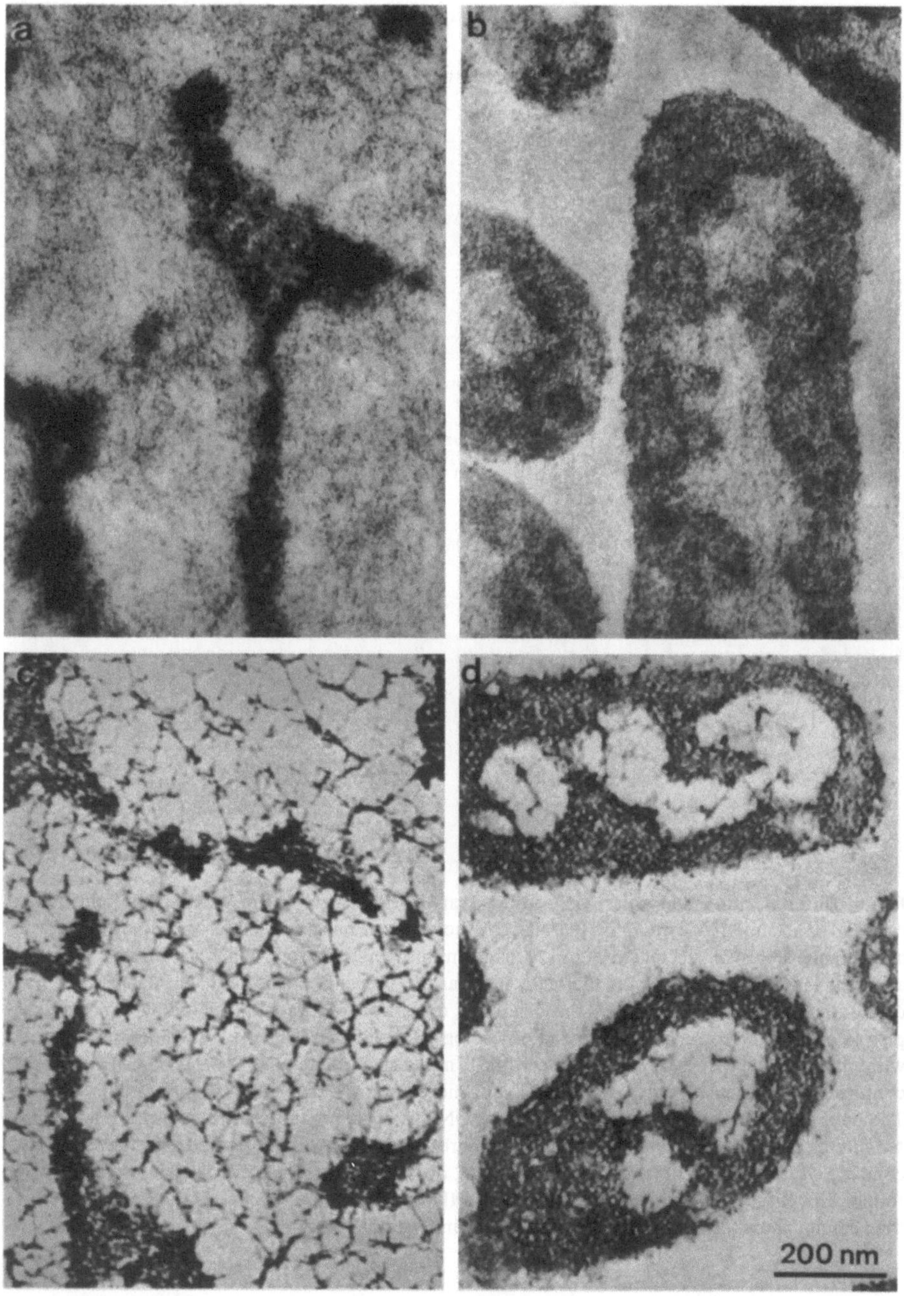

Fig. 7a–d. Chromosomes of *Amphidinium elegans* (decondensed) undergo the same type of aggregation during dehydration as do nucleoids (de Haller et al. 1964). *E. coli* cells of an exponentially growing culture were mixed with a culture of *Amphidinium* and processed together. **a** and **b** show decondensed (expanded) parts of some three chromosomes (as shown in Fig. 5a) and *E. coli* respectively). In both organisms the DNA plasmas are fine-fibrillar. Figures **c** and **d** are from aliquots of the above sample, **a** and **b**, in which the post-fixation with uranyl acetate was replaced by treatment with EDTA. The DNA aggregation is obvious in **c** and **d**, although it is somewhat less coarse than that shown for the phage pool in Fig. 2. The essence is that it is exactly comparable for the two DNA-plasmas (bacteria and dinoflagellates). Processing as in Figs. 5 and 6

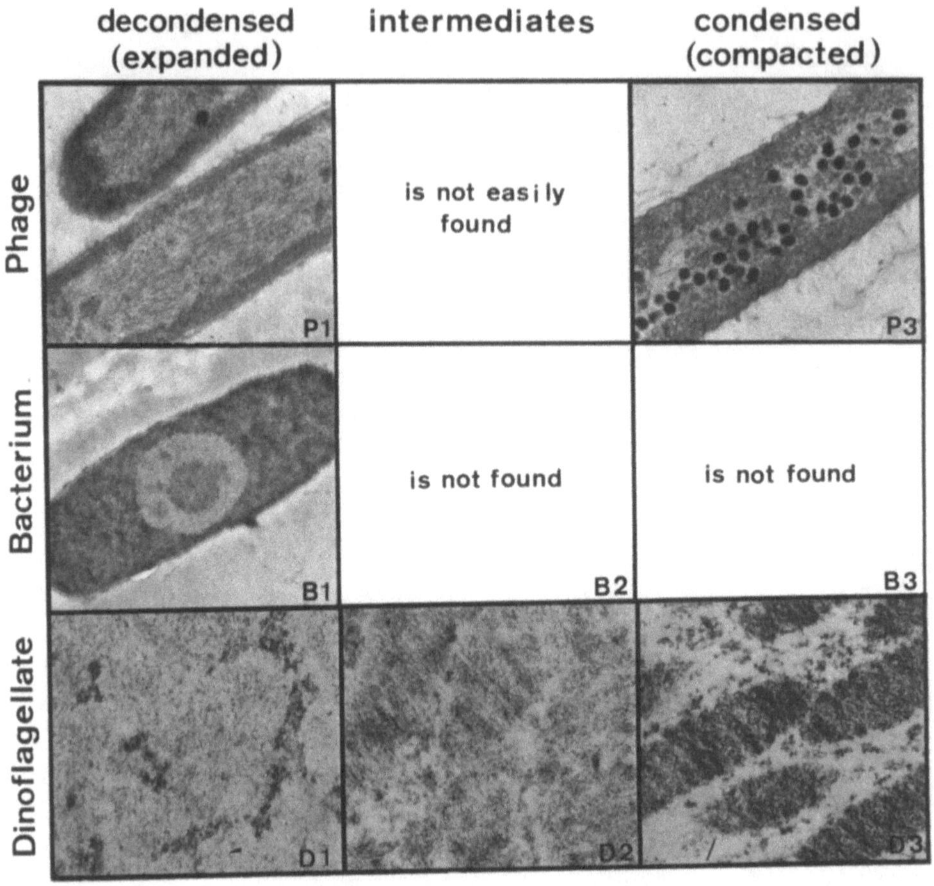

Fig. 8. Different non-eukaryotic organisms compared with respect to the occurrence of physiological condensation. DNA-containing plasmas are shown from phages, bacteria and dinoflagellates in comparable states of decondensation (expansion) or condensation. What we compare are the apparent local concentrations of the DNA plasmas, as discussed in the text. For bacteria, no condensation has been observed, but the degree of dispersion of the DNA plasma within the cytoplasma might vary strongly according to physiological conditions. The present bacterium is treated with chloramphenicol so as to show a confined, non-dispersed state of the nucleoid. Even in this confined, non-dispersed state, the DNA is not more compact than it is in more dispersed states. The pool of replicating and transcribing phage DNA is of about the same compaction as the bacterial nucleoid or the expanded dinoflagellate chromosome. The DNA packed into the virion is as compact as the most compact chromosome of dinoflagellates. In both cases intermediates are found. For phages it is difficult to distinguish partly full heads from partial sections and, therefore, they are not shown. References are found in Kellenberger (1980)

more easily in non-eukaryotic DNA plasmas than with eukaryotic nuclei. (2) OsO_4 fixation under R.K. conditions (i.e. in the presence of amino acids and Ca, pH around 6) and/or treatment with uranyl acetate or indium chloride is necessary in order to avoid consecutive aggregations; this is not the case to any comparable degree with eukaryotic cells. If, instead of a post-fixation in uranyl acetate, the bacterial cells are post-treated with EDTA (called "Versene" in former times), which chelates divalent cations, aggregation occurs only with non-eukaryotic DNA plasmas (Kellenberger et al. 1958; Ris and Singh 1961; de Haller et al. 1964). (3) And last but not least, the ultrastructure of adequately osmium-fixed non-eukaryotic DNA plasmas is very distinct from that of the chromatin of eukaryotic cells, as seen by their fine-fibrillar structure.

Interestingly enough, this behavior is the same as that of purified DNA. This was shown by surrounding bacteria with a solution of DNA (Schreil 1964), and mixing dinoflagellates with bacteria (de Haller et al. 1964). In all these cases, the observed aggregates were of a very similar nature (Fig. 7).

The precise reason for this distinct and particular behavior, observed with all the non-eukaryotic DNA plasmas, is still not known and is open to speculation. The oldest hypothesis was that the non-eukaryotic DNA is not organised in the form of nucleosomes as is the eukaryotic chromatin. The high relative content of protein in the latter would facilitate cross-linking even by aldehydes, while in the former, the DNA would be neutralised by polyamines and divalent cations (Moncany and Kellenberger 1981) and would thus be more difficult to cross-link and gel (Kellenberger et al. 1981). Another, second hypothesis was that the protein partners are destroyed by OsO_4; indeed, it was shown that proteins are cleaved by OsO_4, although the fragments become separated only after SDS treatment (Baschong et al. 1984). A third hypothesis is based on the assumption that OsO_4 can produce single-strand breaks in DNA and thus can release supercoils even without affecting proteins.

It is fortunate that the applications of cryotechniques in electron microscopy have shown that the DNA plasma of non-eukaryotes consists of very finely dispersed material, independent of fixation before freezing or during substitution (Hobot et al. 1985). The existence of dehydration-induced aggregations at 0°–20°C, which were misleading for so many years, is thus demonstrated. By these new techniques, it should now become possible to reinvestigate the in-situ structure, thus complementing the in vitro techniques, which also have their severe limitations. In these cases, one should remember that intracellularly, we face very high concentrations of Mg^{2+} and very likely also of Ca^{2+} (Chang et al. 1984) and that these are reduced to nil by the obligatory presence of EDTA in nearly all in vitro experiments! That means, in particular, that natural protein partners of DNA might specifically bind to the Mg or Ca salt of DNA that is neutral or even positively charged, while all tests of DNA binding are made with the very easily dissociated Na salt of DNA, where all basic proteins bind because they easily compete with sodium (or potassium) (see Bjornsti et al. this volume). The change of the ionic environment upon isolation of nucleoids has also to be kept in mind. Native protein partners might be lost and other basic proteins might become bound. It was also shown that upon fixation by aldehydes and OsO_4, the cells become permeable and potassium and magnesium, among others, start to leak out (experiments with Borle and Moncany, to be published). The intracellular ionic conditions are thus changed, which explains rearrangements of the DNA-containing plasmas, resulting in modifications of their general shape (Hobot et al. 1985).

Acknowledgements. Our own work reported in this paper was subsidized by the States of Geneva and Basel as well as by the Hoffmann La Roche Foundation for team work and by the Swiss National Science Foundation. We are grateful to Dr. Jan Hobot and Werner Villiger for micrographs 2, 3 and 4. Our thanks go also to Marlies Zoller and Margrit Jaeggi for the graphic arts work and Regula Niederhauser for typing partiently the successive versions of this paper.

References

Baschong W, Baschong-Prescianotto C, Wurtz M, Carlemalm E, Kellenberger C, Kellenberger E (1984) Preservation of protein structures for electron microscopy by fixation with aldehydes and/or OsO_4. Eur J Cell Biol 35:21–26

Blank J, Trench RK (1985) Speciation and symbiotic Dinoflagellates. Science 229:656–658

Cairns J (1963) The bacterial chromosome and its manner of replication as seen by autoradiography. J Mol Biol 6:208

Carlemalm E, Kellenberger E (1982) The reproducible observation of unstained embedded cellular material in thin sections: visualisation of an integral membrane protein by a new mode of imaging for STEM. EMBO J 1:63–67

Carlemalm E, Colliex CH, Kellenberger E (1985) Contrast formation in electron microscopy of biological material. In: Hawkes PW (ed) Advances in electronics and electronphysics. Academic Press, New York, pp 269–334

Chang Ch-F, Shuman H, Somlyo AP (1984) Electron probe analysis, X-ray mapping and electron energy loss spectroscopy of elemental distribution of *E. coli* B. In: Bailey GW (ed) Proc 42th Annu Meeting Electron Microsc Soc Am. San Francisco Press, San Francisco, pp 570–571

Earnshaw WC, King J, Eiserling FA (1978) The size of the bacteriophage T4 head in solution with comments about the dimensions of virus particles as visualized by electron microscopy. J Mol Biol 122:247–253

Giesbrecht VP, Drews G (1982) Die „Kernstrukturen" der Bakterien und ihre Beziehungen zu denen der „Mesokaryoten". In: Metzner H (ed) Die Zelle – Struktur und Funktion. Wissenschaftliche Verlagsgesellschaft, Stuttgart, pp 542–597

Grassé PP, Dragesco J (1957) L'ultrastructure du chromosome des Peredeniens et ses consequences genetiques. CR Acad Sci Paris 245:2447–2452

Grell KG, Wohlfarth-Bottermann KE (1957) Licht- und electronenmikroskopische Untersuchungen an dem Dinoflagellaten Amphidinium elegans. Z Zellforsch Mikrosk Anat 47:7–17

de Haller G, Kellenberger E, Rouiller G (1964) Etude au microscope electronique des plasmas contenant de l'acide desoxyribonucleique III. Variations ultra-structurales des chromosomes d'Amphidinium. J Microsc (Paris) 3:627–642

Hearst JE, Vinograd J (1961) A three component theory of sedimentation equilibrium in a density gradient. PNAS 47:999–1004

Herzog M, Soyer MO (1981) Distinctive features of Dinoflagellate chromatin. Absence of nucleosomes in a primitive species (Prorocentrum micans) E. Eur J Cell Biol 23:295–302

Herzog M, von Boletzky S, Soyer MO (1984) Ultrastructural and biochemical nuclear aspects of eukaryotic classification: independent evolution of the Dinoflagellates as a sister group of the actual eukaryotes. Origins Life 13:205–215

Hobot JA, Villiger W, Escaig J, Maeder M, Ryter A, Kellenberger E (1985) The shape and fine structure of the nucleoid observed on sections of ultrarapid frozen and cryosubstituted bacteria. J Bacteriol 162:960–971

Hohn T (1976) Packaging of genomes in bacteriophages: a comparison of ssRNA bacteriophages and dsDNA bacteriophages. Philos Trans R Soc Lond B Biol Sci 176:143–150

Ingraham JL, Maaløe O, Neidhardt B (1983) Growth of the bacterial cell. Sinauer, Sunderland, Mass.

van Iterson W (1984) Inner structures of bacteria. Van Nostrand Reinhold Scientific and academic edns, New York

Kellenberger E (1953) Les formes caracteristiques des nucleoides de *E. coli* et leurs transformations dues a l'action d'agent mutagenes-inducteurs et de bacteriophages. In: Symp Citologia Batterica, Supplemento Rendiconti Ist. Superiore di Sanita, Roma, pp 45–66. Also: thesis Geneva, University Microfilms Int. PO Box 1346, Ann Arbor, Mich. 48106, USA Cust. Ref. 84-03,992

Kellenberger E (1962) The study of natural and artificial DNA-plasms by thin sections. In: Harris JJC (ed) The interpretation of ultrastructure. Academic Press, New York pp 233–249

Kellenberger E (1980) Control mechanisms in the morphogenesis of bacteriophage heads. BioSystems 12:201–223

Kellenberger E, Ryter A (1964) In bacteriology. In: Siegel BM (ed) Modern developments in electron microscopy. Academic Press, New York, pp 335–393

Kellenberger E, Ryter A, Séchaud J (1958) Electron microscopy study of DNA-containing plasms II. Vegetative and mature phage DNA as compared with normal bacterial nucleoids in different physiological states. J Biophys Biochem Cytol 4:671–678

Kellenberger E, Carlemalm E, Stauffer E, Kellenberger C, Wunderli H (1981) In vitro studies of the fixation of DNA, nucleoprotamine, nucleohistone and proteins. Eur J Cell Biol 25:1–4

Kubai DF, Ris H (1969) Division in the Dinoflagellate Gyrodinium colini (Schiller). J Cell Biol 40: 508–518

Livolant F, Giraud MM, Bouligand Y (1978) A goniometric effect observed in sections of twisted fibrous materials. Biol Cell 31:159–168

Michel-Salamin L, Gautier A, Soyer-Gobillard MO, Herzog M, de Billy F, Dubochet J, McDowall AW, Kellenberger E, Carlemalm E (1984) Appearance of "arch-shaped" chromosomes in Dinoflagellates as observed in thin sections and cryofractures following various preparation procedures. Electron Microsc 3:1803–1804

Moncany MLJ, Kellenberger E (1981) High magnesium content of *E. coli B*. Experientia (Basel) 37:846–847

Onorato L, Stirmer B, Showe MK (1978) Isolation and characterization of bacteriophage T4 mutant preheads. J Virol 27:409–426

Reichelt R, Carlemalm E, Villiger W, Engel A (1985) Concentration determination of embedded biological matter by scanning transmission electron microscopy. Ultramicroscopy 16:69–80

Ris H (1962) Interpretation of ultrastructure in the cell nucleus. In: Harris JJC (ed) The interpretation of ultrastructure. Academic Press, New York, pp 69–99

Ris H, Singh RN (1961) Electron microscopy studies on blue green algae. J Biophys Biochem Cytol 9:63–80

Rizzo PJ (1981) Comparative aspects of basic chromatin proteins in Dinoflagellates. BioSystems 14:433–443

Rizzo PJ, Burghart RC (1982) Histone-like protein and chromatin structure in the wall-less Dinoflagellate Gymnodinium nelsoni. BioSystems 15:27–34

Roberts TM, Trittle RC, Allen JR, Loeblich R, Klotz LC (1974) New genetic and physico-chemical data on structure of Dinoflagellate chromosomes. Nature 248:446–447

Schreil WH (1964) Studies on the fixation of artificial and bacterial DNA plasms for the electron microscopy of thin sections. J Cell Biol 22:1–20

Soyer MO (1981) Presence of intranuclear micro cables in a primitive Dinoflagellate protist: morphological description and discussion of their possible evolutionary significance. BioSystems 14:299–304

Spector DL, Vasongelos AC, Triemer RE (1981) DNA-duplication and chromosome structure in the Dinoflagellates. Protoplasma 105:185–194

Thomas RN, Cox ER (1973) Observations on the symbiosis of peridinium balticum and its intracellular alga VI. Ultrastructure. J Phycol 9:304–323

Valkenburg JAC, Woldringh CL, Brakenhoff GJ, van der Voort HTM, Nanninga N (1985) Confocal scanning light microscopy of the *E. coli* nucleoid: comparison with phase contrast and electron microscopy images. J Bacteriol 161:478–483

DNA-Binding Proteins and Their Interactions with Structure-Building Ions

G. GERACI[1], L. TOMEI[1], V. CALABRO[1], G. QUAGLIAROTTI[2], and L. DE PETROCELLIS[2]

The organization of DNA into a chromosome structure is implemented by its interactions with specific classes of protein molecules. In eukaryotes five different proteins, the histones, are found associated with chromosome DNA. The histones can be subdivided into two sets, on the basis of their roles. The first set comprises the so-called core histones, H2A, H2B, H3 and H4, organized in an octamer around which about 200 base pairs of DNA are wound in two loops constituting the fundamental chromatin particle: the nucleosome. The second set of eukaryotic histone molecules comprises those of the H1 type, which combine with DNA independently of the "core" histones, stabilizing the nucleosome and participating in the formation of higher-order chromatin structures. There are a variety of H1 histones in different organisms, cell types and in each cell.

In the prokaryotic nucleoid, the DNA is also – at least in part – organized in nucleosome-like beads formed by a set of prokaryotic histone-like proteins. The detailed molecular structure of one of these, the DNA-binding protein II from *B. stearothermophilus,* has recently been determined. The model proposed to describe the interaction of that protein with DNA has very interesting new aspects in comparison with other models proposed for the interactions of proteins binding to specific DNA sites (Tanaka et al. 1984). These aspects may well be relevant also for the binding of histones to DNA, since both prokaryotic and eukaryotic histones have the specific function of organizing the DNA into a compact structure so as to fit it into the limited cell space in an ordered way.

For these reasons, it is likely that DNA-structuring proteins and proteins that recognize specific DNA sequences and/or structures bind to DNA by different mechanisms. Thus, for the latter type of protein, the initial electrostatic interaction is only a step in the process of formation of a stable complex. RNA polymerase and repressors of gene activity are supposed to slide freely on DNA before reaching the final site of binding because their rates of combination with their binding sites exceed the values expected from the diffusion limit. For histones and histone-like proteins, including protamine, such a mechanism does not seem adequate. In eukaryotic chromatin, the superhelical state of DNA appears completely dependent upon combination with histones, while in contrast, the negative superhelical turns usually observed in protein-depleted bacterial DNA appear at least in part restrained by protein interactions

1 Department of Genetics, General and Molecular Biology, University of Naples, Naples, Italy
2 Institute of Cybernetics, CNR, Arco Felice, Naples, Italy

Bacterial Chromatin
(ed. by C.O. Gualerzi and C.L. Pon)

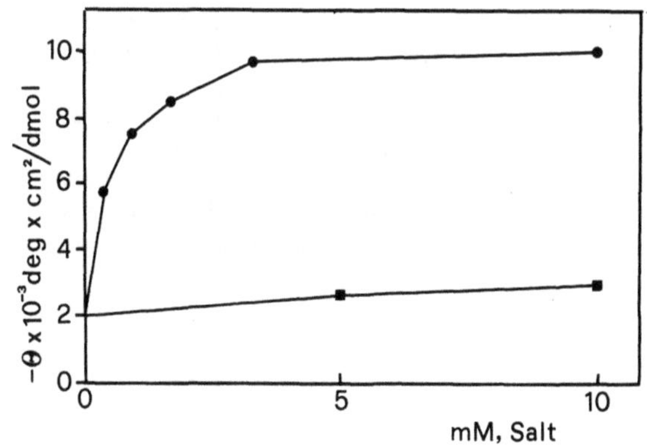

Fig. 1. $[\Theta]_{222\ nm}$ values of the circular dichroism spectra of *Chaetopterus v.* sperm histone H1 as a function of sodium chloride (■–■) and of sodium phosphate (●–●) concentration. Histone 0.1 mg ml^{-1} in 2 mM Tris-acetate, pH 6.0

(Pettijohn 1982). It is difficult to assume that the electrostatic binding of these proteins to DNA can permit free sliding. As a matter of fact, the evidence so far available on this aspect shows that nucleosomes have very limited freedom on the DNA to which they are bound (McGhee and Felsenfeld 1980).

H1 Histones

In studies of the structural organization of the H1 histones obtained from sperm chromatins of the marine annelid worm *Chaetopterus variopedatus* and from sea urchins, strong binding sites have been found for ions like sulfate, phosphate and perchlorate. The presence of sites for such interaction becomes evident when one studies the dependence of the structure of H1 molecules upon pH, ionic strength and ion types, as monitored by circular dichroism measurements in the peptide region of the spectrum.

The structure of the H1 molecule is not dependent on pH in the range pH 3 to pH 8 but, in all cases studied up to now, it is dependent on NaCl concentration. This is expected when one considers that counter-ions shield positive charges on the histone molecule, permitting the peptide to fold. If however sulfate, phosphate or perchlorate ions are present instead of chloride ions, the structuring of the H1 molecule occurs at an ionic strength one or two orders of magnitude lower than that typical of NaCl (Fig. 1).

We have seen that the structuring of H1 by NaCl differs from the structuring by multivalent ions and perchlorate only in that the latter have a greater effect – that is, the difference appears quantitative. However, fluorescence spectroscopy also shows that there is a qualitative difference between the effects of these ions. The binding of phosphate to the H1 molecule changes the fluorescence quantum yield of the two tyrosine groups. This change is quite different from that observed when the NaCl concentration is increased. As shown in Fig 2, the quantum yield of the tyrosine fluorescence of the H1 molecule is increased by NaCl to a value even greater than that expected for free tyrosine. This is an unusual finding, previously reported only for

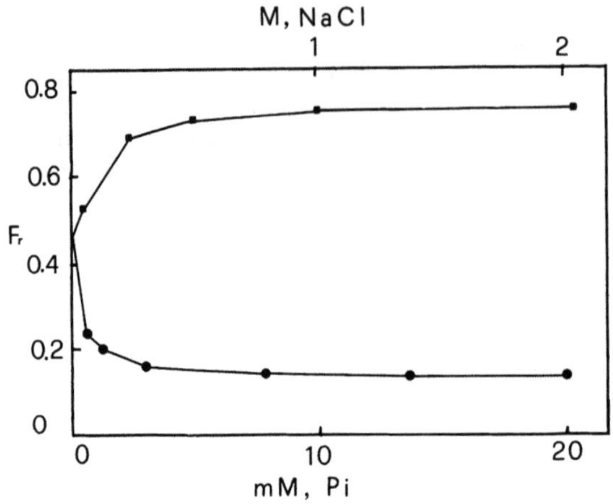

Fig. 2. Perturbation of the fluorescence of tyrosine residues of *Chaetopterus v.* sperm histone H1 by sodium chloride (■–■) and sodium phosphate (●–●). Excitation at 280 nm; emission at 305 nm. Histone 0.1 mg ml^{-1} in 2 mM Tris-acetate, pH 6.0. Values relative to free-tyrosine fluorescence

calf thymus H1 (Giancotti et al. 1977). Correlation between the percentage of induced structure and the increase in quantum yield suggests that structuring of the molecule provides the tyrosine residues with an environment enhancing fluorescence emission. Contrary to what is observed with NaCl, addition of phosphate to the *Chaetopterus v.* H1 histone causes a quenching of the fluorescence. Since the CD spectra show that the protein is structured by the salt, the quenching must derive from local effects due to the binding of phosphate.

The number of phosphate sites per histone molecule has been determined by gel filtration of the histone on Sephadex columns that were first equilibrated with phosphate in the presence of different concentrations of NaCl in order to counteract unspecific, weak binding of phosphate to positively-charged amino-acid residues. Extrapolation of the results to zero NaCl concentration shows the presence of about ten phosphate sites per histone molecule. The interaction between phosphate ions and histones is so strong that, in the absence of NaCl, pyrophosphate acts as a bifunctional reagent causing histone aggregation and eventually histone precipitation.

Protamines

Structural studies of the protamine molecule isolated from the sperm chromatin of *Chaetopterus v.* (De Petrocellis et al. 1983) give clear indications of the presence of cation-cation interactions mediated by a linking anion. The molecule, like other protamines, has a very simple amino-acid composition, shown in Table 1. When analyzed by gel filtration on Sephadex columns at any salt concentration above 10 mM and at pH values between 3 and 9, it invariably shows a symmetrical peak, with an elution volume corresponding to a molecular weight of about 45 kD. Yet in 8 M urea, 0.9 M acetic acid it shows an elution volume corresponding to 12 kD molecular weight. The protamine is immediately precipitated from its solutions by the addition of SDS.

Table 1. Amino-acid composition of *Chaetopterus v.* sperm protamine

Amino-acid residue	Thr	Ser	Pro	Gly	Ala	Val	Lys	Arg
Mol %	2.1	20.8	2.2	2.7	10.9	5.3	14.4	41.7

Reactions of the protamine with reagents modifying arginine or lysine residues have always given partial results. Chloride ions can be removed only if two cycles of treatment with Dowex 1 × 8 charged with OH is carried out on solutions of protamine in 8 M urea. The importance of the linking anion is demonstrated by dissolving the protamine in barbital buffer. Under these conditions, the protamine is eluted from Sephadex gel-filtration columns in a volume corresponding to 12 kD molecular weight. When it is deprived of chloride ions by Dowex treatment, the protamine shows, in barbital buffer, an elution volume corresponding to 15 kD molecular weight. Addition of any salt except barbital reconstitutes the usual elution volume corresponding to 45 kD molecular weight. Barbital ion, due to its structure, is not a suitable linker anion between two positive groups.

Conclusion

These data on histone H1 and on *Chaetopterus v.* protamine show that these two positively-charged molecules can make unusually stable electrostatic bonds. The high stability of those interactions could be based on the formation of triplets of charges, with two positively-charged groups of the protein binding firmly to one negatively-charged group of the ligand ion and thus restricting sliding of the two molecules relative to each other.

References

De Petrocellis B, Parents A, Tomei L, Geraci G (1983) An H1 histone and a protamine molecule organize the sperm chromatin of the marine worm *Chaetopterus variopedatus*. Cell Differ 12: 129–135

Giancotti V, Fonda M, Crane-Robinson C (1977) Tyrosine fluorescence of two tryptophan free proteins: histone H1 and H5. Biophys Chem 6:379–383

McGhee JD, Felsenfeld G (1980) Nucleosome structure. Annu Rev Biochem 49:1115–1156

Pettijohn DE (1982) Structure and properties of the bacterial nucleoid. Cell 30:667–669

Tanaka I, Appelt K, Dijk J, White SW, Wilson KS (1984) 3-Å resolution structure of a protein with histone-like properties in prokaryotes. Nature 310:376–381

The Packaging of DNA in Bacteriophage Lambda

J. WELSH and C.R. CANTOR[1]

1 Introduction

The structure of bacteriophage lambda has been studied extensively by several groups using a variety of physical and chemical techniques. However, many aspects of bacteriophage structure remain unreasolved. It is not known how the DNA is arranged three-dimensionally within the capsid, nor is the mechanism of packaging understood. However, some general features of lambda structure have been discerned (Review: Earnshaw and Casjens 1980). For example, it is now well established that the DNA is somehow pushed into a preformed structure known as the prohead, rather than assembly of the capsid around the condensed DNA molecule (Hohn and Hohn 1974; Kaiser et al. 1975), or a concerted protein-DNA condensation mechanism as has been observed in the animal virus SV40 (Bina et al. 1983).

Much more is known about the assembly of the protein components of bacteriophages (Hendrix and Casjens 1975), but here there are also important elements missing from the picture. Through complementation studies, a general outline of the assembly process of lambda has been established (Murialdo and Becker 1978), but the locations and functions of several minor proteins and fusion-cleavage proteins are still unknown (Murialdo and Ray 1975). Also unknown are the molecular details of the assembly of the scaffold, a postulated protein structure around which the capsid is assembled.

Structural studies on lambda have been partially motivated by the hope that understanding this relatively simple system might lead to a better understanding of more complex DNA-protein complexes, such as bacterial and eukaryotic genomes. To put this into perspective, there are approximately 20 proteins known to be either components of lambda or involved in its assembly (Kellenberger and Edgar 1971), whereas, the typical eukaryotic chromosome has over 500 associated proteins (Lewin 1980). Techniques for localizing proteins in complex structures like lambda remain unrefined. It is also difficult to study the conformation adopted by the DNA in nucleo-protein structures. Since bacteriophage lambda has only two major capsid proteins and a single, small DNA molecule, it serves as a convenient and simple model for developing techniques for studying the proximity of proteins to particular sequences in a DNA molecule.

1 Department of Genetics and Development, Columbia University College of Physicians and Surgeons, 701 West 168th Street, New York, NY 10032, USA

Bacterial Chromatin
(ed. by C.O. Gualerzi and C.L. Pon)

2 General Morphology of Lambda

Electron microscopy of bacteriophage lambda has defined its external dimensions. It consists of an icosahedral protein capsid of approximately 600 Å diameter and a tail of about 1500 Å in length and 90 Å in diameter (Kellenberger and Edgar 1971). The thickness of the capsid wall is about 40 Å (Lickfeld et al. 1976). Analysis of urea-treated heads revealed the presence of one major structural protein, gpE (Buchwald et al. 1970). The gpD has been identified as the second major capsid protein (Casjens and Hendrix 1974). All of the remaining proteins in the structure apparently derive from either the tail or the head-tail connector.

The major structural component of the lambda capsid is gpE, a 38-kD protein present in 420 copies per phage. As discussed below, gpE assembles with the help of several other proteins to form a sacklike structure known as the prohead. At one position on the prohead is a morphologically distinct annulus, which is important for initiation of assembly of the prohead and which later becomes the site of tail attachment. During phage morphogenesis, the prohead undergoes several protein fusion-cleavage and cleavage events, packages a unit length genome and undergoes a conformational change that leads to the characteristic icosahedral shape. In later sections, we discuss some of the details of these changes.

The gpD (11 kD, 420 copies/phage) is the second major protein of the phage capsid, and it associates with the capsid during or after the DNA packaging reaction.

Packaging of bacteriophage lambda DNA into phage requires the preformation of mature, packaging-competent proheads (Hohn and Hohn 1974; Kaiser et al. 1975). Prohead assembly has been studied extensively by complementation analysis and studies of morphologically aberrant mutant phage. Since the formation of this structure is a prerequisite for the packaging of phage DNA, we will briefly consider this assembly process.

3 Structure and Morphogenesis of the Protein Capsid

Early studies of bacteriophage lambda morphogenesis showed that packaging proceeds through preformation of proheads rather than through mechanisms that involve concerted protein-DNA condensation. Though empty phagelike heads can be seen in small numbers in wild type infected cells, it was not clear if these structures were true precursors of mature phage or simply phage that had been lysed and had lost their DNA during the preparation. Hohn and Hohn (1974) and Kaiser et al. (1975) showed that these empty phagelike particles were competent to package DNA in in vitro packaging systems. The gpA^- mutants which accumulate proheads can be complemented in vitro by mutants that serve as a source of intact gpA. This experiment showed that proheads occur prior to the function of gpA in the morphogenic pathway, and are not aberrant structures.

The formation of proheads is a complex process that is not fully understood, but some general mechanistic features have emerged. Of primary importance in studying the mechanism of assembly is identification of proteins involved in both the final structure and those involved transiently in its formation. Morphogenesis of the capsid involves the interactions of eleven genes, six of which are required for the formation of the mature prohead (i.e., proheads that are ready to package DNA), and five of which are required for DNA packaging and completion of the mature phage head. First we will consider the composition and arrangement of subunits in mature proheads, then we will address the functions of the transient factors during assembly.

3.1 Protein Components of the Capsid

Mature proheads are composed primarily of the gene product gpE, prsent in 420 copies, the same number as in the mature phage. There are several minor structural components in both the prohead and the mature virus. The gpB, a 62 kD protein (2–4 copies) and pB*, a cleavage product of gpB with a molecular weight of 56 kD (10–12 copies) have been localized at the proximal vertex of the prohead where the tail eventually becomes attached. The gpB and pB* have been shown to be components of the so-called preconnector (Kochan and Murialdo 1983; Kochan et al. 1984), which is a 15-nm diameter disk of 13 nm thickness, having a 2.3-nm hole passing through the center. The DNA probably passes through the hole during packaging and infection. By optical filtration of electron micrographs of periodic arrays of gpB polymers, it was found that the gpB polymer has 12-fold rotational symmetry, and from the side resembles an assymmetric dumbbell. The groove that gives the structure its dumbbell appearance may be the site where the capsid attaches to the connector. Other components of the prohead are pX1 and pX2, 31 and 29 kD, respectively (six copies each). These two proteins are fusion-cleavage products of gpC and gpE. Their locations in the structure are still uncertain, but they may form an annulus which attaches the major capsid structure containing gpE to the preconnector.

All of the components of the mature prohead are also found in the mature head, in addition to at least two and probably three other proteins. The second major capsid protin, gpD, an 11 kD protein present in 420 copies, is thought to stabilize the capsid by binding to interstices left in the gpE matrix. The gpD deposition is not an obligatory step in packaging since amber mutants in D are viable, although less so than wild-type phage (Sternberg and Weisberg 1977b).

Two events involving proteins gpW and gpFII occur after packaging of phage DNA. The molecular nature of these events is completely unknown, but complementation experiments have shown that gpW must act on the capsid before gpFII acts or a biologically inactive capsid results. Furthermore, gpFII is found in the completed head. Purified FII^- heads can be activated by W^- extracts, but heads from W^- extracts cannot be complemented by F^+ extracts (Casjens et al. 1972). The gpW may be a component of the completed head, but this has not been demonstrated directly. After the action of these two gene products, the capsid is complete and ready for the attachment of tails.

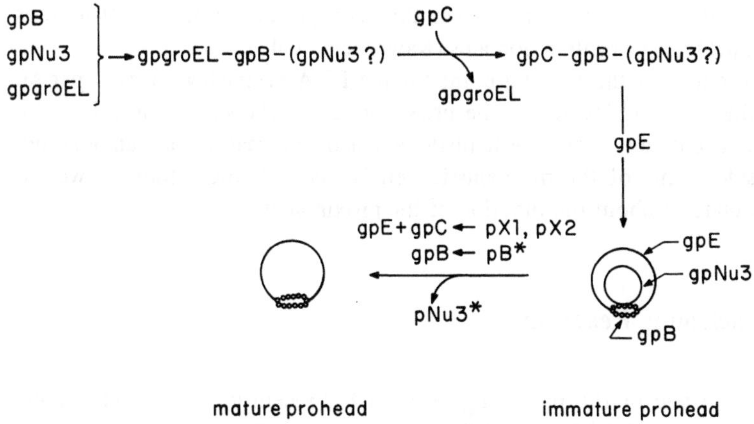

Fig. 1. Outline of the bacteriophage lambda assembly

3.2 Capsid Assembly

The morphogenesis of the mature prohead has been studied by complementation and electron micrographic analysis of mutants and proceeds through a structural intermediate known as the immature prohead. The assembly of the immature capsid is outlined schematically in Fig. 1. The first step involves interaction of gpB, gpNu3, and the host encoded protein gpgroEL. These presumably form a complex containing gpgroEL, gpB, and possibly gpNu3 (although the physical interaction of gpNu3 has not been directly demonstrated). It is thought that gpC interacts with this structure, while gpgroEL detaches, resulting in a structure containing gpB, gpC, and possibly gpNu3. This complex may serve as a nucleus for the proper assembly of the major capsid protein gpE (Murialdo and Becker 1978). Mutants in gpB, gpC, gpgroEL, and gpNu3 all form aberrant structures such as polyheads, monsters and aberrant proheads, implicating the essential role of the initiation complex in proper assembly.

The immature prohead is formed by condensation of gpE with the initiation complex in a reaction that depends on either the prior or coordinate assembly of gpNu3 into a scaffold structure. In the electron microscope, the prohead appears as a roughly spherical structure containing a protein scaffold core. This scaffold is composed of about 70 copies of gpNu3 and it apparently facilitates the efficient organization of gpE into the structure. Maturation of the immature prohead to form the mature prohead involves the loss of the scaffold as well as cleavage of gpB to form pB*, and the fusion-cleavage of gpC and several copies of gpE to form pX1 and pX2. Fewer than five copies of gpNu3 remain in the mature prohead.

During maturation, the prohead undergoes a structural change from the superficially amorphous spherical structure to the familiar icosahedral structure. This structural transition has been studied by circular dichroism which shows that it is not accompanied by large changes in the alpha helix or beta sheet structure of the proteins. However, one or two tyrosine residues move from a polar to a nonpolar environment during the expansion. This suggests that the microscopic structural transitions leading to the change to an icosahedral structure occur through reorganization of intermolecular or interdomain interactions, rather than changes in the internal structures of the individual proteins or protein domains (Kawaguchi et al. 1983).

The mature prohead is the starting point for the DNA packaging reaction per se. It is clear that the structure located at the proximal vertex plays a crucial role in the packaging reaction, but few of the mechanistic details of this reaction are understood. Below, we consider some of the mechanistic details that are understood, as well as some of the speculations about the function of the proximal vertex.

4 The DNA Packaging Reaction

A comprehensive review of the packaging reaction has recently appeared (Feiss and Becker 1983). Here, we highlight some of the most important points.

Bacteriophage lambda is packaged from long linear concatemers that are produced by rolling circle replication (Smith and Skalka 1966). The mechanism of packaging involves the specific recognition of sites at the two extremes of the full-length genome, called cos (for "cohesive" end) sites, by a phage-encoded factor known as the terminase function (Emmons 1974; Feiss and Campbell 1974). Initially, terminase recognizes and binds to cos sites randomly along the linear concatemer. This complex then binds to proheads at the proximal vertex to begin packaging (Feiss and Bublitz 1975). Packaging of the DNA proceeds directionally from the left toward the right hand end of the lambda genetic map, and terminates when a second cos site is encountered. After encapsidation of a full length genome, the terminase endonucleolytically cleaves at the cos site. Two to three consecutive full length genomes from the concatemer are packaged processively after the first terminase-cos recognition event (Feiss and Bublitz 1975). The basis for this processiveness is not known. Interestingly, linear concatemers, linear monomers, and circular multimers can be packaged but circular monomers cannot (Enquist and Skalda 1973).

The likelihood of terminase cutting at the second cos site is determined by the length of the genome. For example, if consecutive cos sites are separated by 77% of the normal full-length genome, the efficiency of cutting at the second site is only 75% of that for the wild type arrangement, whereas for a separation of 90%, the packaging efficiency goes up to 100%.

There are several experiments that suggest that packaging occurs directionally from the initial cos site toward the right-hand end of the genome. Hohn (1975) has shown that only the left-hand-most restriction fragment from the lambda genome is packaged in an in vitro packaging extract. Also, when this in vitro packaging extract is presented with a restriction fragment containing an intact cos site, only the sequences to the right of the cos site are packaged. Additional evidence for the directionality of lambda DNA packaging comes from particles known as doc L and doc R (Sternberg and Weisberg 1977a). Doc particles arise from the packaging of a lambda genome that has integrated into the host genome. If cutting at the second cos site is imprecise, it is possible for the particle to pick up a host marker. Doc L particles, which contain a host marker proximal to the right-hand end of the integrated genome, occur 100 times more frequently than doc R particles. It is thought that doc L particles initiate packaging at the left-hand cos site, and proceed through the right-most cos site into the host genome, thereby suggesting a polarity of packaging.

The terminase function appears to correlate with two proteins, Nul and gpA, that are coded for at the extreme left-hand end of the lambda genome (Echols and Murialdo 1978). Nul and gpA mutants are phenotypically identical. These proteins may interact with the cos site as hetero-oligomers. In fact, highly purified terminase preparations complement both Nul^- and A^- mutations, indicating that the proteins copurify under a variety of chromatographic conditions and further suggesting that they may interact strongly (see Feiss and Becker 1983). The gpA has been shown to interact with DNA in vitro, where it places double stranded cuts into lambda DNA at the cos sites (Becker et al. 1977). Sternberg and Weisberg (1977a) have shown that gpA is important in the sequence specificity of DNA packaging. Occasionally, the packaging machinery of lambda erroneously packages bacterial DNA. Sternberg and Weisberg showed that missense mutations in gpA resulted in a change in the ratio of lambda DNA to bacterial DNA packaged. Missense mutations in other proteins did not affect this ratio.

Several features of the terminase-cos interaction have been sorted out. The stoichiometry of interaction appears to be three to four terminase protomers per cos site for packaging concatemeric DNA, and half that number for packaging a linear monomer, suggesting that two protomers are involved in cutting per se, and two are involved in the actual packaging reaction (Hohn 1975; Becker et al. 1977; Feiss and Becker 1983). The binding of terminase to DNA, and its endonucleolytic activity, are ATP requiring processes, and result in ATP hydrolysis. Binding and cutting in vitro also depend on magnesium, spermidine and a 22 kD *E. coli* basic protein. These are also among the requirements for in vitro lambda packaging systems.

Mechanistically, the packaging of the lambda genome is not understood. It is possible that the proteins that comprise the terminase function are able to translocate along the DNA in an ATP-dependent manner. If this is true, it might serve as a molecular pump, holding on to the capsid and feeding in the DNA (Becker et al. 1977). Polarity would then be established during the initial terminase-cos recognition event. Other proteins at the proximal vertex are also candidates for such a molecular pump and some interesting models have been suggested. One model suggests that the symmetry mismatch between the proximal vertex and the head-tail connector would result in weak interactions, so that the connector can rotate with respect to the head (Hendrix 1978). If such rotation occurs and is ATP-dependent, models can be envisioned where rotation "screws" the DNA into the head.

Polyamines and other counterions are clearly important for DNA packaging into bacteriophage, since polyelectrolyte repulsion must be overcome (Earnshaw and Casjens 1980). Some have suggested that polycation-induced condensation is the major mechanistic feature of packaging, and this idea is supported by the fact that polyamine-condensed DNA has about the same density as packaged DNA (Laemmli 1975; Lerman 1973). The major objection to this view is that once condensed by polycations a mechanism must be found for decondensation upon infection. One possible mechanism would involve encapsidated polyamines diffusing out of the phage head in response to the lower concentration of polyamines outside the cell, allowing decondensation during ejection (Laemmli 1970).

Another model for packaging suggests that prohead expansion is mechanistically responsible for internalizing the DNA (Hohn and Hohn 1974; Serwer 1975). This expansion is 20% in diameter, corresponding to a 100% increase in volume, and occurs sometime after packaging has begun. This expansion is thought to result from a change in the array of protein-protein contacts among the capsid proteins (Kawaguchi et al. 1983). Expansion only occurs, however when a sufficiently large DNA molecule is packaged. For example, the packaging of a molecule of 11.4% total genome length does not result in prohead expansion, whereas a molecule of 44.5% length does promote expansion (see Earnshaw and Casjens 1980). Therefore, expansion per se cannot be mechanistically responsible for the entire packaging reaction.

5 The Structure of Packaged DNA

Lambda DNA is 48.5 kb long, and is located almost entirely within the capsid. It is a linear, noncircularly permuted molecule with twelve base pair cohesive ends (Hershey et al. 1963). The right-hand end of the molecule is inserted into the tail of the phage (Thomas 1974; Chattoraj and Inman 1974). The location of the left-hand end is unknown.

One of the most remarkable features of bacteriophage lambda structure is the apparent economy of DNA packaging. The lambda capsid of about 540 Å inside diameter would have a volume of 8.2×10^7 Å^3, while the 48.5 kb DNA molecule has a calculated volume of about 8.1×10^7 Å^3. Therefore, virtually 100% of the internal volume of the phage should be occupied by DNA. A more precise estimate of the fraction of the volume of the head taken up by DNA is difficult to make. Thin sections of lambda examined by electron microscopy reveal that not all of the capsid volume is occupied by DNA (Cummings et al. 1965). This extra volume appears as a "hole" in the center of the phage head. One argument for this geometry is that the DNA will tend to form the largest circle possible to minimize torsional strain, thereby avoiding the center of the particle and expanding out to the interior walls (Earnshaw and Harrison 1977).

Despite these uncertainties, it is clear that lambda DNA is packaged with great economy, and this fact imposes several constraints on models of the structure of the packaged DNA one might construct. The maximum estimate for the area of the internal wall of the capsid with an interior diameter 540 Å is 9.2×10^5 Å^2. Since DNA has a diameter of about 25 Å, a uniform layer of DNA against this sruface would define a sphere of diameter 515 Å with area, 8.0×10^5 Å^2. This corresponds to 9×10^3 base pairs that can be accommodated in the first packing shell. It follows that 80% of the DNA molecule is excluded from contact with the protein walls of the capsid. The fraction of DNA in contact with the capsid walls has been determined by low angle X-ray scattering for morphologically similar bacteriophage T4, P22, and 29 to be 16%, 24%, and 35–40% respectively (Earnshaw et al. 1976; Earnshaw and Harrison 1977; Subirana et al. 1979). The correct model for the structure of packaged phage DNA would describe which sequences are in contact with the capsid and which sequences are more internal.

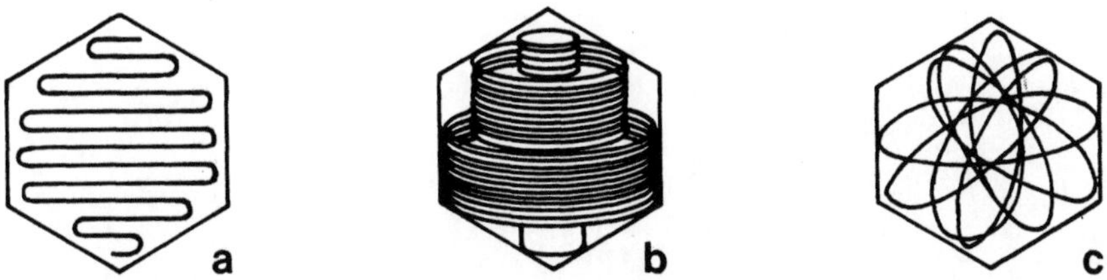

Fig. 2a–c. Possible DNA packaging arrangements

How is the DNA arranged within the capsid? X-ray diffraction studies on lambda and related viruses P22, T2, and T7 reveal a high degree of long range order in the packaged DNA (Earnshaw and Harrison 1977; North and Rich 1961). Low angle X-ray diffraction from unoriented samples of lambda and P22 indicates that the DNA must be packed in parallel arrays in hexagonal closest packing with an apparent domain size of 125 Å. The apparent domain size is essentially a measurement of the number of layers of DNA contributing to a well-defined diffraction pattern. A 125 Å domain size suggests about five layers of well-ordered DNA. For wild-type lambda, the interhelical distance is 27.4 Å, and the DNA appears to be in normal B-form. Furthermore, the diffraction at about 25 Å is modulated in intensity at intervals of approximately 1/540 Å, the reciprocal of the internal capsid diameter. This, it is argued, indicates that ordered DNA segments on one side of the phage must be diffracting with fixed phase relative to ordered segments on the other side of the phage, indicating long-range order over the entire phage. It is interesting that this order persists even though the capsid may not be entirely filled with DNA (Earnshaw and Harrison 1977).

Without additional information, the X-ray diffraction results cannot be synthesized into a unique model for the packing arrangement. In fact, there are a great many candidate models, all of which have parallel arrays of DNA. For example, the DNA could be arranged such that strands make sharp turns and fold back on themselves (Fig. 2a). Alternatively, Fig. 2b illustrates a coaxial spool model, where the DNA winds in concentric circles forming layers of large radius near the capsid and circles of progressively smaller radius approaching the center.

Richards et al. (1973) examined electron micrographs of several bacteriophage prepared by drying in thin films of negative stain. These phage were partially disrupted and flattened on the grid, and the DNA was observed in various degrees of liberation. In all cases, the strands of DNA traced out circular patterns. In favorable cases, concentric circles or tight spirals were indicated. These authors suggested models corresponding to those illustrated in Fig. 2b and c.

The X-ray diffraction data of Earnshaw and Harrison and those of North and Rich seem to be inconsistent with the model in Fig. 2c. In this model, the DNA is wrapped concentrically without parallel disposition of the strands, that is, it is not hexagonally packed. Since parallel packing of the strands is indicated by the diffraction data, Earnshaw and Harrison preferred the coaxial spool model.

In some cases, structural insight can be gained through an understanding of biological function. For example, since bacteriophages inject their DNA through a small hole in the capsid or through a narrow channel in the tail, as is the case with lambda,

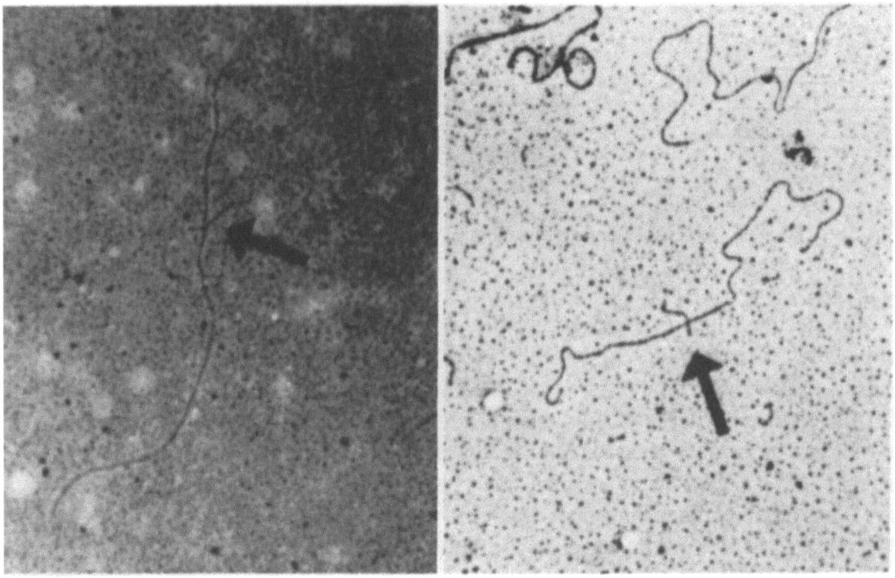

Fig. 3. Bis-psoralen cross-linking of bacteriophage lambda DNA. (R. Haas, Thesis, Columbia University, 1981)

the DNA molecule must remain linear during injection. Structures that are knotted might be difficult to resolve into linear structures fast enough to inject, and therefore, models for intraphage DNA packing are biased against knotted structures (however, see Liu et al. 1981). Another example is that the concentric winding models generally assume that the DNA which is in contact with the capsid walls exits the head last, thereby minimizing the amount of motion the entire mass of DNA must go through to unwind. In these models, DNA exits from the center of the particle first. Finally, since the ultimate goal of the bacteriophage is to eject its DNA, interaction between the calsid and the DNA must not be very great compared to the energy released upon DNA expulsion.

Structural models for intraphage DNA have been tested by DNA-DNA cross-linking studies. Haas et al. (1982) used a bis-psoralen to fix points of contact between different points on the DNA while packaged inside the capsid. By then purifying the DNA and electron microscopically examining the "X's" which result when the cross-linked molecules were digested with a restriction endonuclease (Fig. 3), they found that each restriction fragment in a Bgl II digest could be cross-linked to any other fragment. Furthermore, the frequency of cross-linking between fragments was apparently largely target size dependent. In the simpler versions of the coaxial spool and solenoid models, each DNA layer is composed of a specific portion of the genome, such that the same sequences would be neighbors from phage to phage. In these models, cross-links should only be possible between defined pairs of sequences. Since cross-links can occur between any two restriction fragments, and in an apparently random pattern, these experiments rule out the simple solenoid and coaxial spool models. They also rule out any model that suggests a bias in the nearest neighbors of any fragment favoring any

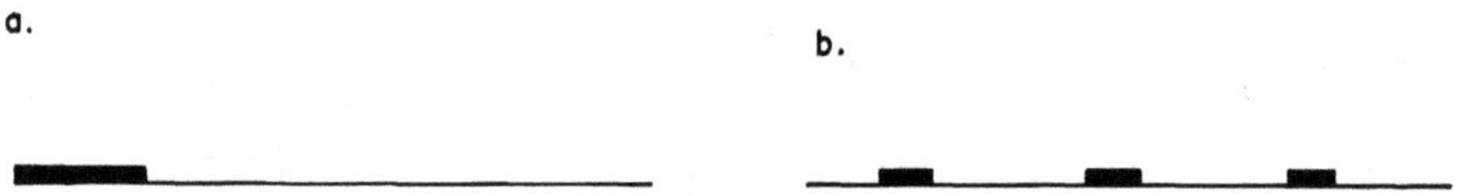

Fig. 4a,b. Array of contacts between the lambda genome and its capsid predicted from different packaging arrangements

other fragment. The cross-linking behavior of one of the fragments, however, suggested that the packaging was not totally random. Subsequent bis-cross-linking studies using a cleavable cross-linker and diagonal electrophoresis supported this finding (Welsh and Cantor, manuscript in prep.). While these bis-cross-linking studies seemed to be inconsistent with all of the existing models for intraphage DNA packaging, the data did not suggest a unique alternative.

Current models for bacteriophage structure make predictions as to the nature of the contacts between the genome and the capsid protein. The coaxial spool model, for example, predits that the left-hand end of the lambda genome is in contact with the protein, to the exclusion of all other parts of the DNA (Fig. 4a). The solenoid model depicted in Fig. 2d predicts contacts such as those mapped in Fig. 4b.

Two laboratories have addressed the structure of lambda using protein-DNA cross-linking techniques. One approach to protein-DNA corss-linking in bacteriophage lambda was to incorporate BrdU into the phage DNA, irradiate to produce cross-links, and analyze the positions of the cross-links by lysing the phage and looking for the position of the capsid ghost along the DNA by electron microscopy (Widom and Baldwin 1983). These results also show reactivity along the entire length of the genome, but the data was folded back on itself since the two different DNA ends could not be distinguished in the electron micrographs. Examination of these same molecules by density gradient sedimentation revealed that all restriction fragments in a Eco Rl digest were represented in the cross-linked fraction.

In another approach, cross-links were formed between the capsid and the DNA by using a bifunctional psoralen derivative having a succinimide ester moiety. Psoralen was tethered to the capsid via reaction of capsid lysines with the succinimide easter, and the psoralen was then photoreacted with the DNA by UV irradiation. Only the DNA proximal to the capsid received a covalently bound psoralen. Psoralens can react photochemically with DNA to produce an interstrand cross-link, and these cross-links are easily located. In these experiments, psoralen cross-links were found throughout the lambda genome in frequencies that suggested random protein-DNA contacts (Schwartz et al. 1983; Welsh and Cantor 1984 and manuscript in prep.).

These results favor models in which a constant fraction of each restriction fragment is in contact with the capsid. One possibility is that each phage is packaged differently and every position on the genome has equal probability of contacting the capsid only when the entire ensemble is considered. Alternatively, the packaging may be such that every restriction fragment has a length-dependent probability of contacting the capsid within a single phage. This could result if the contact points were closely spaced along the DNA, or if the DNA were free to diffuse within the capsid.

The coaxial spool model represents an extreme in the spectrum of highly ordered, parallel array models, and although it is ruled out by cross-linking experiments, it can

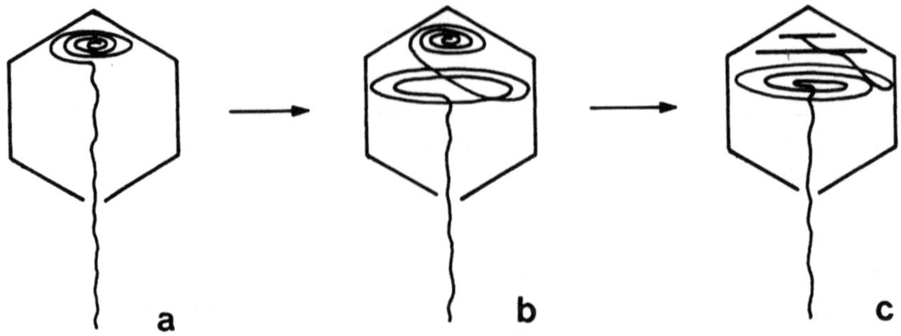

Fig. 5a–c. Stacked disk model for bacteriophage lambda DNA packaging. a Formation of the first disk. b The second disk would stack against the first. c Subsequent disks are predicted to stack against previously formed disks

help to delineate the range of possible parallel array models. Another extreme parallel array model, a stacked disk model can be constructed as shown in Fig. 5. One end of the DNA enters the capsid and forms a spiral disk against the opposite wall. When the first disk is complete, the DNA goes on to form a second disk stacked against the first, and so on until the packaging is complete. In the construction of a disk, as in the construction of a layer in the coaxial spool model, each new loop of DNA is nearest-neighbor to the previous loop, and the immediately following loop.

These models delineate the extreme solenoid-type models, and are inconsistent with DNA-DNA and protein-DNA cross-linking data. DNA-DNA cross-linking data argue against these models, since both models predict preferential cross-linking between certain regions. The coaxial spool is ruled out by protein-DNA cross-linking, since no preference is shown for cross-linking to the capsid by any fragment. The stacked disk model is consistent with protein-DNA cross-linking, but only if special structural criteria are met. The key element in this argument is the periodicity of contact between the DNA and capsid, and how this pattern overlaps the restriction map. If the frequency of DNA-protein contacts is a linear function of the fragment size, the probability of cross-linking any given fragment to the capsid will also be a linear function of fragment size. For example, consider the case where each disk contains 2500 bp. Approximately one-fifth, or 500 bp would be in continuous contact with the capsid. A 20-kb fragment would come into contact with the capsid eight times, for a total of 4000 bp, while a 5-kb fragment would contact the capsid two times, for a total of 1000 bp. This would lead to the same Poisson statistics of cross-linking as a totally random model.

Although the DNA-DNA cross-linking and protein-DNA cross-linking results are largely inconsistent with the coaxial spool and stacked disk models per se, these experiments can be reconciled to a hybrid of the two models, similar to the model proposed by Harrison (1983). In the hybrid model, DNA can randomly partition between the concentric, layered configuration of the coaxial spool model and the stacked disk configuration, with each new wind of DNA adding to a growing layer or disk. Implicit to this model is that each individual phage would have a unique structure.

We emphasize that this is only one example of a specific, highly organized model that would give an apparently random cross-linking pattern. In fact, the constraint of variation by one DNA thickness which occurs when new winds are added to growing layers or disks may not be warranted by the available data. It is possible that the DNA trajectory may jump from one end of the capsid to the other. However, the spontaneous stacking of random sequences into hexagonal arrays has often been observed, and is a reasonable element of any model.

It is not yet possible to argue for one or another packaging model on the basis of purely energetic grounds, because we do not know whether the packaged DNA is in a torsional ground state. If it is, one might postulate a scheme where all of the loop sizes are intermediate in size rather than varying from very large to very tight. Wang and colleagues, and Shore and Baldwin have demonstrated the nonlinear dependence of torsional energy on DNA loop radius, and this becomes an important factor for loops smaller than 100 bp (Horowitz and Wang 1984). The largest possible loop within lambda is about 500 bp.

It is important to keep in mind that intraphage DNA is not in a solid state. It was shown by ^{31}P-NMR that intraphage DNA is restricted in motion relative to DNA in solution, but not as restricted as DNA in the solid state. In another study, Ashikawa et al. (1984) found by fluorescence anisotropy depolarization of intercalated ethidium bromide, that the (presumably) torsional mobility of intraphage DNA reflected the interhelix distance in wild type and deletion mutants of lambda and glycosylation of the bases in T4. Deletion mutants had previously been shown to have greater interhelix distances than wild-type phage by X-ray diffraction (Earnshaw and Harrison 1977).

Some of the most intriguing data concerning the orientation of DNA within bacteriophage comes from linear dichroism and birefringence studies (Hall and Schellman 1982; Kosturko et al. 1979). Both of these methods indicate that the overall orientation of the DNA in bacteriophage is not isotropic. Linear dichroism indicates that the axis of the DNA in bacteriophage lambda is preferentially oriented perpendicularly to the axis of the phage as defined by the line of the tail, whereas the DNA of phage T_4 is oriented parallel to the axis. Hall and Schellman also found that the magnitude of the linear dichroism does not correspond to that predicted by any of the simple solenoid models.

On the basis of these results and arguments, we propose the following model for the packaging of bacteriophage lambda DNA in its capsid. The DNA follows roughly circular to spiral paths, which, on average form hexagonal arrays. Nearest neighbors in a given hexagonal array are not likely to be nearest neighbors in primary sequence, nor are the same subsets of neighbors likely to appear from one phage to the next, as would be predicted by the parallel array models. Disorder of this type may be caused by the DNA axis crossing over itself rather than lying in a perfect parallel array. Such overlaps may not be energetically very costly, or may be trapped kinetically during packaging. Since there is no apparent bias against the right-hand-end cross-linking to the capsid, and since DNA-DNA cross-linking data suggests that there is no progressive stacking of the DNA from the right-hand-end to the left-hand-end, it is unlikely that the right-most part of the DNA is located in the center of the capsid. Thus it is also unlikely that exit of DNA from the capsid during infection proceeds by unraveling of the DNA from the inside of a spool-like structure. Since there is evidence of extra space within the capsid by thin section electron microscopy, the packing and torsional interactions which hold the DNA in hexagonal arrays in the static particle must be small enough to allow sliding of one loop past another during infection. It is not clear whether this type of sliding can occur in the static particle by thermal motion. Because of this uncertainty, it is impossible to assess models wherein structural bias is introduced as a consequence of the packaging mechanism.

References

Ashikawa I, Furuno T, Kinosita K Jr, Ikegami A, Takahashi H, Akutsu H (1984) Internal motion of DNA in bacteriophages. J Biol Chem 259:8338–8344

Becker A, Gold M (1978) Enzymatic breakage of the cohesive end site of phage lambda DNA: terminase (ter) reaction. Proc Natl Acad Sci USA 75:4199–4203

Becker A, Marko M, Gold M (1977) Early events in the packaging of bacteriophage lambda DNA. Virology 78:291–305

Bina M, Ng S-C, Blasquez V (1983) Simian virus 40 chromatin interaction with capsid proteins. J Biomolecular Structure and Dynamics 1:689–704

Buchwald M, Murialdo H, Siminovitch L (1970) The morphogenesis of bacteriophage lambda. II. Identification of the principal structural proteins. Virology 42:390–400

Casjens S (1973) Bacteriophage lambda FII gene protein: role in head assembly. J Mol Biol 90: 1–23

Casjens S, Hendrix RW (1974) Locations and amounts of the major structural proteins in bacteriophage lambda. J Mol Biol 88:535–545

Casjens S, Hohn T, Kaiser AD (1972) Head assembly steps controlled by genes F and W in bacteriophage lambda. J Mol Biol 64:551–563

Chattoraj D, Inman RB (1974) Location of DNA ends in P2, 186, P4, and lambda bacteriophage heads. J Mol Biol 87:11–22

Cummings DJ, Chapman VA, Delong SS (1965) An electron microscopic study of lambda and lambda dg bacteriophage in thin sections. J Mol Biol 14:418–422

Earnshaw WC, Casjens SR (1980) DNA packaging by the double-standed DNA bacteriophages. Cell 21:319–331

Earnshaw WC, Harrison SC (1977) DNA arrangement in isometric phage heads. Nature 268:598–602

Earnshaw WC, Casjens S, Harrison SC (1976) Assembly of the head of bacteriophage P22: X-ray diffraction from heads, proheads and related structures. J Mol Biol 104:387–410

Echols H, Murialdo H (1978) Genetic map of bacteriophage lambda. Microbiol Rev 42:577–591

Emmons SW (1974) Bacteriophage lambda derivatives carrying two copies of the cohesive end site. J Mol Biol 83:511–526

Enquist LW, Skalka A (1973) Replication of bacteriophage lambda DNA dependent on the function of host and viral genes. I. Interaction of red, gam and rec. J Mol Biol 75:185–212

Feiss M, Becker A (1983) DNA packaging and cutting. In: Hendrix RW, Roberts JW, Stahl FW, Weisberg RA (eds) Lambda II. Cold Spring Harbor Laboratory, Cold Spring Harbor, New York, pp 305–330

Feiss M, Bublitz A (1975) Polarized packaging of bacteriophage lambda chromosomes. J Mol Biol 94:583–594

Feiss M, Campbell A (1974) Duplication of the bacteriophage lambda cohesive end site: genetic studies. J Mol Biol 83:527–540

Haas R, Murphy RF, Cantor CR (1982) Testing models of the arrangement of DNA inside bacteriophage lambda by crosslinking the packaged DNA. J Mol Biol 159:71–92

Hall SB, Schellman JA (1982) Flow dichroism of capsid DNA phages. II. Effect of DNA deletions and intercalating dyes. Biopolymers 21:2011–2031

Harrison SJ (1983) Packaging of DNA into bacteriophage heads: A model. J Mol Biol 171:577–580

Hendrix RW (1978) Symmetry mismatch and DNA packaging in large bacteriophages. Proc Natl Acad Sci USA 75:4779–4793

Hendrix RW, Casjens SR (1975) Assembly of bacteriophage lambda heads: Protein processing and its genetic control in petit lambda assembly. J Mol Biol 91:187–199

Hershey AD, Burgi E, Ingraham L (1963) Cohesion of DNA molecules from phage lambda. Proc Natl Acad Sci USA 49:748–755

Hohn B (1975) DNA as a substrate for packaging into bacteriophage lambda in vitro. J Mol Biol 98:93–106

Hohn B, Hohn T (1974) Activity of empty, headlike particles for packaging of DNA of bacteriophage lamda in vitro. Proc Natl Acad Sci USA 71:2372–2376

Horowitz DS, Wang JC (1984) Torsional rigidity of DNA and length dependence of the free energy of DNA supercoiling. J Mol Biol 173:75–91

Kaiser AD, Syvanen M, Masuda T (1974) Processing and assembly of the head of bacteriophage lambda. J Supramol Structure 2:318–328

Kaiser D, Syvanen M, Masuda T (1975) DNA packaging steps in bacteriophage lambda head assembly. J Mol Biol 91:175–186

Kawaguchi K, Noda H, Katsura I (1983) Structure and inherent properties of the bacteriophage lambda head shell III. Spectroscopic studies on the expansion of the prohead. J Mol Biol 164: 573–587

Kellenberger E, Edgar R (1971) Structure and assembly of phage particles. In: Hershey AD (ed) The bacteriophage lambda. Cold Spring Harbor Laboratory, Cold Spring Harbor, New York, p 271

Kochan J, Murialdo H (1983) Early intermediates in bacteriophage lambda prohead assembly II. Identification of biologically active intermediates. Virology 131:100–115

Kochan J, Carrascosa JL, Murialdo H (1984) Bacteriophage lambda preconnectors: purification and structure. J Mol Biol 174:433–447

Kosturko LD, Hogan M, Dattagupta N (1979) Structure of DNA within three isometric bacteriophages. Cell 16(3):515–522

Laemmli UK (1970) Cleavage of structural proteins during the assembly of the head of bacteriophage T4. Nature 227:680–685

Laemmli UK (1975) Characterization of DNA condensates induced by poly(ethylene oxide) and polylysine. Proc Natl Acad Sci USA 72:4288–4292

Lerman L (1973) Cold Spring Harbor Symp. Quant Biol 38:59

Lewin B (1980) In: Gene expression 2. Eucaryotic Chromosomes, 2nd ed. John Wiley and Sons, New York

Lickfeld KG, Menge B, Hohn B, Hohn T (1976) Morphogenesis of bacteriophage lambda: electron microscopy of thin sections. J Mol Biol 103:299–318

Liu LF, Perkocha L, Calendar R, Wang JC (1981) Knotted DNA from bacteriophage capsids. Proc Natl Acad Sci USA 78:5498–5502

Murialdo H, Becker A (1978) Head morphogenesis of complex double-stranded deoxyribonucleic acid bacteriophages. Microbiol Rev 42:530–575

Murialdo H, Ray PN (1975) Model for arrangement of minor structural proteins in head of bacteriophage lambda. Nature 257:815–817

North ACT, Rich A (1961) X-ray diffraction studies of bacterial viruses. Nature 191:1242–1245

Richards KE, Williams RC, Calendar R (1973) Mode of DNA packaging within bacteriophage heads. J Mol Biol 78:255–259

Schwartz DC, Saffran W, Welsh J, Haas R, Goldenberg M, Cantor CR (1983) New techniques for purifying large DNAs and studying their properties and packaging. Cold Spring Harbor Quant Biol 47:189–195

Serwer P (1975) Bouyant density sedimentation of macromolecules in sodium iothalamate density gradients. J Mol Biol 92:433–448

Smith M, Skalka A (1966) Some properties of DNA from phage infected bacteria. In: Proceedings of a symposium on macromolecular metabolism. Little, Brown and Company, Boston, Mass, pp 127–142

Sternberg N, Weisberg R (1977a) Packaging of coliphage lambda DNA. I. The role of the cohesive end site and the A protein. J Mol Biol 117:717–731

Sternberg N, Weisberg R (1977b) Packaging of coliphage lambda DNA. II. The role of the gene D protein. J Mol Biol 177:733–759

Subirana JA, Lloveras J, Lombardero M, Vinuela E (1979) X-ray scattering of the non-isometric Bacillus subtilis phage φ29. J Mol Biol 128(1):101–106

Thomas JO (1974) Chemical linkage of the tail to the right-hand end of bacteriophage lambda DNA. J Mol Biol 87:1–9

Welsh J, Cantor CR (1984) Protein-DNA crosslinking. Trends in Biochem 9(12):505–508

Widom J, Waldwin RL (1983) Tests of spool models for DNA packaging in phage lambda. J Mol Biol 171:419–437

The Association of the *Escherichia coli* Chromosome with the Cell Membrane

G.B. OGDEN and M. SCHAECHTER [1]

The notion that the bacterial chromosome is attached to the membrane is by now firmly embedded in the folklore of bacterial structure. It is derived from a series of experiments on the segregation of the *Escherichia coli* chromosome and its plasmids, which Jacob et al. (1963) used for their original postulation of the replicon model. Unfortunately, in the subsequent years this notion remains more compelling for its logical appeal than for the rigorousness of the experimental evidence. In any case, this aspect of the replicon model has not contributed very much to our understanding of genome segregation or of the regulation of its synthesis. Despite this, no one so far has presented a serious alternative for the physical basis for chromosome segregation.

We will review here the salient work done in this field and some of the reasons why progress has been slow. We will also present some of the considerable work done on the partition of bacterial plasmids, since we believe that these experiments represent the most useful models for understanding the association of bacterial replicons with the membrane.

The facts can be summarized as follows: Fractionation of *E. coli, Bacillus subtilis,* or several other bacterial species leads to the isolation of a subcellular complex containing both DNA and membrane (see Moyer 1979; Ogden and Schaechter 1985, for references). The particulars depend on the methods used. If the cells are opened by a harsh method, such as sonication or French pressure cell treatment, the chromosome will be sheared and only a small proportion of the DNA will be found in the membrane fraction. On the other hand, if the cells are opened gently by lysis of spheroplasts, virtually all the DNA is found associated with membrane material. Two methods in the latter class have been used extensively. One consists of preparing so-called membrane-associated nucleoids, that is, folded DNA obtained with concentrations of detergent that solubilize only part of the membrane (Drlica et al. 1978). The other relies on the afinity of membranes for hydrophobic crystals of magnesium-sarkosyl to obtain a DNA-membrane fraction (the so-called M-band; Tremblay et al. 1969). More generally, the repertoire of separation techniques encompasses velocity centrifugation, equilibrium centrifugation, and phase partition (Leibowitz and Schaechter 1975). In each case, DNA added in vitro to controls does not bind to the membranes. In several instances, it has been shown that fragments of chromosomal DNA generated

1 Department of Molecular Biology and Microbiology, Tufts University, Schools of Medicine, Veterinary Medicine and Dental Medicine, 136 Harrison Avenue, Boston, MA 02111, USA

Bacterial Chromatin
(ed. by C.O. Gualerzi and C.L. Pon)

in vivo or in vitro do not attach to the membrane (e.g., Ivarie and Pene 1973; Dworsky and Schaechter 1973; Parks et al. 1981; Sargent et al. 1983). Thus, these controls make it seem reasonable that the association of the chromosome to the membrane seen in cell fractionation is not due to trivial artifacts. Unfortunately, this is not the same as showing that this association exists in vivo, even in the absence of evidence to the contrary.

Why is this considerable body of data not more definitive? The most obvious reason is that cell fractionation studies must rely on independent confirmation obtained with different methods. In this field, however, we cannot invoke suitable morphological or biochemical studies. More importantly, none of the genetic work done on chromosome metabolism in bacteria is particularly relevant. Genetic studies done with the chromosome are of little value because we know too little about the nature of the DNA-membrane attachment to predict the phenotype of relevant mutants and to work with relevant proteins. Luckily, some work has been carried out with plasmids in this connection, and therein lies the current hope for a genetic approach.

The most useful information on the attachment of the chromosome to the membrane deals with the specificity of the interactions involved. Several authors have reported that the origin of chromosomal replication is involved in this attachment (see Leibowitz and Schaechter 1975, for older references). However, this region is not unique in this respect and other portions of the chromosome are also found bound to the membrane. Their number can be estimated by a combination of in vivo DNA scissions and cell fractionation. The number of attachment sites has been found to vary between 20 and 80 per genome (Abe et al. 1977; Drlica et al. 1978; Dworsky and Schaechter 1973; Parks et al. 1981; Sargent et al. 1983). The sites of attachment that do not involve the origin seem to be randomly distributed on the chromosome (Drlica et al. 1978), vaguely suggesting that the chromosome may move along the membrane. These sites are sensitive to rifampin (Dworsky and Schaechter 1973), and in cells treated with this inhibitor of RNA polymerase the chromosome appears to be bound to the membrane only at its origin. In addition, these sites are selectively cleaved in vitro by single-strand specific nucleases (Abe et al. 1977). These facts are open to various interpretations, but they suggest that the chromosome is attached to the membrane for reasons other than genome segregation. It should be pointed out that the number of attachment sites reported for *E. coli* is similar to the number of supercoiled domains (Worcel and Burgi 1972). However, this relationship has not been studied further.

In recent years we have found it more constructive to turn away from cell fractionation techniques and to attempt in vitro reconstruction experiments (Hendrickson et al. 1982; Kusano et al. 1984). A similar approach has proven useful with plasmids in *B. subtilis* (Tanaka and Sueoka 1983). In *E. coli* we have shown that fragments carrying the origin of DNA replication bind to the membrane with startling selectivity. The origin of replication possesses a unique degree of affinity for the membrane and, under appropriate conditions, is the only fragment that binds to the membrane (Fig. 1). In order to demonstrate this binding, nonspecific sites on the membrane must be saturated with calf thymus or other heterologous competitor DNA. The reaction requires divalent cations, especially magnesium. Components of this binding reaction appear to be sensitive to steps in fractionation and to be labile on storage. Binding takes place

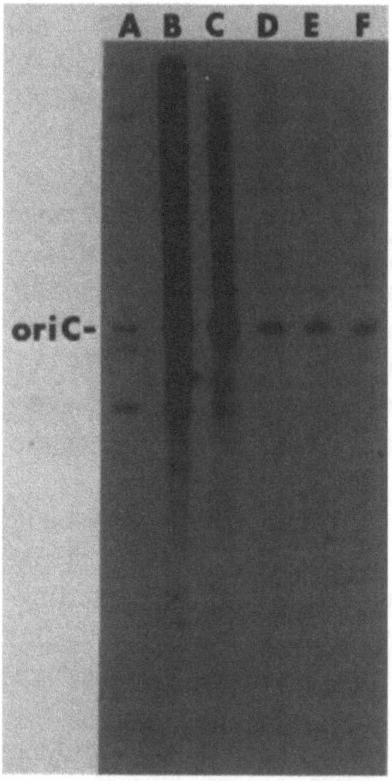

Fig. 1. Filter binding of *Ava*I restriction fragments from the entire *E. coli* chromosome. *E. coli* strain pC2 (*dna*C^{ts}) cells were lysed by passage through a French pressure cell and the released DNA was purified by sucrose velocity and then CsCl density centrifugation (the DNA banded at a density of 1.70–1.65 g cm^{-2}). An outer membrane fraction which separated from the DNA during the CsCl centrifugation was dialyzed separately from the DNA (for details see Hendrickson et al. 1982). The DNA was restricted with *Ava*I and its 5′ termini labeled. It was then incubated for 30 min at 4°C with 5 μg of protein equivalents of outer membrane in the presence of different amounts of calf thymus DNA. The DNA and membrane were passed through a nitrocellulose filter and the bound DNA was eluted and applied to a 5% acrylamide gel and subjected to electrophoresis (according to the procedure of Hendrickson et al. 1982). *Lane A*, pCM959 DNA cut with *Ava*I and end labeled (463 bp *Ava*I fragment containing *oriC* is indicated). *Lanes B–F:* Filter binding in the presence of calf thymus DNA: (*B*) 0, (*C*) 5, (*D*) 25, (*E*) 50, (*F*) 75 μg ml^{-1}, respectively. (Photograph kindly provided by Dr. Zvi Gross)

at two sites in the *oriC* region; in the accepted notation for this region, one site includes bases +91 to +151, the other includes bases +151 to +253. The apparent affinity of this DNA for the membrane is due to the presence of two proteins (of molecular weight 55 kD and 75 kD). These proteins were found fortuitously because they remain attached to DNA even after centrifugation through cesium chloride. They are the only proteins in *E. coli* with these properties. At present, we do not know how these proteins function. Analogous proteins have been found bound to the DNA at the nuclear matrix attachment sites of higher cells (Bodnar et al. 1983). Others have found that a membrane protein, called B′, binds specifically to single-stranded origin DNA from *E. coli* (Jacq et al. 1983). It seems conceivable, therefore, that the role of the "CsCl-resistant proteins" is to distort the DNA to allow it to bind to specific membrane "receptor" proteins.

The earliest proposal that plasmid DNA is bound to the cell membrane came from the observation of Jacob et al. (1963) that certain F′ lac plasmids co-segregated with the bacterial chromosome. There is little experimental evidence that attachment to the membrane is involved in the partitioning of plasmids. On the other hand, there is considerable evidence for a specific mechanism to ensure precise segregation. Naturally occurring plasmids are present in defined numbers in exponentially growing cells (Clowes 1972); this stability is maintained even when this number is as small as one or two plasmid copies per chromosome (Rosner 1972). Obviously, if the segregation mechanism were passive, most progency cells would contain plasmids only if the copy numbers were very large. However, plasmid-free cells are ordinarily not observed even for single-copy plasmids (Frame and Bishop 1971).

Active partition mechanisms have been suggested for several low copy-number *E. coli* plasmids, including P1 (Austin and Abeles 1983a,b), F (Ogura and Hiraga 1983), and R1 (Nordstrom et al. 1980). Deletions in these plasmids often result in derivatives which are lost from cells at a frequency expected from random partitioning of the plasmids during cell division. Deletions in a high copy-number plasmid (ColEl) can result in instability, but this does not appear to be due to the loss of any *par* function. Rather, it reflects an increase in concatemer formation, which effectively reduces the number of ColEl segregants per cell (Summers and Sheratt 1984). Another mechanism (as yet unproven) for ensuring precise plasmid segregation would be a clustering or compartmentalization of the plasmid with the nucleoid (Kline and Miller 1975). For now, plasmids containing a legitimate *par* function would appear to be the best candidates for establishing a link between the bacterial envelopes and a genome during the segregation process.

As yet, attempts to demonstrate a physical association between plasmid DNA and membrane have fared neither better nor worse than those involving the bacterial chromosome. Various techniques have been used for the isolation of membrane fragments containing DNA from *E. coli* plasmids R (Levy 1971), ColEl (Sparks and Helinski 1979), R6K (Archibald et al. 1983), RK2 (Firshein et al. 1982), and plasmids in *B. subtilis* (Sueoka et al. 1984).

Early attempts at demonstrating a link between plasmid DNA and the cell envelope often suffered from an inability to distinguish phenotypic differences between associations involving plasmid replication and those involving partitioning. ColEl is an interesting case. Electron microscopy has demonstrated the association of a particular region of the DNA with purported membrane fragments (Sparks and Helinski 1979). This DNA-membrane complex was recovered from cesium chloride-sucrose gradients and involved one-quarter of all the plasmid DNA molecules examined. The portion of the DNA associated with the membrane encompassed the origin/terminus region and the relaxation nick site of ColEl. However, the significance of this association is unclear, as ColEl does not have a *par* function (Summers and Sheratt 1984).

Novick et al. (1980) attempted to demonstrate indirectly an association between plasmids and cell membranes by examining plasmid stability before and after making spheroplasts of *Staphylococcus aureus.* They found that spheroplasting results in the loss of some plasmids, but not of others; plasmid loss was not due to low copy number since plasmid replication appeared to be normal in the spheroplasts. Thus, two mechanisms of segregation apparently exist; spheroplasting disrupts the segregation of plasmids that require the participation of the cell envelope, but has no effect on the others. Firshein et al. (1982) found that a miniplasmid derived from the low copy number plasmid RK2 can be isolated as a DNA-membrane complex (M-bands) from minicells. The M-band associated DNA was capable of rifampin-sensitive synthesis of nascent plasmid DNA without the addition of exogenous template or enzymes. However, attachment at a specific region of the DNA was not demonstrated.

Our understanding of the mechanism of plasmid segregation has made considerable progress because it has become possible to distinguish between mutations in replication (*rep*) and partitioning (*par*). It was shown that deletion of a restriction fragment from the *E. coli* plasmid Rl*drd*-19, far removed from the origin of replication, results in a rate of plasmid-curing consistent with random segregation. This instability was found to be independent of copy number, leading the authors to conclude that the region deleted codes for a *par* function (Nordstrom et al. 1980). Analogous results have been produced using transposon mutagenesis of mini-F derivatives (Bergquist et al. 1981) and by producing partition-defective plasmids in pSC101 (Meacock and Cohen 1980). The *par* locus identified in these plasmids acts in *cis*, as expected. More recent deletion analysis has shown that three distinct, but adjacent, fragments of pSC101 are responsible for the *par* phenotype (Tucker et al. 1984).

Gustafson et al. (1983) coupled genetic studies on the *par* locus to cell fractionation. They looked for differential binding of *par*$^+$ and *par*$^-$ derivatives of plasmids Rl and pSC101 to the inner and outer membrane of *E. coli*. While *par*$^+$ plasmids were found to be associated with both inner and outer membrane material, the *par*$^-$ derivatives were associated only with the inner membrane. This suggests a specific association between the Rl and pSC101 *par* loci and the outer membrane. It is interesting to note that the low copy-number plasmids F and Pl also contain a *cis*-acting site encoding a *par* function. The *par* region is complex, as shown by the fact that it also encodes one or two *trans*-acting proteins essential for plasmid maintenance (Austin 1984; Austin and Abeles 1983a,b; Austin and Wieizbicki 1983; Ogura and Hiraga 1983). It has been suggested that these proteins may be required for membrane binding (Austin 1984). However, direct evidence is lacking in this regard.

The in vitro binding of plasmid DNA to membranes has also been studied by Korn et al. (1983). The plasmid used (pSL103) is chimeric, containing the replicative origin of an *S. aureus* plasmid and selectable genes (*trp*C$^+$) from *Bacillus pumilus*. The binding of pSL103 in vitro was prevented only by addition of DNA containing the pSL103 origin. Binding was not affected by heterologous DNA nor by the *trp*C$^+$ region of *B. pumilus*. In similar experiments using total cell lysates, Tanaka and Sueoka (1983) determined that the binding of the origin fragment is sensitive to high salt and takes place at four sites, two of which flank the origin itself.

There are only a few studies on the effect of host mutations on DNA-membrane associations and they are exemplified by work on *B. subtilis* (Winston and Sueoka 1980; Sueoka et al. 1984). These authors used known *B. subtilis* initiation mutants to seek a connection between membrane attachment to the replicative origin and the initiation of DNA replication. Temperature-sensitive mutants in initiation of DNA replication (*dna*-1 and *dna*B19) showed decreased association of the chromosome with the membrane at the nonpermissive temperature. At the restrictive temperature, the chimeric plasmid pSL103 (mentioned above) was also released from the membrane and failed to initiate DNA replication in *dna*-1 temperature-sensitive bacteria. In a *dna*B19 mutant, however, membrane attachment and DNA initiation both occurred at the nonpermissive temperature. In all cases, DNA detachment from the membrane was reversed when the temperature was lowered. Results similar to these were obtained when temperature shifts were carried out in vitro using fractions containing both chromosome and plasmid DNA-membrane complexes. Thus, *dna*-1 appears to be involved in the attachment of both the chromosome and the plasmid to the membrane, while *dna*B19 is involved with the chromosome only. The *dna*-1 mutant is of particular interest because in these mutants a membrane protein (35 kD) is not found at the restrictive temperature (Imada et al. 1976).

In conclusion, the subject of DNA-membrane binding has proven to be technically complex and progress in analyzing it has been distressingly slow. However, the earlier cell fractionation studies have been extended recently to include more critical experiments involving reconstitution in vitro and genetic studies of plasmid segregation. In view of these important advances, it is likely that significant progress in this field will be made.

References

Abe M, Brown C, Hendrickson WG, Boyd DH, Clifford P, Cote RH, Schaechter M (1977) Release of *Escherichia coli* DNA from membrane complexes by single-stranded endonucleases. Proc Natl Acad Sci USA 74:2756–2760

Archibald ER, Wilson JD, Allison DP, Sheehy RJ (1983) Membrane-bound fractions of R6K plasmid DNA in *Escherichia coli.* J Bacteriol 156:414–418

Austin SJ (1984) Bacterial plasmids that carry two functional centromere analogs are stable and are partitioned faithfully. J Bacteriol 158:742–745

Austin S, Abeles A (1983a) Partition of unit-copy miniplasmids to daughter cells I. Pl and F miniplasmids contain discrete, interchangeable sequences sufficient to promote equipartition. J Mol Biol 169:353–372

Austin S, Abeles A (1983b) The partition of unit-copy miniplasmids to daughter cells II. The partition region of miniplasmid Pl encodes an essential protein and a centromere-like site at which it acts. J Mol Biol 169:373–387

Austin S, Wierzbicki A (1983) Two mini-F encoded proteins are essential for equipartition. Plasmid 10:73–81

Bergquist PL, Downhard RA, Coughey PA, Gardner, Lane HED (1981) Analysis of mini-F plasmid replication by transposition mutagenesis. J Bacteriol 147:888–899

Bodnar JW, Jones CJ, Coombs DH, Pearson GD, Ward DC (1983) Proteins tightly bound to HeLa cell DNA at nuclear matrix attachment sites. Mol Cell Biol 3:1567–1579

Clowes RC (1972) Molecular structure of bacterial plasmids. Bacteriol Rev 36:361–405

Drlica K, Burgi E, Worcel A (1978) Association of the folded chromosome with the cell envelope of *Escherichia coli:* nature of the membrane-associated DNA. J Bacteriol 134:1108–1116

Dworsky P, Schaechter M (1973) Effect of rifampin on the structure and membrane attachment of the nucleoid of *Escherichia coli.* J Bacteriol 116:1364–1374

Firshein W, Strumph P, Benjamin P, Burnstein K, Kornacki J (1982) Replication of a low-copy-number plasmid by a plasmid DNA-membrane complex extracted from minicells of *Escherichia coli.* J Bacteriol 150:1234–1243

Frame R, Bishop JO (1971) The number of sex factors per chromosome in *E. coli.* Biochem J 121:93–103

Gustafson P, Wolf-Watz H, Lind L, Johansson K, Nordstrom K (1983) Binding between the *par* region of plasmids Rl and pSC101 and the outer membrane fraction of the host bacteria. EMBO J 2:27–32

Hendrickson WG, Kusano T, Yamaki H, Balakrishnan R, King M, Murchie J, Schaechter M (1982) Binding of the origin of replication of *Escherichia coli* to the outer membrane. Cell 30:915–923

Imada S, Carroll LE, Sueoka N (1976) DNA-membrane complex in *Bacillus subtilis.* In: Schlessinger D (ed) Microbiology-1976. Am Soc Microbiol Wash DC, pp 116–122

Ivarie RD, Pene JJ (1973) Association of many regions of the *Bacillus subtilis* chromosome with the cell membrane. J Bacteriol 114:571–576

Jacob F, Brenner S, Cuzin F (1963) On the regulation of DNA replication in bacteria. Cold Spring Harbor Symp Quant Biol 28:329–340

Jacq A, Kohiyama M, Lothar H, Messer W (19830 Recognition sites for a membrane-derived DNA binding protein preparation in the *E. coli* replication origin. Mol Gen Genet 191:460–465

Kline BC, Miller JR (1975) Detection of non-integrated plasmid deoxyribonucleic acid in the folded chromosome of *Escherichia coli:* physiological approach to studying the unit of segregation. J Bacteriol 121:165–172

Korn R, Winston S, Tanaka T, Sueoka N (1983) Specific *in vitro* binding of a plasmid to a membrane fraction of *Bacillus subtilis.* Proc Natl Acad Sci USA 80:574–578

Kusano T, Steinmetz D, Hendrickson WG, Murchie J, King M, Benson A, Schaechter M (1984) Direct evidence for specific binding of the replicative origin of the *Escherichia coli* chromosome to the membrane. J Bacteriol 158:313–316

Leibowitz PJ, Schaechter M (1975) The attachment of the bacterial chromosome to the cell membrane. Int Rev Cytol 41:1–28

Levy SB (1971) Physiological and functional characteristics of R-factor deoxyribonucleic acid segregated into *Escherichia coli* minicells. J Bacteriol 108:300–308

Meacock PA, Cohen SN (1980) Partition of bacterial plasmids during cell division: a *cis*-acting locus that accomplishes stable plasmid inheritance. Cell 20:529–542

Moyer MP (1979) The association of DNA and RNA with membranes. Int Rev Cytol 61:1–61

Nordstrom K, Molin S, Aagaard-Hansen (1980) Partitioning of plasmids Rl in *Escherichia coli* I. Kinetics of loss of plasmid derivatives deleted of the *par* region. Plasmid 4:215–227

Novic R, Sanchez-Rivaz C, Gruss A, Edelman I (1980) Involvement of the cell envelope in plasmid maintenance: plasmid curing during the regeneration of protoplasts. Plasmid 3:348–358

Ogden GB, Schaechter MS (1985) Chromosomes, plasmids and the bacterial cell envelopes. In: Levine L (ed) Microbiology-1985. Am Soc Microbiol Wash DC, pp 282–286

Ogura T, Hiraga S (1983) Partition mechanism of F plasmid: two plasmid gene-encoded products and a *cis*-acting region are involved in partition. Cell 32:351–360

Parks LC, Dicker DT, Conger AD, Daneo-Moore L, Higgins ML (1981) Effect of chromosomal breaks induced by X-irradiation on the number of mesosomes and the cytoplasmic organization of *Streptococcus faecalis.* J Mol Biol 146:413–431

Rosner JS (1972) Formation, induction and curing of bacteriophage Pl lysogens. Virology 49: 679–689

Sargent MG, Bennett MF, Burdett IDJ (1983) Identification of specific restriction fragments associated with a membrane subparticle from *Bacillus subtilis.* J Bacteriol 154:1389–1396

Sparks R, Helinski D (1979) Association of cellular membrane of *E. coli* minicells with the original terminus of replication of plasmid Col El DNA. Nature 277:572–575

Sueoka N, Korn R, McKenzie T, Tanaka T, Winston S (1984) Two types of binding of pUB110 to *Bacillus subtilis* membrane. In: Ganesan AT, Hoch JA (ed) Genetics and biotechnology of bacilli. Academic Press, New York, pp 79–88

Summers DK, Sherratt DJ (1984) Multimerization of high copy number plasmids cause instability: ColEl encodes a determinant essential for plasmid monomerization and stability. Cell 36: 1097–1103

Tanaka T, Sueoka N (1983) Site-specific *in vitro* binding of plasmid pUB110 to *Bacillus subtilis* membrane fraction. J Bacteriol 154:1184–1194

Tremblay GY, Daniels TJ, Schaechter M (1969) Isolation of a cell membrane-DNA-nascent RNA complex from bacteria. J Mol Biol 40:65–76

Tucker WT, Miller CA, Cohen SN (1984) Structural and functional analysis of the *par* region of pSC101 plasmid. Cell 38:191–201

Winston S, Sueoka N (1980) DNA-membrane association is necessary for initiation of chromosomal and plasmid replication in *Bacillus subtilis.* Proc Natl Acad Sci USA 77:2834–2838

Worcel A, Burgi E (1972) On the structure of the folded chromosome of *Escherichia coli.* J Mol Biol 71:127–147

DNA Topoisomerase Mutations in Bacteria

K. DRLICA[1,2], G.J. PRUSS[1], S.H. MANES[1], and S.G. CHEVALIER[1,2]

1 Introduction

The chromosome of *Escherichia coli* is a DNA molecule having a length more than a thousand times that of the cell in which it resides (Cairns 1963). How this DNA is compacted and packaged is poorly understood. It appears that the DNA is arranged into about 50 large (100 kbp), topologically independent domains (Worcel and Burgi 1972; Sinden and Pettijohn 1981), and each of these domains, or loops, is probably under negative superhelical tension (Worcel and Burgi 1972; Sinden et al. 1980). While it is becoming increasingly clear that the superhelical tension is a result of topoisomerase action, how the loops are established and maintained is still a mystery. Another level of compaction appears to be the packaging of DNA into nucleosomelike structures (Griffith 1976; Varshavsky et al. 1977). Unfortunately, bacterial chromatin, unlike its eukaryotic counterpact, has been difficult to isolate and study.

Our own work has focused on topoisomerases and supercoiling, and below we briefly describe some of the characteristics of topoisomerase mutations. These mutations have been the primary tool for studying the physiology of DNA supercoiling in bacteria. If chromatin structure, supercoiling, and topoisomerase activity are related, as several eukaryotic studies indicate (Ryoji and Worcel 1984; Glikin et al. 1984; Villeponteau et al. 1984), topoisomerase mutations may become valuable tools for studying chromatin structure in vivo.

Topoisomerase mutations are now available in three genes, *gyrA* and *gyrB*, which encode DNA gyrase (Gellert et al. 1976a), and *topA*, which encodes topoisomerase I (Wang 1971). A third enzyme, topoisomerase III, has been extracted from bacterial cells (Dean et al. 1983), but no mutations are yet available for physiological studies. The topoisomerase mutations can be divided into three categories, those that confer resistance to specific antibiotics, those that make cells temperature-sensitive, and those that alter steady-state levels of supercoiling. Each category is described below, and mutants that are generally available are listed in Table 1.

Gyrase is inactivated by two classes of antibiotic, each specific for a different subunit of the enzyme. The A subunit is sensitive to nalidixic acid, oxolinic acid, and a

1 Department of Biology, University of Rochester, Rochester, NY 14627, USA
2 Present address: Public Health Research Institute of the City of New York, 455 First Avenue, New York, NY 10016, USA

Bacterial Chromatin
(ed. by C.O. Gualerzi and C.L. Pon)

Table 1. Topoisomerase mutations

Mutation	Phenotype[a]	Source/Reference
gyrA(nal^R)	Resistance to nalidixic, oxolinic acids	Gellert et al. (1977), Sugino et al. (1977)
gyrA43	Thermal sensitivity for growth; at restrictive temperature supercoiling, RNA synthesis, DNA synthesis, and growth of certain bacteriophages decrease	Kreuzer and Cozzarelli (1979), Steck et al. (1984)
gyrA224	Steady-state levels of supercoiling are about 25% below wild type	DiNardo et al. (1982), Pruss et al. (1982)
gyrB(cou^R)	Resistance to coumarin antibiotics	Gellert et al. (1976b)
gyrB41	Resistance to coumarin antibiotics; thermal sensitivity for growth; at restrictive temperature supercoiling, RNA synthesis decrease, DNA synthesis decreases slowly; isolated nucleoids have doublet morphology	Gellert et al. (1979), Filutowicz and Jonczyk (1981), Steck and Drlica (1984), Steck et al. (1984)
gyrB(Ts)	Resistance to coumarin antibiotics; thermal sensitivity for growth; at restrictive temperature supercoiling, RNA synthesis decrease, DNA synthesis decreases slowly	Orr et al. (1979), Steck et al. (1984)
gyrB402	Characteristics similar to gyrB(Ts) except that DNA synthesis is rapidly inhibited at restrictive temperature; growth of phage T7 lower at restrictive temperature	Filutowicz and Jonczyk (1983), Steck et al. (1984), Steck and Drlica (1985)
gyrB225	Steady-state levels of supercoiling about 25% below wild type	DiNardo et al. (1982), Pruss et al. (1982)
gyrB226	Steady-state levels of supercoiling about 25% below wild type	DiNardo et al. (1982), Pruss et al. (1982)
gyrB-him(Ts)	Resistance to coumarin antibiotics; temperature-sensitive for phage lambda site-specific recombination, growth of bacteriophage Mu, plasmid maintenance, introduction of supercoils into superinfecting phage lambda	Friedman et al. (1984a,b)
topA10	Steady-state levels of supercoiling about 15% above wild type, heterogeneous topoisomer distribution	Sternglanz et al. (1981), Pruss et al. (1982), Pruss (1985)

Table 1 (Continued)

Mutation	Phenotype[a]	Source/Reference
ΔtopA	Slow growth in absence of gyrase compensatory mutation, such as gyrB225; heterogeneous plasmid topoisomer distribution	DiNardo et al. (1982), Pruss et al. (1982), Sternglanz et al.
ΔtopA(supX)[b]	Steady-state supercoiling of plasmids above normal, heterogeneous plasmid topoisomer distribution; suppresses leu-500 promoter mutation	Mukai and Margolin (1963), Overbye and Margolin (1981), Overbye et al. (1983), Pruss (1985)
amber topA(supX)[b]	Similar to ΔtopA(supX)	same as ΔtopA(supX)
topA(supX) point mutants[b]	Similar to ΔtopA(supX)	Richardson et al. (1984)

[a] The list of phenotypes is only partial. For additional phenotypes, see Drlica (1984) and the references cited above.

[b] Mutations are in *Salmonella typhimurium;* all others are in *Escherichia coli.*

series of related compounds (Sugino et al. 1977; Gellert et al. 1977). Novobiocin, chlorobiocin, and coumermycin A_1 are inhibitors of the B subunit (Gellert et al. 1976b; Fairweather et al. 1980). Both types of antibiotic affect a number of chromosomal functions in living cells, including DNA replication and transcription. Cells with drug-resistant mutations, which occur in *gyrA* and *gyrB,* serve as controls for nonspecific drug effects and allow inhibitor studies to be interpreted in terms of gyrase action. No antibiotics have been discovered that inactivate topoisomerases I or III.

Temperature-sensitive mutations have been obtained in *gyrA* and in *gyrB*. Experiments using temperature-sensitive gyrase mutants complement studies using antibiotic inhibitors of gyrase. In both cases a shift experiment can be performed to examine the physiological consequences of rapidly inactivating gyrase. The temperature-sensitive mutants are especially useful for this purpose because the inhibitors may create drug-protein-DNA complexes that cause effects beyond simple inactivation of gyrase. It is likely that this is the case with nalidixic and oxolinic acids (for review see Peebles et al. 1979).

The third category of mutation alters the steady-state level of supercoiling. Such mutations have been obtained in *gyrA*, *gyrB*, and *topA*, and they generally result in reduced topoisomerase activity. Cells containing these mutations are viable, and they are useful for studying the net consequences of small alterations in supercoiling.

It is important to note that at present we cannot change levels of supercoiling without perturbing topoisomerase activities. Consequently, we are unable to distinguish between physiological effects of the two.

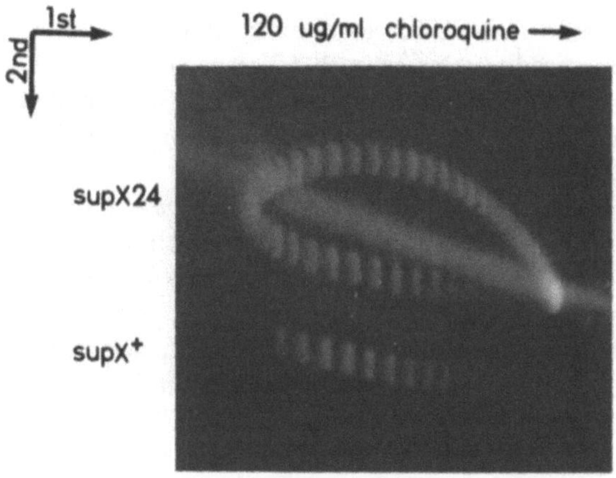

Fig. 1. Two-dimensional gel electrophoresis of pBR322 DNA isolated from the supX24 mutant and from its $supX^+$ parental strain. Plasmid DNA was isolated from strains pBR322-PM233 (*supX*24) and pBR322-PM596 (*$supX^+$*) and subjected to electrophoresis in two dimensions. The direction of migration in the first dimension is from *left to right;* the direction of migration in the second dimension is from *top to bottom.* The chloroquine concentrations were 120 $\mu g\ ml^{-1}$ in the first dimension and 600 $\mu g\ ml^{-1}$ in the second dimension. The *bright diagonal streak* is probably due to fragments of chromosomal DNA. Before electrophoresis in the gel shown, plasmid DNA was heated in boiling water for 60 s to reduce the amount of chromosomal DNA in the region of the gel to which supercoiled pBR322 DNA migrates. This heat treatment is responsible for the absence of open-circular (and linear) pBR322 DNA. The center of the supercoiling distribution of pBR322 DNA isolated from the *$supX^+$* strain corresponds to a linking difference of about −25. Figure 1 is adapted from a figure in Pruss (1985)

2 Control of Supercoiling

Mutant and inhibitor studies indicate that at least two topoisomerases are involved in controling the level of intracellular superhelical tension. Gyrase appears to be responsible for introducing and maintaining DNA supercoiling: inactivation of this enzyme blocks the introduction of supercoils into bacteriophage lambda DNA during superinfection of a lysogen (Gellert et al. 1976b, 1977); it also causes a loss of supercoiling from the bacterial chromosome (Drlica and Snyder 1978) and from plasmid DNA (Pettijohn and Pfenninger 1980). Topoisomerase I (Wang 1971) appears to modulate the effects of gyrase by relaxing DNA: *topA* mutants (Sternglanz et al. 1981; Mukai and Margolin 1963; Trucksis et al. 1981) have higher-than-normal levels of DNA supercoiling (Pruss et al. 1982; Richardson et al. 1984; Pruss 1985; also, see Fig. 1).

The level of DNA supercoiling in bacteria appears to be a regulated parameter because perturbation of topoisomerase activity elicits corrective responses. For example, inhibition of gyrase by inhibitors or by shift of a temperature-sensitive gyrase mutant to nonpermissive temperature increases expression from gyrase genes (Menzel and Gellert 1983). Likewise, the presence of topoisomerase I causes gyrase expression to increase (Gellert et al. 1983). In another example, mutations which inactivate *topA* (Sternglanz et al. 1981) and which we expect to raise levels of supercoiling, are compensated by gyrase mutations (DiNardo et al. 1982; Pruss et al. 1982) that lower supercoiling (Pruss et al. 1982). Even mild treatments, such as temperature changes, which are expected to slightly alter the helical pitch of DNA and thus supercoiling, may elicit topoisomerase action in a wild-type strain to restore the normal level of supercoiling (Goldstein and Drlica 1984).

Some of the control over supercoiling apparently breaks down in *topA* mutants, causing supercoiling distributions to become more heterogeneous than normal. We first noticed this phenomenon as a broadening of the dye-titration curve of nucleoids isolated from a *topA* mutant (Pruss et al. 1982). A more striking case can be seen with plasmid pBR322 DNA extracted from *topA* mutants (Pruss 1985). This phenomenon is illustrated in Fig. 1 where plasmid DNA isolated from a *topA* (supX) mutant is displayed after two-dimensional gel electrophoresis in the presence of the intercalating dye chloroquine. Some of the pBR322 DNA from *topA* mutants is so highly supercoiled that it cannot be titrated even with very high concentrations of chloroquine. Density gradient analysis of the most highly supercoiled DNA from one mutant indicates that it has more than twice the superhelical density than DNA from wild-type cells (Pruss 1985). We do not yet have a molecular explanation for the unusual supercoiling distribution of pBR322 DNA isolated from *topA* mutants. However, the degree of heterogeneity is replicon-specific (Pruss 1985), a feature that will enable us to carry out comparative studies that may provide insights into the nature of the phenomenon.

3 Oxolinic Acid Anomalies

Biochemical studies of gyrase inhibitors indicate that nalidixic and oxolinic acids trap a gyrase-DNA reaction intermediate in such a way that subsequent treatment with detergents leads to DNA cleavage (for review, see Peebles et al. 1979). We reasoned that the drug-gyrase-DNA complexes might mark the sites of gyrase action. If this is the case, then the distribution of the enzyme on the chromosome could be determined by measuring the size of chromosomal DNA fragments isolated from cells treated with oxolinic acid (Hill and Fangman 1972; Piesetsky et al. 1972). Our measurements indicated that on average the spacing among the sites is about 100 kbp (Snyder and Drlica 1979), corresponding to about one gyrase site per topological domain (Sinden and Pettijohn 1981).

The availability of topoisomerase mutations that change DNA supercoiling provided us with the opportunity to determine whether the number of oxolinic acid-induced DNA cleavage events corresponds with chromosomal DNA supercoiling. Experiments using strains RS2 (*topA*10) and SD104 (*gyrB*225) obtained from R. Sternglanz and S. DiNardo indicated that small (20%) increases or decreases in steady-state supercoiling do not affect the size of the DNA fragments generated by oxolinic acid. However, these studies revealed two unusual properties of strain DM800 (*gyrB225 ΔtopA;* Sternglanz et al. 1981). The first property is that the number of cleavage events is more than twice that observed in a wild-type strain (see Fig. 2 and Table 2). This number is unaffected by transduction of a *topA*$^+$ allele into DM800 (data not shown). Thus, the altered fragment size, although obtained using a strain that once contained a

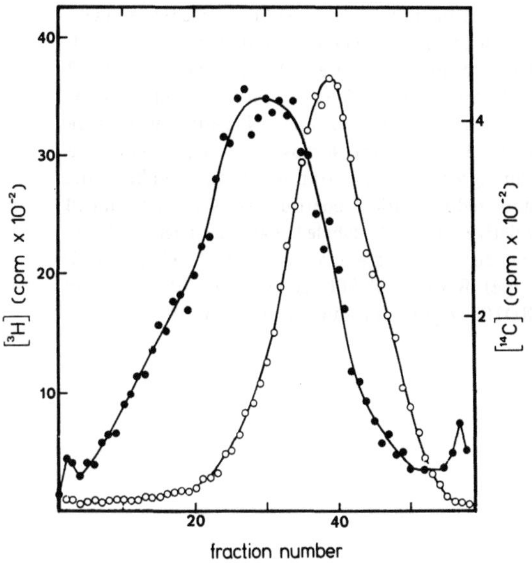

Fig. 2. Effect of oxolinic acid on size of DNA isolated from strain DM800. *E. coli* K-12 strains DM800 (Δ*topA*, *gyrB*225 o—o) and DM4100 (wild-type •—•) were grown in the presence of [^{3}H]- or [^{14}C]-labeled thymidine, respectively, treated with oxolinic acid (2 μg ml^{-1} for 10 min), and harvested by centrifugation. The resuspended cells from the two strains were pooled, lysed, and treated with 0.5% SDS as described in Snyder and Drlica (1979). Centrifugation was carried out in a Beckman SW41 rotor for 65 h at 8100 rpm. The gradient contained 0.68 μg DNA. Sedimentation is from *right to left*

Table 2. Number-average molecular weights of chromosomal DNA after oxolinic acid treatment[a]

Bacterial strains	DM4100	DM800
Genotype	wt	gyrB225 ΔtopA
M_n/M_n(T4B)[b]	0.48 ± 0.02 (9)	0.21 ± 0.05 (11)
DNA size (kbp)	100	40
M_n/M_n(DM800)	2.3	1
Fragments/genome-equivalent[c]	44	110

[a] Oxolinic acid was added to cultures at 2 μg/ml for 10 min prior to cell lysis.

[b] Number-average molecular weights are determined from sedimentation of DNA into sucrose density gradients (see Snyder and Drlica 1979 for method). Error estimates represent one standard deviation from the mean. Numbers in parentheses are the number of determinations used to calculate means. The molecular weight of T4 DNA is taken as 132 × 10^6 daltons.

[c] The size of the *E. coli* genome is taken as 4400 kbp.

deletion in *topA*, does not require the presence of that mutation. The second observation is that in strain DM800, the oxolinic acid-gyrase-DNA complexes are abnormally labile: some of the cleavage events occur in the absence of sodium dodecyl sulfate. This lability is most easily observed during titration of chromosomal DNA supercoiling by ethidium bromide. The titration curve is flattened relative to that observed with nucleoids isolated from wild-type cells treated with oxolinic acid (cf. Fig. 3A and B). As above, continued presence of the *topA* deletion is not required to observe the lability of the oxolinic acid-gyrase-DNA complexes (Fig. 3C).

The genetic basis for the two unusual effects of oxolinic acid described above is still unknown. When the *gyrB*225 allele from DM800 is transduced into a wild-type

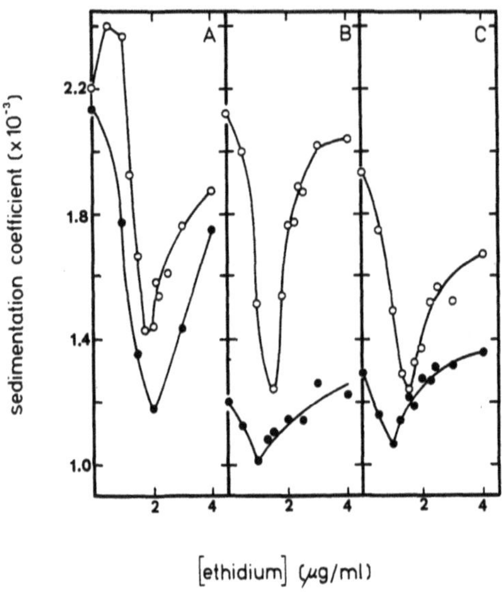

Fig. 3. Effect of oxolinic acid on nucleoid sedimentation in the presence of ethidium bromide. [^{3}H]-Labeled nucleoids were isolated from *E. coli* K-12 cells treated with 0 (○–○) or 2 μg ml^{-1} (●–●) oxolinic acid for 10 min. Nucleoids were sedimented into neutral sucrose density gradients containing various concentrations of ethidium bromide, and sedimentation coefficients were determined relative to [^{14}C]-labeled bacteriophage T4B included in each gradient. **A** Strain DM4100 (wild-type). **B** Strain DM800 (Δ*topA*, *gyrB*225). **C** Strain SD108 (*topA*$^+$ transductant of DM800)

strain, the resulting transductant has a lower-than-normal level of chromosomal supercoiling, but it exhibits neither abnormally labile drug-gyrase-DNA complexes nor increased DNA cleavage induced by oxolinic acid (data not shown). Our working hypothesis is that strain DM800 contains as yet unidentified mutations.

4 DNA Relaxation by Coumermycin A_1

Inhibition of gyrase by coumermycin A_1 leads to a loss of supercoiling from bacterial chromosomes (Drlica and Snyder 1978). Although it seemed reasonable that relaxation arises from the competing activity topoisomerase I, we have long been puzzled by the observation that low, but saturating doses of oxolonic acid cause little or no DNA relaxation (Snyder and Drlica 1979; Manes et al. 1983; also see Fig. 3). Since gyrase itself is able to relax DNA in vitro when inhibited by coumermycin A_1 (Gellert et al. 1976b), most of the relaxation in vivo may be due to reversing the action of gyrase rather than to topoisomerase I activity. We have initiated a study of this phenomenon by measuring chromosomal DNA supercoiling in nucleoids isolated from Δ*topA* mutants following treatment of cells with coumermycin A_1. Figure 4 shows that chromosomal DNA supercoiling is relaxed by the drug. Thus, topoisomerase I cannot be responsible for the relaxation, leading us to favor gyrase as the source of the relaxation. Whether topoisomerase III (Dean et al. 1983) is involved in DNA relaxation in vivo is still unknown.

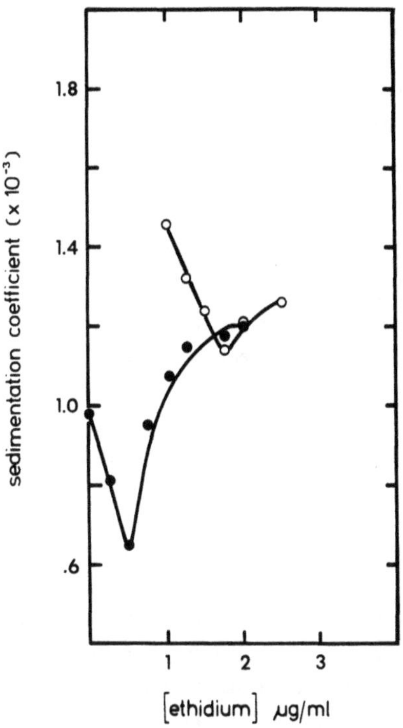

Fig. 4. Effect of coumermycin A_1 on chromosomal DNA relaxation in a ΔtopA mutant. [^{3}H]-Labeled nucleoids were isolated from *E. coli* K-12 strain DM800 (Δ*topA*) treated with 0 (○–○) or 50 (●–●) µg ml^{-1} coumermycin A_1 for 25 min. Nucleoids were sedimented into neutral sucrose density gradients containing various concentrations of ethidium bromide, and sedimentation coefficients were determined relative to [^{14}C]-labeled bacteriophage T4B included in each gradient. For experimental methods, see Drlica and Snyder (1978)

5 Perspective

The rapid progress in understanding the biochemistry of topoisomerases (reviewed by Cozzarelli 1980; Gellert 1981) and the development of genetic tools has made possible the investigation of how these enzymes affect chromosome function. Gyrase and topoisomerase I appear to regulate the level of superhelical tension in the cell and thereby contribute to the structure of the DNA component of chromatin. In addition, the decatenating activity of gyrase appears to be important for proper segregation of replicated daughter chromosomes: nucleoids isolated from a temperature-sensitive *gyrB* mutant have an unusual doublet morphology if the cells are incubated at restrictive temperature prior to nucleoid isolation, and the doublet nucleoids can be resolved into singlets by purified gyrase (Steck and Drlica 1984).

It is occasionally important to stress that physiological experiments are often difficult to interpret rigorously. This is especially true for those that involve cell fractionation, for macromolecular rearrangements can occur during cell lysis and fractionation. Fortunately, rearrangements do not pose a problem in the study of DNA topology as long as precautions are taken to quickly inactivate topoisomerases and other enzymes that cleave or ligate DNA. In the case of temperature-sensitive gyrase mutants, we found that rapid chilling of cells was important to prevent the reintroduction of supertwists by gyrase, an event that occurs if cells are cooled slowly (Steck et al. 1984; Steck and Drlica 1984). A second area where caution must be exercised is the study of suppressor mutations. For example, the events that generate gyrase mutations to compensate for a deletion in *topA* may also lead to acquisition of additional mutations. It is possible that an additional unknown mutation(s) is involved in the abnormal oxolinic acid-induced cleavage that we described above for strain DM800. A third area requiring the careful construction of control experiments is the use of gyrase inhibitors, especially oxolinic acid. For example, one cannot interpret physiological responses in terms of supercoiling without measuring changes in supercoiling – this drug is a potent inhibitor even when little change in supercoiling occurs (Snyder and Drlica 1979; Manes et al. 1983). Indeed, some combinations of topoisomerase mutations even enable oxolinic acid to *raise* supercoiling and at the same time inhibit the incorporation of nucleotides into RNA (Manes et al. 1983). Another aspect of oxolinic acid and related drugs also merits attention. Use of these drugs to mark the sites of gyrase action depends on the drug trapping gyrase on the DNA at its normal site of action. The observation that oxolinic acid does not change the location of gyrase binding to DNA in vitro, as determined by nuclease protection studies (Fisher et al. 1981; Morrison and Cozzarelli 1979), supports this assumption.

Several observations lead us to suspect that relationships exist among chromatin structure, DNA supercoiling, and topoisomerases. Three points arise from eukaryotic studies. First, nucleosome formation involves wrapping DNA around histones (for review, see McGhee and Felsenfeld 1980), a process that is favored by negative supercoiling (Germond et al. 1975). Indeed, nucleosome assembly in crude extracts is accompanied by striking changes in DNA-linking numbers (Glikin et al. 1984). Second, novobiocin, which inhibits topoisomerase II activity in vitro, blocks nucleosome assembly onto circular DNA injected into frog oocytes (Ryoji and Worcel 1984). Third, novobiocin alters the nuclease protection pattern, and presumably the chromatin structure, around active globin genes in chicken cells (Villeponteau et al. 1984). Support derived from bacterial cells is less direct because chromatin structure is so poorly understood. Here, the data centers around the histonelike protein HU (Rouviere-Yaniv and Gross 1975). This protein, which appears to be a component of bacterial chromatin (Varshavsky et al. 1977), acts with topoisomerase I as a specificity factor in a *dnaA, oriC*-dependent DNA replication initiation system (Dixon and Kornberg 1984). In addition, topoisomerase I and HU physically interact in vitro (Rouviere-Yaniv, unpublished observations).

As methods develop for isolation of plasmids in the form of chromatin (Busby et al. 1979; Wu et al. 1982), we expect the structural characterization to follow the lead set by prior studies of eukaryotic chromatin. But once the general outline of bacterial chromatin structure is understood, it will be possible to use the topoisomerase mutations to perturb DNA structure inside living cells so we can begin to unify our thinking about supercoiling and chromatin structure.

Acknowledgments. This work has been supported by Public Health Service grants GM32005 and GM24320, a Career Development award to K.D. from the National Cancer Institute, and an EMBO long-term fellowship to K.D.

Notes added in proof: The *gyrB41* mutation consists of two mutations (Menzel and Gellert 1983). Some mutations which compensate for the absence of *topA* map outside the gyrase genes (Raji et al. 1985, J. Bacteriol. 162:1173–1179).

References

Busby S, Kolb A, Buc Henri (1979) Isolation of plasmid-protein complexes from *Escherichia coli.* Eur J Biochem 99:105–111

Cairns J (1963) The chromosome of *Escherichia coli.* Cold Spring Harbor Symp Quant Biol 28: 43–46

Cozzarelli N (1980) DNA gyrase and the supercoiling of DNA. Science 207:953–960

Dean F, Krasnow M, Otter R, Matzuk M, Spangler S, Cozzarelli N (1983) *Escherichia coli* type I topoisomerases: identification, mechanism, and role in recombination. Cold Spring Harbor Symp Quant Biol 47:769–777

DiNardo S, Voelkel K, Sternglanz R, Reynolds A, Wright A (1982) *Escherichia coli* DNA topoisomerase I mutants have compensatory mutations in DNA gyrase genes. Cell 31:43–51

Dixon N, Kornberg A (1984) Protein HU in the enzymatic replication of the chromosomal origin of *Escherichia coli.* Proc Natl Acad Sci USA 81:424–428

Drlica K (1984) Biology of bacterial DNA topoisomerases. Microbiol Rev 48:273–289

Drlica K, Snyder M (1978) Superhelical *Escherichia coli* DNA: relaxation by coumermycin. J Mol Biol 120:145–154

Fairweather N, Orr E, Holland I (1980) Inhibition of deoxyribonucleic acid gyrase: effects on nucleic acid synthesis and cell division in *Escherichia coli* K-12. J Bacteriol 142:153–161

Filutowicz M, Jonczyk P (1981) Essential role of *gyrB* gene product in the transcriptional event coupled to *dnaA*-dependent initiation of *Escherichia coli* chromosome replication. Mol Gen Genet 183:134–138

Filutowicz M, Jonczyk P (1983) The *gyrB* gene product functions in both initiation and chain polymerization of *Escherichia coli* chromosome replication: suppression of the initiation deficiency in *gyrB*-ts mutants by a class of the *rpoB* mutations. Mol Gen Genet 191:282–287

Fisher L, Mizuuchi K, O'Dea M, Ohmori H, Gellert M (1981) Site-specific interaction of DNA gyrase with DNA. Proc Natl Acad Sci USA 78:4165–4169

Friedman D, Olson E, Carver D, Gellert M (1984a) Synergistic effect of *himA* and *gyrB* mutations: evidence that Him functions control expression of *ilv* and *xyl* genes. J Bacteriol 157:484–489

Friedman D, Pantefaber L, Olson E, Carver D, O'Dea M, Gellert M (1984b) Mutations in the DNA *gyrB* gene that are temperature sensitive for lambda site-specific recombination, Mu growth, and plasmid maintenance. J Bacteriol 157:490–497

Gellert M (1981) DNA topoisomerases. Annu Rev Biochem 50:879–910

Gellert M, Mizuuchi K, O'Dea M, Nash H (1976a) DNA gyrase: an enzyme that introduces superhelical turns into DNA. Proc Natl Acad Sci USA 73:3872–3876

Gellert M, O'Dea M, Itoh T, Tomizawa J (1976b) Novobiocin and coumermycin inhibit DNA supercoiling catalyzed by DNA gyrase. Proc Natl Acad Sci USA 73:4474–4478

Gellert M, Mizuuchi K, O'Dea M, Itoh T, Tomizawa J (1977) Nalidixic acid resistance: a second genetic character involved in DNA gyrase activity. Proc Natl Acad Sci USA 74:4772–4776

Gellert M, Fisher L, O'Dea M (1979) DNA gyrase: purification and catalytic properties of a fragment of gyrase B protein. Proc Natl Acad Sci USA 76:6289–6293

Gellert M, Menzel R, Mizuuchi K, O'Dea M, Friedman D (1983) Regulation of DNA supercoiling in *E. coli*. Cold Spring Harbor Symp Quant Biol 47:763–767

Germond J, Hirt B, Oudet P, Gross-Bellard M, Chambon P (1975) Folding of the DNA double helix in chromatin-like structures from simian virus 40. Proc Natl Acad Sci USA 72:1843–1847

Glikin G, Ruberti I, Worcel A (1984) Chromatin assembly in *Xenopus* oocytes: *in vitro* studies. Cell 37:33–41

Goldstein E, Drlica K (1984) Regulation of bacterial DNA supercoiling: plasmid linking numbers vary with growth temperature. Proc Natl Acad Sci USA 81:4046–4050

Griffith J (1976) Visualization of prokaryotic DNA in a regularly condensed chromatin-like fiber. Proc Natl Acad Sci USA 73:563–567

Hill W, Fangman W (1972) Single-strand breaks in deoxyribonucleic acid and viability loss during deoxyribonucleic acid synthesis inhibition in *Escherichia coli.* J Bacteriol 116:1329–1335

Kreuzer K, Cozzarelli N (1979) *Escherichia coli* mutants thermosensitive for deoxyribonucleic acid gyrase subunit A: effects on deoxyribonucleic acid replication, transcription, and bacteriophage growth. J Bacteriol 140:424–435

Manes S, Pruss G, Drlica K (1983) Inhibition of RNA synthesis by oxolinic acid is unrelated to average DNA supercoiling. J Bacteriol 155:420–423

McGhee J, Felsenfeld G (1980) Nucleosome structure. Annu Rev Biochem 49:1115–1156

Menzel R, Gellert M (1983) Regulation of the genes for *E. coli* gyrase: homeostatic control of DNA supercoiling. Cell 34:105–113

Morrison A, Cozzarelli N (1979) Site-specific cleavage of DNA by *E. coli* DNA gyrase. Cell 17: 175–184

Mukai F, Margolin P (1963) Analysis of unlinked suppressors of an 0° mutation in *Salmonella*. Proc Natl Acad Sci USA 50:140–148

Orr E, Fairweather N, Holland I, Pritchard R (1979) Isolation and characterization of a strain carrying a conditional lethal mutation in the *cou* gene of *Escherichia coli* K-12. Mol Gen Genet 177:103–112

Overbye K, Margolin P (1981) Role of the *supX* gene in ultraviolet light induced mutagenesis in *Salmonella typhimurium*. J Bacteriol 146:170–178

Overbye K, Basu S, Margolin P (1983) Loss of DNA topoisomerase I activity alters many cellular functions in *Salmonella typhimurium*. Cold Spring Harbor Symp Quant Biol 47:785–791

Peebles C, Higgins N, Kreuzer K, Morrison A, Brown P, Sugino A, Cozzarelli N (1979) Structure and activities of *Escherichia coli* DNA gyrase. Cold Spring Harbor Symp Quant Biol 43:41–52

Pettijohn D, Pfenninger O (1980) Supercoils in prokaryotic DNA restrained *in vivo*. Proc Natl Acad Sci USA 77:1331–1335

Pisetsky D, Berkower I, Wickner P, Hurwitz J (1972) Role of ATP in DNA synthesis in *Escherichia coli*. J Mol Biol 71:557–571

Pruss G (1985) DNA topoisomerase I mutants: increased heterogeneity in linking number and other replicon-dependent changes in DNA supercoiling. J Mol Biol 185:51–63

Pruss G, Manes S, Drlica K (1982) *Escherichia coli* DNA topoisomerase I mutants: increased supercoiling is corrected by mutations near gyrase genes. Cell 31:35–42

Richardson SM, Higgins L, Lilley D (1984) The genetic control of DNA supercoiling in *Salmonella typhimurium*. EMBO J 3:1745–1752

Rouviere-Yaniv J, Gros F (1975) Characterization of a novel, low molecular-weight DNA-binding protein from *Escherichia coli*. Proc Natl Acad Sci USA 72:3428–3432

Ryoji M, Worcel A (1984) Chromatin assembly in *Xenopus* oocytes: *in vivo* studies. Cell 37:21–32

Sinden R, Pettijohn D (1981) Chromosomes in living *Escherichia coli* cells are segregated into domains of supercoiling. Proc Natl Acad Sci USA 78:224–228

Sinden R, Carlson J, Pettijohn D (1980) Torsional tension in the DNA double helix measured with trimethylpsoralen in living *E. coli* cells. Cell 21:773–783

Snyder M, Drlica K (1979) DNA gyrase on the bacterial chromosome: DNA cleavage induced by oxolinic acid. J Mol Biol 131:287–302

Steck T, Drlica K (1984) Bacterial chromosome segregation: evidence for DNA gyrase involvement in decatenation. Cell 36:1081–1088

Steck T, Pruss G, Manes S, Burg L, Drlica K (1984) DNA supercoiling in gyrase mutants. J Bacteriol 158:397–403

Sternglanz R, DiNardo S, Voelkel K, Nishimura Y, Hirota Y, Becherer K, Zumstein L, Wang J (1981) Mutations in the gene coding for *Escherichia coli* DNA topoisomerase I affecting transcription and transposition. Proc Natl Acad Sci USA 78:2747–2751

Sugino A, Peebles C, Kreuzer K, Cozzarelli N (1977) Mechanism of action of nalidixic acid: purification of *Escherichia coli nalA* gene product and its relationship to DNA gyrase and a novel nicking-closing enzyme. Proc Natl Acad Sci USA 74:4767–4771

Trucksis M, Golub E, Zabel D, DePew R (1981) *Escherichia coli* and *Salmonella typhimurium supX* genes specify deoxyribonucleic acid topoisomerase I. J Bacteriol 147:679–681

Varshavsy A, Nedospasov V, Bakeayev V, Bakeayev T, Georgiev G (1977) Histone-like proteins in the purified *Escherichia coli* deoxyribonucleoprotein. Nucleic Acids Res 4:2725–2745

Villeponteau B, Lundell M, Martinson H (1984) Torsional stress promotes the DNase I sensitivity of active genes. Cell 39:469–478

Wang J (1971) Interaction between DNA and an *Escherichia coli* protein. J Mol biol 55:523–533

Worcel A, Burgi E (1972) On the structure of the folded chromosome of *Escherichia coli.* J Mol Biol 71:127–147

Wu F, Kolb A, Buc H (1982) A transcriptionally active plasmid-protein complex isolated from *Escherichia coli.* Biochim Biophys Acta 696:231–238

New Electron Microscopic Data on the Structure of the Nucleoid and Their Functional Consequences*

M.A. BJORNSTI[1], J.A. HOBOT[1], A.S. KELUS[2], W. VILLIGER[1], and E. KELLENBERGER[1]

Introduction

The DNA of the bacterial genome is organised quite differently from that of eukaryotic cells: Since only one linkage group exists, the "genome" of the bacterium is considered to be a single chromosome. A simple mechanism of division would allow an equipartition of the bacterial genome, in contrast to organisms with multiple chromosomes, where the chromosomes, condensed in metaphase, are presumably separated by the mitotic spindle fibers. In the condensed, compact form, the DNA is neither replicating nor transcriptionally active. In prokaryotes, replication and transcription are continuous when the cells are under optimal growth conditions (see Ingraham et al. 1983). Furthermore, Miller et al. (1970) showed that initiation of translation occurs as soon as a piece of the messenger RNA becomes available on the transcribing DNA.

The DNA in vivo is not in the form of its sodium or potassium salt, but is neutralised by other cations abundant in the bacterial cell, such as magnesium, putrescine and spermidine, as well as being associated with some proteins. The latter are either structurally necessary partners of DNA, comparable to the histones of eukaryotes, or functional proteins involved in the metabolic activities of DNA. It is clear that these proteins might also be involved in the supercoiling of the bacterial DNA (Pettijohn 1982).

Considering the coupling of transcription and translation, one is tempted to propose that the DNA is more or less dispersed throughout the cell, intermingled with the polysomes. Classical cytochemical work, however, suggested that even for growing cells, the DNA is not completely dispersed and local concentrations were observed, the "chromatinic bodies", "nuclear equivalents" or "nucleoids". These localised nucleoids are seen in growing cells by optical microscopy (Mason and Powelson 1956; Robinow 1956; Valkenburg et al. 1985) as regions of lower density, and are comparable to those observed by electron microscopy after fixation (Kellenberger 1953; Fig. 1). These nucleoids were similarly shown after hydrolysis and Feulgen-type staining (Piekarski 1937; Robinow 1944, 1956), after RNAse digestion (Tulasne and Vendrely 1947) or by acridine orange fluorescence (Séchaud and Anderson 1960,

* Dedicated to Karl Lickfeld in honor of his 60th birthday.

1 Department of Microbiology, Biozentrum of the University of Basel, 4056 Basel, Switzerland
2 Basel Institute for Immunology, 4005 Basel, Switzerland

Bacterial Chromatin
(ed. by C.O. Gualerzi and C.L. Pon)

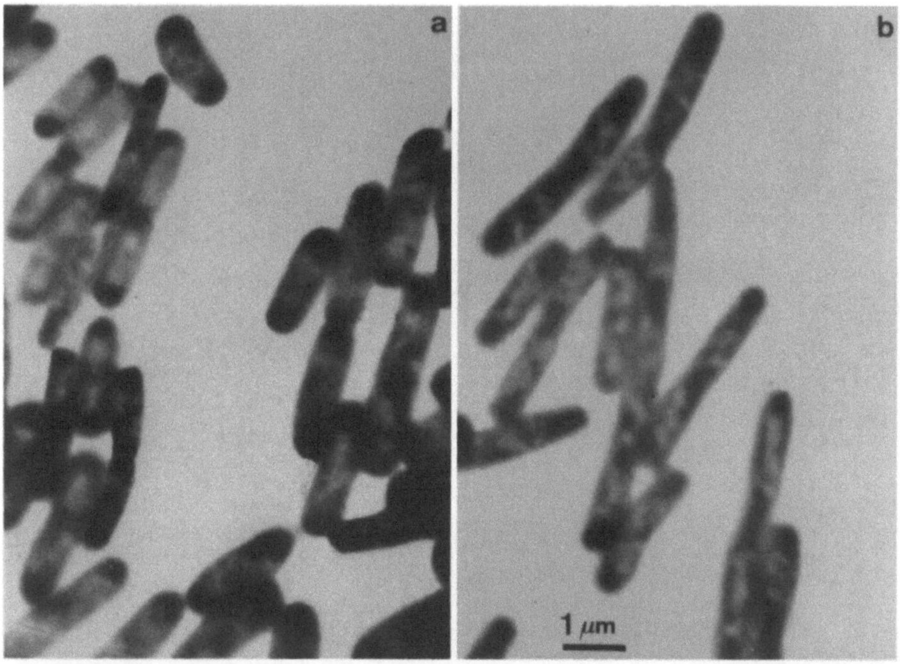

Fig. 1a,b. *E. coli* K12 cells from an exponentially growing culture (2×10^8 cells/ml) prepared by the agar filtration method (Kellenberger and Bitterli 1976) following (**a**) fixation for 10 min by adding 4% formaldehyde to the culture and (**b**) fixation for 10 min by adding 0.2% OsO_4 to the culture. After filtration onto the supporting film, both samples were postfixed with vapors of either formaldehyde or OsO_4. Observed unsectioned (whole cells), the differences in the results obtained with the two fixatives are not overwhelming, although with aldehyde the nucleoid seems to be more dispersed, thus leading to a more diffuse aspect [Micrograph out of the archives (Nos. 16862 and 16812) from A. Ryter and E. Kellenberger 1955]

unpublished; for methods, see Séchaud 1960; Anderson et al. 1959). Antibiotics, phage infection and radiation produce specific changes in the shape of these nucleoids visible by at least several of the different techniques enumerated above (Kellenberger 1953; Kellenberger et al. 1958; Ryter 1960; Daneo-Moore and Higgins 1972).

Fixation by aldehydes and OsO_4 promote a permeabilisation of the cells, evidenced by the immediate leakage of potassium and then of magnesium from the cells (Moncany 1982; Moncany et al., in preparation). These effects, per se and in association with differnt salt concentrations of the medium, induce changes in nucleoid shapes (Whitfield and Murray 1956; Woldringh 1973) rendering a precise evaluation of the degree of dispersion rather difficult (Fig. 2). The observation of frozen-hydrated cryosections (Dubochet et al. 1983) and of cryosubstituted resin sections (Hobot et al. 1985) confirmed that the DNA of growing cells is not completely dispersed throughout the cells. In the following sections, we will show thin sections of cells obtained after cryosubstitution. The new technique also allows us to use immunocytochemical techniques to localise single-stranded and double-stranded nucleic acids as well as putative partners of the DNA. We show the first results obtained on the intracellular localisation of the protein HU (Rouviere-Yaniv and Gros 1975).

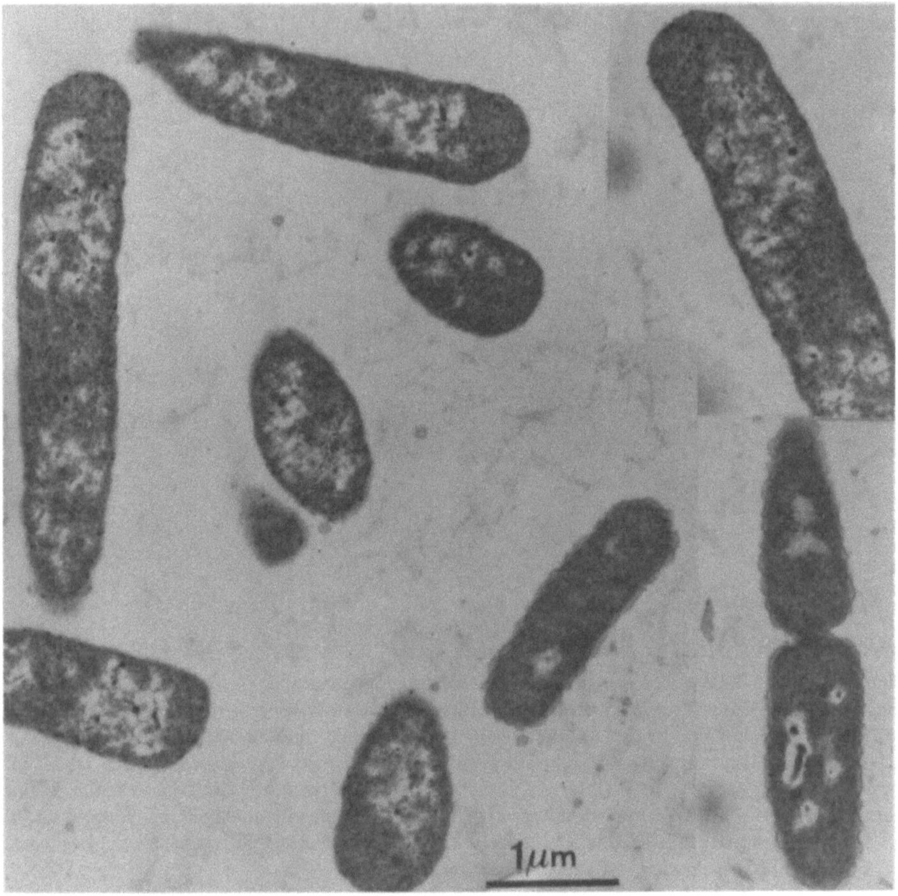

Fig. 2. Two aliquots of the same exponentially growing culture of *E. coli* B cells were fixed with either (1) 1% glutaraldehyde for 120 min or (2) 1% OsO_4 for 120 min. The two cultures were then washed and mixed together in liquid agar. The mixture was dehydrated in ethanol at room temperature and further processed for embedding in Epon. On the *left*, with low magnification, we see a uranyl acetate/lead citrate-stained section of the mixture: the cells fixed with the two fixatives show clear differences in the dispersion of the nucleoid. The OsO_4-treated cells exhibited more shrinkage and thus a different aspect of the ribosomes. In both cases the DNA plasms aggregated during room temperature dehydration, resulting in "black" precipitates in an "empty nuclear vacuole", as is easily visible in the two *insets* on the *right* (*top:* glutaraldehyde fixed cell; *bottom:* OsO_4 fixed cells). Aggregation of DNA is prevented by OsO_4 fixation under RK conditions (Kellenberger and Ryter 1964). Under these conditions, as well as with InCl, uranyl acetate and $KMnO_4$, a viscous sol of DNA is transformed by cross-linking into a gel (Kellenberger et al. 1981). Both micrographs were made by E. Carlemalm and R. Gyalog from our laboratory

In normally growing *E. coli* B cells, the intracellular concentrations of potassium can vary from 100 to 600 mM, according to the outside osmotic pressure (Epstein and Schultz 1965). Magnesium is always high (70–150 mM) as shown by several authors (Moncany and Kellenberger 1981; Günther and Dorn 1969; Kung et al. 1976; Lusk et al. 1968; Süssmuth and Widmann 1979; Chang et al. 1984). Equilibrium dialysis experiments performed in our laboratory with M. Moncany (thesis 1982, and publication in preparation) have shown that Mg, putrescine and spermidine bind preferentially to DNA. We have characterised the proteins (in SDS polyacrylamide gels) which bind to DNA neutralised with these cations and compared them with those bound to the usual Na-salt. The spectra of binding proteins, obtained in these two ways, show mainly quantitative differences, but also some qualitative ones. The significance of the results from these approaches (DNA-cellulose chromatography and immunolabelling of embedded cryosubstituted cell sections) are discussed with respect to the structural organisation of DNA.

1 Nucleoid Morphology and Fine Structure

The fine structure of the bacterial nucleoid has been difficult to characterise in situ with conventional electron-microscopical techniques for thin sectioning, owing not only to the observed differences in the nucleoid shape following OsO_4 or aldehyde fixation, but also because of the influence of the external salt concentration in the growth medium (Whitfield and Murray 1956; Kellenberger and Ryter 1964; Woldringh 1973). Both adlehydes and OsO_4 induce a rapid leakage of small cellular solutes (Moncany 1982; Woldringh 1973), which equilibrates with the external concentration. This can affect the degree of dispersion of the nucleoid. Centrifugation causes ionic leakage in bacteria, which can be much reduced when filtration (with the correct maintenance of temperature and aeration) is used instead (Epstein and Schultz 1965; Moncany and Kellenberger 1981). We used the principles of ultra-rapid freezing (thus avoiding the use of any chemical fixatives beforehand), with prior harvesting of the bacterial cells by filtration. By coupling this to the method of cryosubstituation we hoped to obtain a clearer picture of the nucleoid shape (Hobot et al. 1985).

Unfixed, metabolically active bacterial cells (*Escherichia coli* B) were frozen by projection onto a copper block cooled via liquid helium (Heuser et al. 1979; Escaig 1982) followed by substitution at –90°C in a 2.5% OsO_4/acetone mixture and subsequent embedding in Epon (Hobot et al. 1985). The same results were later obtained when frozen cells were substituted in 3% glutaraldehyde/acetone (–90°C) and embedded in Lowicryl K4M at –35°C (Fig. 3A). The cryosubstituted bacterial cells show ribosome-free areas filled with rather grainy structures intermingled with fibrous elements. Grains of up to 15 nm and segments of fibers in the range of 4–6 nm can be seen (Fig. 3A). This holds not only for the larger, ribosome-free areas, but also for the smaller spaces between the ribosomes. The ribosome-free area (shaded for easier observation in Fig. 3B) appears to be localised within the cytoplasm as a specific region having a cleft (very indented leaf shape) outline. Varying the external salt concentration had no effect on the morphological appearance of the ribosome-free area. *Bacillus subtilis* cells after cryosubstitution revealed a similar aspect of the ribosome-free area as seen in *E. coli* B, and also lacked mesosomes (Hobot et al. 1985; Dubochet et al. 1983; Ebersold et al. 1981), suggesting them to be artefacts of preparation procedures in confirmation of many previous observations with freeze fracture. We further demonstrated (Hobot et al. 1985) that the ribosome-free areas contain the

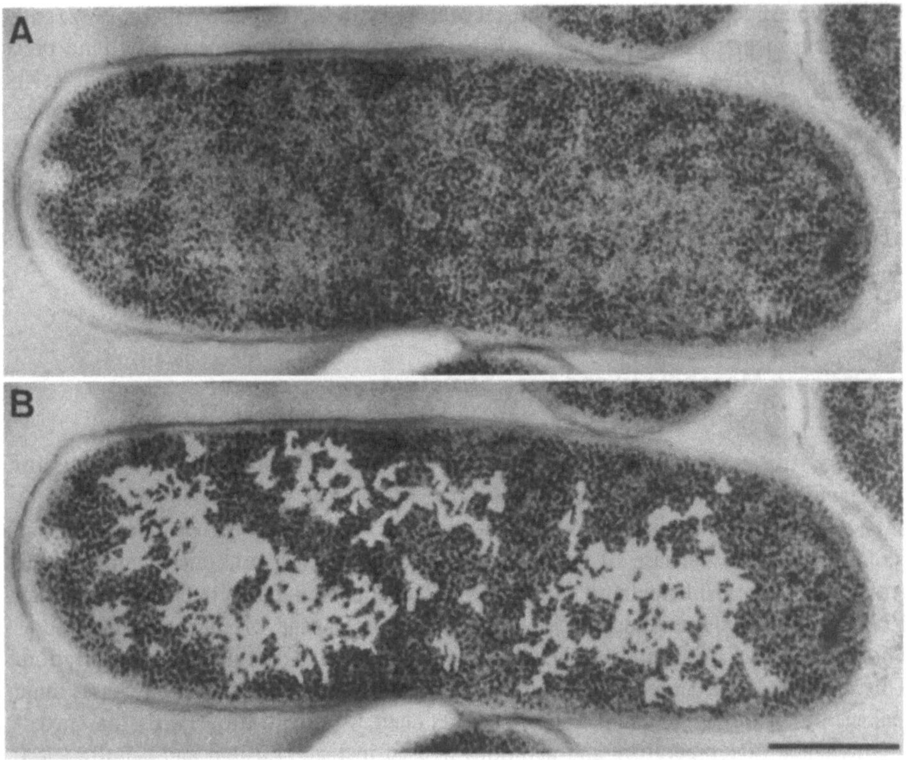

Fig. 3A,B. Morphology of the bacterial nucleoid in cryosubstituted *E. coli* B cells. Exponentially growing cells were rapidly frozen on a helium-cooled copper block and substituted at low temperature (−90° C) in 3% glutaraldehyde in acetone with subsequent infiltration and polymerisation in Lowicryl K4M at −35° C. **A** The ribosome-free area appears localised, having a cleft outline, and is filled with a fibrillar and fine-grained plasm. **B** In this micrograph, the ribosome-free area has been *shaded* for clarity. (The sections in these and subsequent micrographs, Figs. 5–8, have been stained with 4% uranyl acetate and Millonig's lead acetate. The *bar* marked in Figs. 3–8 is 0.5 μm)

bulk of the DNA by using Gautier's (1976) osmium ammine stain which is specific for DNA. In this procedure, following H_2O_2 treatment of the thin sections to remove any OsO_4, which might interfere with the staining reaction, the thin sections are subjected to hydrolysis with 5 N HCl for 15 min at 20°C. This selectively hydrolyses RNA, leaving the DNA alone to react with the osmium ammine reagent. To hydrolyse the DNA, much longer incubation times are needed (A. Gautier, personal communication). The net result of this combined H_2O_2 and HCl hydrolysis treatment is presented in Fig. 4, where the darker regions are analogous to the ribosome-free areas and contain DNA. The lighter region, bleached of OsO_4, corresponds to the cytoplasm/ribosomal RNA. However, this technique is not sensitive enough to reveal DNA, which could be dispersed among the ribosomes, corresponding to the grainy/fibrous areas seen between the ribosomes in Fig. 3.

Fig. 4. Thin section of cryosubstituted *E. coli* B frozen as in Fig. 3, but substituted in 1% OsO_4 in acetone and embedded in Epon. Sections were treated with 2% H_2O_2 (20 min, 20°C) to bleach away the OsO_4, and with 5 N HCl (15 min, 20°C) to hydrolyse RNA selectively [basis of Gautier's (1976) osmium-ammines staining]: The darker regions of the cell correspond now to the ribosome-free area described in Fig. 3. No section staining used

2 Localisation of DNA

The technique of cryosubstitution has given us not only new insights into the overall morphology of the nucleoid/ribosome-free area itself, but has also given a novel aspect for the fine structure of the DNA-containing plasma. The already mentioned grainy/fibrillar appearance of the DNA-containing plasma is different from the fine fibrillar aspect obtained after OsO_4 fixation under RK conditions by a much higher content of "grainy" elements. These differences could be due either to modifications of supercoiling by OsO_4-RK (Lilley and Palecek 1984) and/or by a destruction of associated protein partners (Baschong et al. 1984). Further experiments are needed to reach an understanding. From these comparisons, it becomes clear that the compact aspects of "black particles" within an empty "nuclear vacuole" (Fig. 2) is purely an artefact of aggregation occurring during the dehydration procedures at room temperature.

The DNA-containing plasma of the nucleoid is, therefore, a highly hydrated gel, of which the organisational details are still unknown. It is much less compact than, e.g., the DNA of a phage head with its 50–75% (v/v) of water (Earnshaw et al. 1978). Embedding of the cryosubstituted specimens in Lowicryl K4M (a low-temperature resin for use down to –35°C, which gives, at these temperatures, improved structural preservation; Carlemalm et al. 1982) made on-section immunolabelling with the protein-A gold technique (Roth et al. 1981) possible.

We used sera from autoimmune mice (gift of Prof. R. Franklin, Biozentrum, Basel), for our DNA localisation studies. MRL/Mp mice (Olac 1976 Ltd.) are a model system for lupus erythematosus, apparently related to the production of autoimmune antibodies. The serum used in our labelling of *E. coli* B cells exhibits a high affinity for double-stranded DNA in ELISA tests (Munns et al. 1984; R. Franklin, personal communication). Following incubations of thin sections of *E. coli* B with this particular serum, and subsequent treatment with a protein-A gold (14 nm diameter) solution, the distribution of double-stranded DNA was found to be over regions corresponding to the ribosome-free area (Fig. 5A). In this micrograph, obtained with sections that

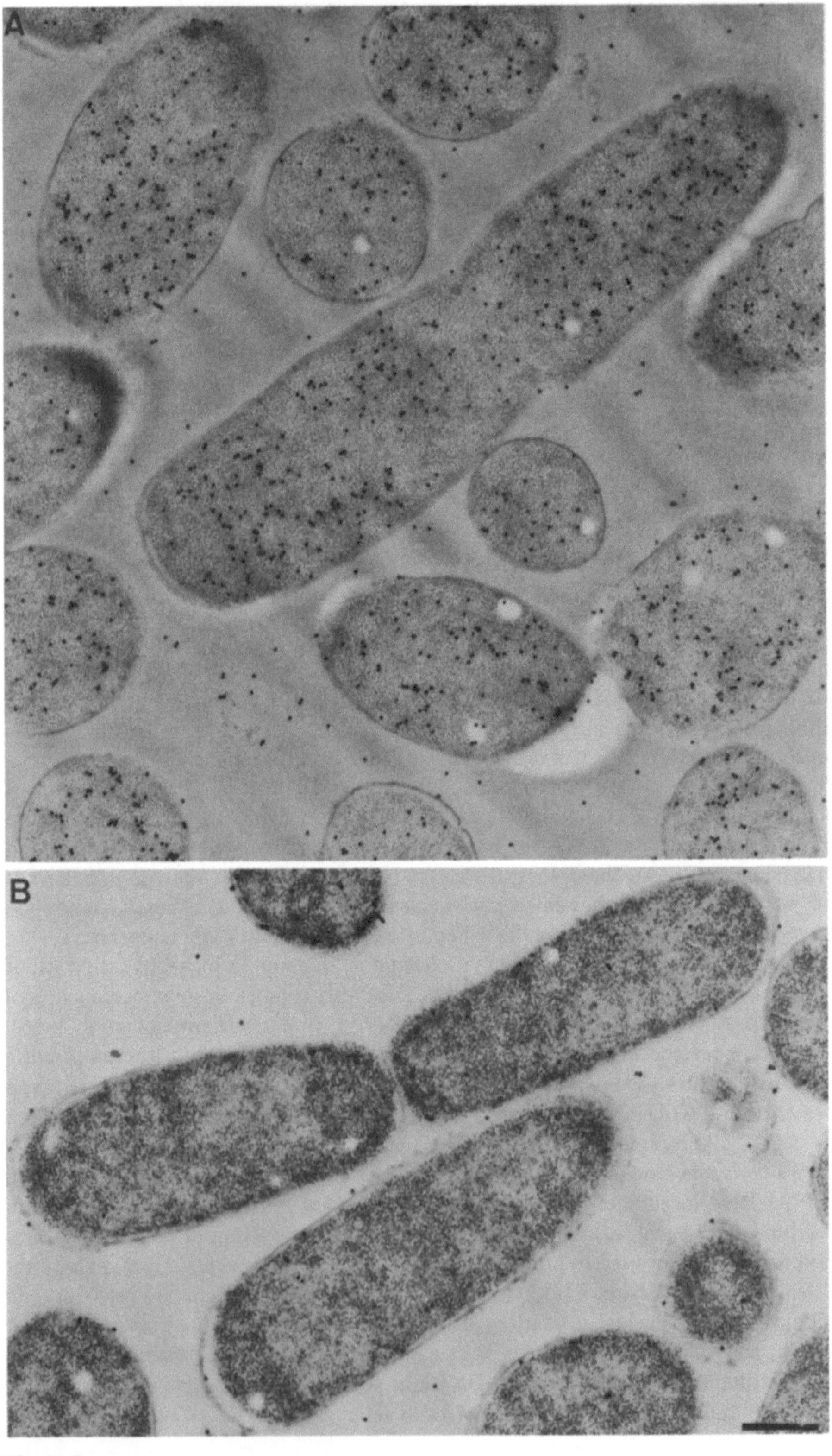

Fig. 5A,B

Table 1. Intracellular distribution of colloidal gold particles in cells immunolabelled with MRL/Mp sera

Number of gold particles per μm^2 [a] Ribosome-free area	Cytoplasm	Background
59.6 ± 10.2[b]	4.7 ± 1.3	2.5 ± 0.9
128.3 ± 17.9[c]	13.1 ± 5.3	10.6 ± 0.8

[a] The areas of the respective regions of the micrographs were measured with a Numonics Corp, electronic graphics calculator coupled to an Apple computer.
[b] Average values from 13 micrographs.
[c] Average values from 3 micrographs.

have been contrasted with uranyl and lead actate staining (Reynolds 1963; Millonig 1961), the distinction between ribosome-free areas (dark grey areas) and cytoplasm is easily visible when comparing with Fig. 3A, over which 86% of the gold particles are located in these ribosome-free areas (Table 1). The strong staining of the DNA-containing plasma with this serum, in the absence of gold label is currently being investigated. If, for example, the serum is preincubated with calf thymus DNA cellulose prior to immunolabelling, neither specific gold labelling nor direct staining of the ribosome-free area is observed (Fig. 5B). The direct staining of the ribosome-free areas by the serum of this autoimmune mouse is observed with the protein-A gold incubation step being omitted. This very specific staining effect is even more striking in chloramphenicol-treated cells incubated with this same serum (Fig. 8, see below), where the "ring" of DNA is clearly stained grey. In both cases, the grainy/fibrillar aspect of the ribosome-free area is hidden under the stain.

3 Nucleoid Components

The bacterial chromosome is negatively supercoiled and is segregated into domains of supercoiling (Pettijohn 1982; Kornberg et al. 1974; Sinden and Pettijohn 1981). The

Fig. 5A,B. Immunocytochemical localisation of bacterial DNA within the ribosome-free area. **A** Thin sections of cryosubstituted *E. coli* cells in K4M were immunolabelled utilising the protein-A-colloidal gold technique (Roth et al. 1981) with sera from MRL/Mp mice exhibiting an affinity for double-stranded (ds) DNA. Sections were incubated with antibody following a preincubation step with 0.5% ovalbumin in physiologically-buffered saline (PBS). After washing with PBS, sections were treated with a protein-A-colloidal gold solution. Protein-A-coated gold particles now attached to the Fc piece of the specific IgG antibody fractions and give an indication of the distribution/localisation of the antigen in question. The slightly darker grey aspect of the ribosome-free areas, within which the gold label is distributed, showing DNA localisation, is observed when immunolabelling with whole sera (cf. Fig. 8a). **B** *E. coli* B treated with the above serum, but which had been preincubated with calf-thymus DNA-cellulose prior to immunolabelling. No specific gold labelling is seen. Also, the grainy/fibrillar aspect of the ribosome-free area is the same as with untreated sections

negatively supercoiled state of the DNA is apparently maintained by the antagonistic actions of topoisomerase I and DNA gyrase (Menzel and Gellert 1983) and is important for a number of functions, for example, DNA replication and transcription (reviewed in Gellert 1981; Vosberg 1985). The proteins involved in maintaining the domains and structure of the nucleoid (what we call protein partners) have not yet been characterised, owing primarily to the instability of nucleoid complexes during isolation. Griffith reported beaded fibers in *E. coli* DNA when cells were lysed on electron microscopic films and immediately dehydrated (Griffith 1976). The structures, unlike eukaryotic nucleosomes, were extremely labile; they could not be isolated and their protein composition was not determined. In vivo labelling of bacterial DNA with trimethylpsoralen indicated that a significant portion of the supercoils are unrestrained and in a state of torsional tension (Sinden et al. 1980). This is in marked contrast to the nucleosomic packing of eukaryotic DNA.

Bacterial DNA-binding proteins, including HU, have been identified in cell extracts, which may play a role in the organisation of the nucleoid (Geider and Hoffmann-Berling 1981; Rouviere-Yaniv and Gros 1975; Laine et al. 1984). However, whether these proteins are structural partners of the DNA in a manner analogous to eukaryotic histones or are associated with DNA transcription or replication is unclear. In vitro the HU protein, for example, forms beaded structures with DNA (Rouviere-Yaniv et al. 1979); however, it is also involved in DNA replication (Dixon and Kornberg 1984) and inhibits DNA transcription (Losso et al. 1982).

The preservation of the DNA plasm in cryosubstituted and K4M embedded cells coupled with the technique of immunolabelling with protein-A colloidal gold has enabled us to begin a detailed study of the structure of the DNA plasm. Thin sections of *E. coli* cells were labelled with monoclonal IgG's exhibiting a high affinity for single-stranded DNA, but also with a lower affinity for RNA (Eilat et al. 1984). The protein-A colloidal gold particles, subsequently bound to the IgG's, are distributed along the DNA-plasm ribosome border or just within the ribosome area, close to the border (Fig. 6). This is in striking contrast to the uniform distribution of the double-stranded DNA in the DNA plasm seen in Fig. 5A. The localisation of single-stranded nucleic acids, accessible to the IgG's, within this DNA plasm/ribosome border zone is consistent with the coupling of DNA transcription and RNA translation in bacteria. A comparison of these observations (Figs. 5A and 6) might serve to distinguish functional regions of the bacterial nucleoid (transcriptional activity) from inactive regions.

The identification of different regions of the nucleoid in this manner encouraged us to look at the intracellular distribution of a putative protein partner of the DNA. Protein HU was purified essentially as described by Imber et al. (1982) and polyclonal antibodies were raised in rabbits. IgG's were purified by DEAE ion exchange chromatography or affinity purified with HU protein on strips of nitrocellulose filters using principles described by Olmsted (1981). The results of immunolabelling experiments

Fig. 6A,B. Distribution of single-stranded DNA along the ribosome-nucleoid border zone in cryosubstituted *E. coli* B. **A,B** As described in Fig. 5, thin sections were immunolabelled with monoclonal IgG's exhibiting a high specificity for single-stranded (ss) DNA. The pattern of gold labelling is along the ribosome-nucleoid border, perhaps tending to be in the more dispersed parts of the ribosome-free areas (i.e. nucleoid)

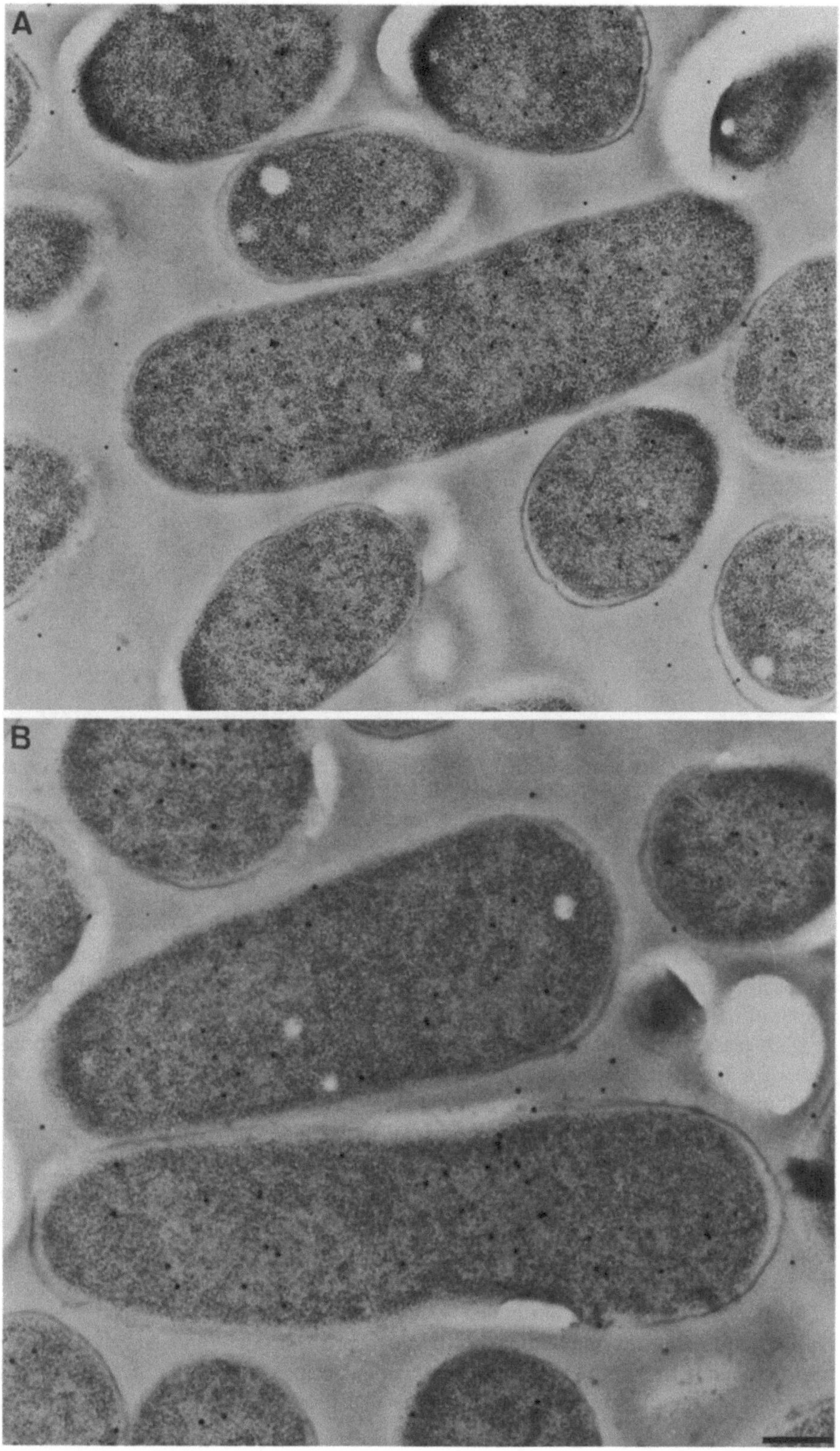

Fig. 6A,B

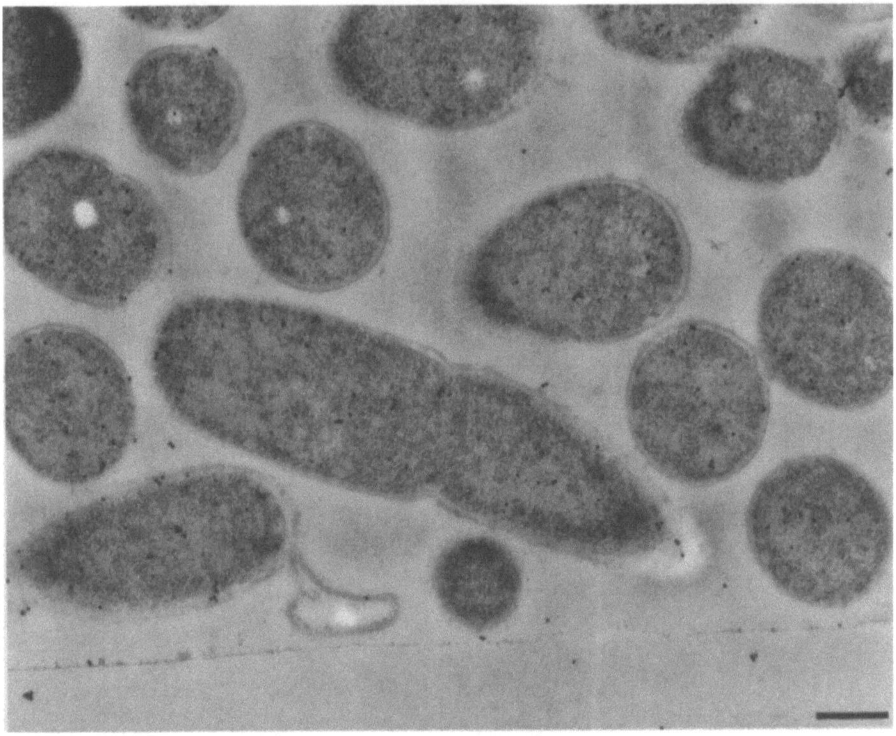

Fig. 7. Intracellular localisation of the HU protein along the ribosome-nucleoid border zone in cryosubstituted *E. coli* B cells. As described in Fig. 5, thin sections were immunolabelled with DEAE-purified rabbit IgG's specific for the protein HU by the protein-A-colloidal gold technique. The pattern of labelling along the ribosome-nucleoid border is the same as in Fig. 6

with these IgG's are represented by Fig. 7. The gold particles are clearly not contained within the DNA plasm, which would be expected for protein partners involved in compacting inactive DNA. Rather, they appear within, or close to, a zone defining the border of the DNA plasm with ribosomes. It is difficult to distinguish the smaller clefts of DNA plasm within ribosome areas in some cells, and this could explain the appearance of a few particles in the cytoplasm. This is true for Fig. 6 as well. The overall distribution of HU is similar to that observed with monoclonal IgG's specific for single-stranded nucleic acids (Fig. 6) and might be representative of an association between HU and transcriptionally active or replicating DNA. We have to keep in mind the possibility that wrapping or coiling of DNA around an HU core in a "nucleosome-like" structure might block antigenic determinants, resulting in no labelling of these complexes. While we cannot exclude this argument, the use of polyclonal IgG's in these experiments reduces the likelihood of epitope inaccessibility.

Preliminary results with antibodies specific for topoisomerase I indicate a similar localisation of this enzyme within the DNA-cytoplasm border zone. At this point, we might speculate that in the absence of a restrained packing of the DNA, transcriptional and replicative complexes may be necessary and sufficient for maintaining the domains

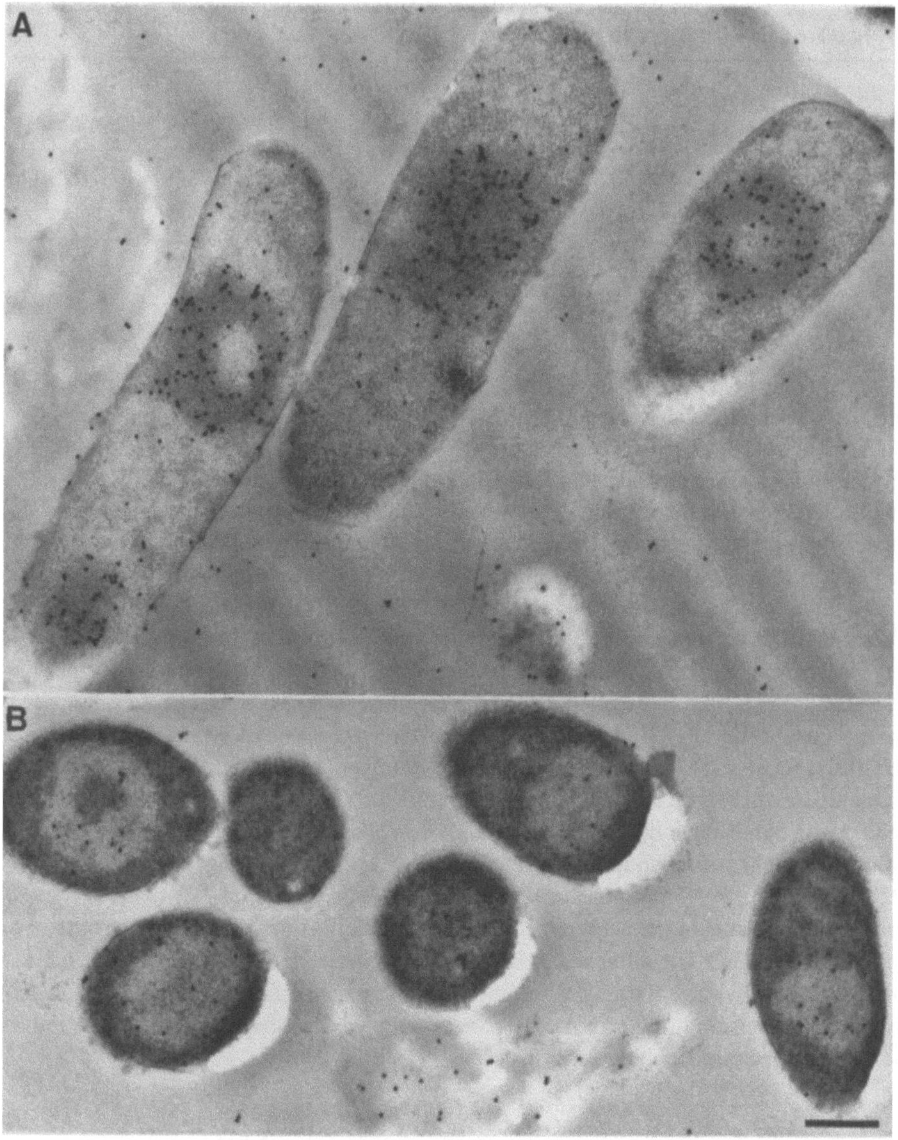

Fig. 8A,B. Immunolabelling (see legend, Fig. 5) of ds- and ss-DNA in thin sections of chloramphenicol (CAM)-treated cells. Exponentially growing cells were treated with 25 mg ml^{-1} CAM for 60 min prior to freezing and substitution. Thin sections were labelled with MRL/Mp sera specific for ds DNA (**A**) and monoclonal IgG's exhibiting higher specificity for ss DNA (**B**) as described above. In both, the gold label is distributed over the ribosome-free area, which appears as a ring-like, doughnut structure. No label appears in the centre of the ring. The darker aspect of the nucleoid ring is seen only with the sera for MRL/Mp mice (**A** and Fig. 5A), and not with the monoclonal IgG's (**B** and Fig. 6)

Table 2. Intracellular distribution of colloidal gold particles in chloramphenicol (25 $\mu g\ ml^{-1}$) treated cells immunolabelled with MRL/Mp sera

Length of chloramphenicol treatment (min)	Number of gold particles per μm^2		
	Ribosome-free area	Cytoplasm	Background
30	103.0 ± 13.3[a]	16.2 ± 2.8	6.1 ± 5.4
60	115.6 ± 8.1[a]	13.7 ± 2.9	3.2 ± 0.4

[a] Average values for 5 micrographs.

of supercoiling of the *E. coli* chromosome, by virtue of their interaction with ribosomes and the cytoplasmic membrane.

We plan to continue our study of nucleoid structure in vivo by determining the intracellular location and distribution of other DNA-binding proteins and RNA polymerase in normal and antibiotic-treated cells, and in bacteriophage-infected cells. The morphology of the bacterial nucleoid changes dramatically when cells are treated with sublethal doses of chloramphenicol (25 $\mu g\ ml^{-1}$) (Kellenberger et al. 1958). The hollow sphere shape of the DNA plasm (Fig. 8A and B) observed following chloramphenicol CAM-induced inhibition of protein synthesis may result from the disruption of polysome association with transcriptionally active DNA. The hollow spherical nucleoid in Fig. 8A is clearly stained a darker grey with the MRL/Mp sera described in the previous section (Fig. 5). The gold particles are evenly distributed throughout the structure and are not found in significant number in the rest of the cell (Fig. 8A, Table 2). A more typical uranyl-lead staining of thin sections of CAM-treated cells is seen with monoclonal IgG labelling in Fig. 8B. Surprisingly, single-stranded nucleic acids are found throughout the entire nucleoid, and are no longer confined to the DNA-cytoplasm border zone as in untreated cells (Fig. 6). In contrast to these results, initial immunolabelling experiments with HU-specific IgG's may indicate that the protein is localised near the outer ribosome border of the nucleoid sphere. Further analysis of cells treated with CAM and other antibiotics (such as Rifamycin) by immunolabelling, might give some insight into the role of coupled transcription and translation in defining nucleoid morphology and structure.

4 Perspectives

The identification in vitro of DNA-binding proteins possibly involved in the structural organisation of the bacterial nucleoid typically includes the isolation of intact nucleoids and/or DNA cellulose chromatography. However, in most cases, the binding of these proteins to DNA is carried out under conditions far from those found in vivo; at salt concentrations of 100 mM or less and in the presence of EDTA (Pettijohn 1982; Kornberg et al. 1974; Yamazaki et al. 1984). The intracellular concentrations of K vary between 100 mM and 650 mM, depending on the osmolality of the medium

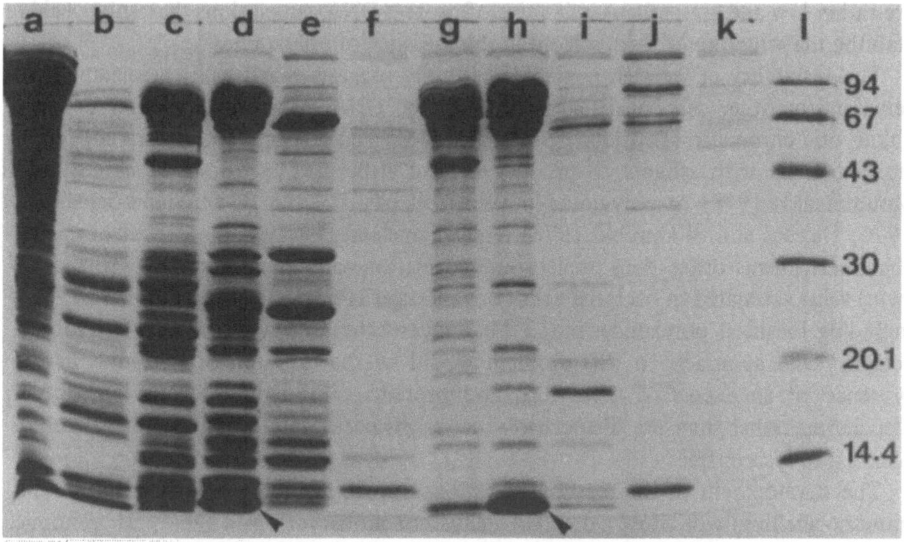

Fig. 9. *E. coli* proteins bound to DNA-cellulose in the presence of Na-EDTA or Mg and polyamines. Successive fractionations with ammonium sulfate were made with a bacterial cell extract following sonication in high salt buffer and a high-speed centrifugation step to remove cellular debris. The pellet resulting from 30% to 50% saturation with ammonium sulfate was resuspended and dialysed against Na-EDTA or Mg, spermidine and putrescine-containing buffers and applied to calf-thymus DNA-cellulose columns equilibrated with the same buffer. Proteins were eluted with steps of increasing salt concentration, lyophilised and analysed following SDS polyacrylamide gel electrophoresis and staining with Coomassie blue. *Lanes b–f* represent fractions obtained from Na-EDTA saturated DNA cellulose, and *lanes g–k* contain fractions from Mg-polyamines saturated DNA cellulose. In both cases, the proteins were eluted, from *left to right* with 0.1, 0.2, 0.3, 0.5, and 1 M NaCl. *Lane a* contains the flow through from the Na-EDTA DNA column. The molecular weights ($\times 10^3$) of the protein standard (*lane l*) are indicated on the *right*. The position of protein HU is designated by *arrows* in *lanes d* and *h*

(Epstein and Schultz 1965), while Mg, putrescine and spermidine are normally found at levels of approximately 70–150 mM, 50 mM and 20 mM, respectively (Moncany 1982; Moncany and Kellenberger 1981). Equilibrium dialysis experiments indicate that most of the intracellular Mg, putrescine and spermidine is bound to the DNA, thus presenting a rather different substrate for protein binding than the Na-salt formed with Na-EDTA.

In our attempt to more clearly define the nucleoid morphology and components in vivo, we have also begun a detailed comparison of proteins binding to DNA saturated with Na (in the presence of EDTA) or Mg and polyamines. Some preliminary results are presented in Fig. 9. Fewer proteins bind to the Mg-polyamine DNA columns as compared with the Na-EDTA DNA column, especially in the lower molecular weight range. Many proteins that bind to both do so with lower affinity in the presence of Mg-polyamines (for example, HU binding in lanes d and h); however, other proteins bind with greater affinity (see lane j). We are pursuing this type of analysis in order to identify additional putative partners which bind to DNA under physiological conditions. Although the results are preliminary, the binding affinities of many proteins are rather low and are perhaps indicative of protein-DNA associations that only weakly stabilise the structure of the nucleoid in the presence of polyamines.

The instability of isolated nucleoids certainly suggests that weakly associated proteins and/or other cell constituents may gel the DNA in a way very different from eukaryotic chromatin (Pettijohn 1982). The biological relevance of in vitro condensation of DNA with ethanol to produce beaded structures or toroids (Eickbush and Moudrianakis 1978) or polyamine-induced collapse of DNA (Gosule and Schellman 1976; Thomas and Bloomfield 1978) remains unclear. However, these studies indicate that mechanisms other than nucleosome formation can concentrate DNA to the 5% (w/v) value estimated in bacterial cells (Kellenberger 1962; Kellenberger et al., this vol.); including localised polyamine and/or Mg condensation and weak protein-DNA interactions. One approach to this problem would be the isolation of nucleoids in the presence of an excess of these ions and proteins. This would favour ion/protein association rather than the dissociation that might occur due to dilution of these complexes into lysis buffer.

The development of the methods described above (including cryosubstitution and immunolabelling) will allow a detailed analysis of prokaryotic and eukaryotic genomic organisation and will provide an opportunity to relate studies of protein-DNA interactions in vitro to structure in vivo. In particular, comparisons of antibiotic-treated cells and viral-infected cells with normal cells might define the effects of specific cellular functions and/or components on chromatin and nucleoid structure. Specific examples might include (1) the CAM-treated bacterial cells described in Sect. 3; (2) bacteriophage T4-infected *E. coli* cells, where the nucleoid and the host DNA is degraded and used for synthesising the vegetative phage DNA; and (3) adenovirus-infected HeLa cells, where viral-encoded proteins replace histones in organising viral DNA late in infection (Burg et al. 1983; Marcus-Sekura and Carter 1983).

Acknowledgements. We are particularly indepted towards Profs. Richard M. Franklin and Thomas A. Bickle who have helped us with many suggestions and R. Dan Eilat and Dr. James C. Wang for the generous gifts of A52 monoclonal IgG's and *E. coli* topoisomerase I IgG's, respectively. We thank Dr. Eric Carlemalm and Renate Gyalog for contributing Fig. 2, Cornelia Kellenberger for excellent technical assistance and Dr. Michel Wurtz and Lilo Müller for efficient help in preparing the illustrations. We are most grateful to Marianne Schäfer for patiently typing and modifying our manuscript.

This work was supported by the state and canton of Basel Stadt and the Swiss National Science Foundation through a grant to E.K. and a postdoctoral exchange fellowship to M.A.B. The Basel Institute for Immunology (A.S.K.) is founded and supported by F. Hoffmann-La Roche & Co. Ltd., Basel, Switzerland.

References

Anderson ES, Armstrong JA, Niven SF (1959) Fluorescence microscopy: observation of virus growth with aminoacridines. In: Virus growth and variation. University Press Cambridge, pp 224–255

Baschong W, Baschong-Prescianotto C, Wurtz M, Carlemalm E, Kellenberger C, Kellenberger E (1984) Preservation of protein structures for electron microscopy by fixation with aldehydes and/or OsO_4. Eur J Cell Biol 35:21–26

Burg Jl, Schweitzer J, Danniell E (1983) Introduction of superhelical turns into DNA by adenoviral core proteins and chromatin assembly factors. J Virol 46:749–755
Carlemalm E, Garavito RM, Villiger W (1982) Resin development for electron microscopy and an analysis of embedding at low temperature. J Microsc (Oxf) 126:123–143
Chang C-F, Shuman H, Somlyo AP (1984) Electron probe analysis, X-ray mapping and electron energy loss spectroscopy of elemental distribution of *E. coli* B. In: Bailey GW (ed) Proc 42th Annu Meet Electron Microsc Soc Am. San Francisco Press, San Francisco, pp 570–571
Daneo-Moore L, Higgins ML (1972) Morphokinetic reaction of *Streptococcus faecalis* (ATOC 9790) cells to the specific inhibition of macromolecular synthesis: nucleoid condensation on the inhibition of protein synthesis. J Bacteriol 109:1210–1220
Dixon N, Kornberg A (1984) Protein HU in the enzymatic replication of the chromosomal origin of *Escherichia coli.* Proc Natl Acad Sci USA 81:424–428
Dubochet J, McDowall AW, Menge B, Schmid EN, Lickfeld KG (1983) Electron microscopy of frozen-hydrated bacteria. J Bacteriol 155:381–390
Earnshaw WC, King J, Eiserling FA (1978) The size of the bacteriophage T4 head in solution with comments about the dimension of virus particles as visualized by electron microscopy. J Mol Biol 122:247–253
Ebersold HR, Cordier JL, Lüthy P (1981) Bacterial mesosomes: method dependent artifacts. Arch Microbiol 130:19–22
Eickbush TH, Moudrianakis EN (1978) The compaction of DNA helices into either continuous supercoils or folded fiber rods and toroids. Cell 13:295–301
Eilat D, Hochberg M, Pumphrey J, Rudikoff S (1984) Monoclonal antibodies to DNA and RNA from NZB/NZW F_1 mice: antigenic specificities and NH_2 terminal amino acid sequences. J Immunol 133:489–494
Epstein W, Schultz SG (1965) Cation transport in *Escherichia coli* V. Regulation of cation content. J Gen Physiol 49:221–234
Escaig J (1982) New instruments which facilitate rapid freezing at 83K and 6K. J Microsc (Oxf) 126:221–229
Gautier A (1976) Ultrastructural localisation of DNA in ultrathin tissue sections. Int Rev Cytol 44: 113–191
Geider K, Hoffmann-Berling H (1981) Proteins controlling the helical structure of DNA. Annu Rev Biochem 50:233–260
Gellert M (1981) DNA topoisomerases. Annu Rev Biochem 50:879–910
Gosule LC, Schellman IA (1976) Compact form of DNA induced by spermidine. Nature 259:333–335
Griffith JD (1976) Visualization of prokaryotic DNA in a regularly condensed chromatin-like fiber. Proc Natl Acad Sci USA 73:563–567
Günther T, Dorn F (1969) Über die intrazelluläre Mg-Ioneninaktivität von *E. coli* Zellen. Z Naturforsch Sect B 24:713–717
Heuser JE, Reese TS, Dennis MJ, Jan Y, Jan L, Evans L (1979) Synaptic vesicle exocytosis captured by quick freezing and correlated with quantal transmitter release. J Cell Biol 81:275–300
Hobot JA, Villiger W, Escaig J, Maeder M, Ryter A, Kellenberger E (1985) The shape and fine structure of the nucleoid observed on sections of ultra-rapid frozen and cryosubstituted bacteria. J Bacteriol 162:960–971
Imber R, Bächinger H, Bickle TA (1982) Purification and characterization of a small DNA-binding protein, HB, from *Bacillus globigii.* Eur J Biochem 122:627–632
Ingraham JL, Maaloe O, Neidhardt B (1983) Growth of the bacterial cell. Sinauer, Sunderland, Mass
Kellenberger E (1953) Les formes caracteristiques des nucleoides de *E. coli* et leurs transformations dues a l'action d'agent mutagenes-inducteurs et de bacteriophages. In: Symp Citologia Batterica, Supplemento Rendiconti Ist. Superiore di Sanita, Roma, pp 45–66. Also: thesis Geneva, University Microfilms Int. P.O Box 1346, Ann Arbor, MI 48106, USA Cust. Ref: 84-03,992
Kellenberger E (1962) The study of natural and artificial DNA-plasms by thin sections. In: The interpretation of ultrastructure, vol 1. Symp Soc Cell Biol, Berne, pp 233–249

Kellenberger E, Bitterli D (1976) Preparation and counts of particles in electron microscopy: Application of negative stain in the agarfiltration method. Microsc Acta 78:131–148

Kellenberger E, Ryter A (1964) In bacteriology. In: Siegel BM (ed) Modern developments in electron microscopy. Academic Press, New York, pp 335–393

Kellenberger E, Ryter A, Séchaud J (1958) Electron microscopy study of DNA-containing plasms II. Vegetative and mature phage DNA as compared with normal bacterial nucleoids in different physiological states. J Biophys Biochem Cytol 4:671–678

Kellenberger E, Carlemalm E, Stauffer E, Kellenberger C, Wunderli H (1981) In vitro studies of the fixation of DNA, nucleoprotamine, nucleohistone and proteins. Eur J Cell Biol 25:1–4

Kornberg T, Lockwood A, Worcel A (1974) Replication of the *Escherichia coli* chromosome with a soluble enzyme system. Proc Natl Acad Sci USA 71:3189–3193

Kung FC, Raymond J, Glaser DA (1976) Metal ion content of *E. coli* versus cell age. J Bacteriol 126:1089–1095

Laine B, Sautiere P, Spassky A, Rimsky S (1984) A DNA-binding protein from *E. coli.* Isolation, characterization and its relationship with proteins H1 and B1. Biochem Biophys Res Commun 119:1147–1153

Lilley DMJ, Palecek E (1984) The supercoil-stabilised cruciform of ColE1 is hyper-reactive to OsO_4. EMBO J 3:1187–1192

Losso A, Miano, Gianfranceschi GL, Gualerzi C (1982) Proteins from the prokaryotic nucleoid II. Inhibition of DNA transcription by NS1 and NS2 (HU proteins). Biochem Int 5:423–427

Lusk J, Williams RJP, Kennedy EP (1968) Magnesium and the growth of *E. coli.* J Biol Chem 243: 2618–2624

Marcus-Sekura CJ, Carter BJ (1983) Chromatin-like structure of adeno-associated virus DNA in infected cells. J Virol 48:79–87

Mason DJ, Powelson DM (1956) Nuclear division as observed in live bacteria by a new technique. J Bacteriol 71:474–479

Menzel R, Gellert M (1983) Regulation of the genes for *E. coli* DNA gyrase: homeostatic control of DNA supercoiling. Cell 34:105–113

Miller OL Jr, Hamkalo BA, Thomas CA Jr (1970) Visualisation of bacterial genes in action. Science 169:392–395

Millonig G (1961) A modified procedure for lead staining of thin sections. J Biophys Biochem Cytol 11:736–739

Moncany MLJ (1982) Determination des conditions intracellulaires chez *E. coli.* Consequences biologiques de leur modification. Thesis Docteur d'Etat, l'Universite de Paris VII

Moncany MLJ, Kellenberger E (1981) High magnesium content of *Escherichia coli* B. Experientia (Basel) 37:846–847

Munns TW, Liszewski MK, Hahn BH (1984) Antibody-nucleic acid complexes. Antigenic domains within nucleosides as defined by solid-phase immunoassay. Biochemistry 23:2958–2964

Olmsted JB (1981) Affinity purification of antibodies from diazotized paper blots of heterogeneous protein samples. J Biol Chem 256:11955–11957

Pettijohn DE (1982) Structure and properties of the bacterial nucleoid. Cell 30:667–669

Piekarski G (1937) Cytologische Untersuchungen an Parathyphus und Coli Bakterien. Arch Mikrobiol 8:428–439

Reynolds ES (1963) The use of lead citrate at high pH as an electron-opaque stain in electron microscopy. J Cell Biol 17:208–213

Robinow CF (1944) Cytological observations on bact. coli, proteus vulgaris and various acrobic spore-forming bacteria with special reference to the nuclear structures. J Hyg 43:413–423

Robinow CF (1956) The chromatin bodies of bacteria. Bacteriol Rev 20:207–242

Roth J, Bendayan M, Carlemalm E, Villiger W, Garavito RM (1981) Enhancement of structural preservation and immunocytochemical staining in low temperature embedded pancreatic tissue. J Histochem Cytochem 29:663–671

Rouvière-Yaniv J, Gros F (1975) Characterization of a novel low molecular weight DNA-binding protein from *Escherichia coli.* Proc Natl Acad Sci USA 72:3428–3432

Rouvière-Yaniv J, Yaniv M, Germond JE (1979) *E. coli* binding protein HU forms nucleosome-like structures with circular double-stranded DNA. Cell 17:265–274

Ryter A (1940) Etude au microscope electronique des transformations nucleaires de E. coli K12S et K12S(λ26) apres irradiation aux rayons ultraviolets et aux rayons X. J Biophys Biochem Cytol 8:399–412

Ryter A, Kellenberger E, Birch-Andersen A, Maaloe O (1958) Etude au microscope electronique de plasma contenant de l'acide desoxyribonucleique I. Les nucleoides des bacteries en croissance active. Z Naturforsch Sect B 13:597–605

Séchaud J (1960) Developpement intracellulaire du coliphage Lambda. Arch Sci (Geneva) 13: 427–474

Sinden RR, Pettijohn DE (1981) Chromosomes in living Escherichia are segregated into domains of supercoiling. Proc Natl Acad Sci USA 78:224–228

Sinden RR, Carlson JD, Pettijohn DE (1980) Torsional tension in the DNA double helix measured with trimethylpsoralen in living *E. coli* cells: analogous measurements in insect and human cells. Cell 21:773–783

Süssmuth R, Widmann A (1979) γ-irradiated ribosomes from Micrococcus radiodurans in a cell-free protein synthesizing system. Z Naturforsch Sect C 34:565–569

Thomas T, Bloomfield V (1978) Collapse of DNA caused by trivalent cations: pH and ionic specificity effects. Biopolymers 22:1097–1106

Tulasne R, Vendrely R (1947) Demonstration of bacterial nuclei with ribonuclease. Nature 160: 225

Valkenburg JAC, Woldringh CL, Brakenhoff GJ, van der Voort HTM, Nanninga N (1985) Confocal scanning light microscopy of the E. coli nucleoid: comparison with phase contrast and electron microscopy images. J Bacteriol 161:478–483

Vosberg HP (1985) DNA topoisomerasis: Enzymes that control DNA conformation. Curr Top Microbiol Immunol 114:19–102

Whitfield JF, Murray RGE (1956) The effects of the ionic environment on the chromatin structures of bacteria. Can J Microbiol 2:245–260

Woldringh CL (1973) Effect of cations on the organisation of the nucleoplasm in E. coli prefixed with osmiumtetroxyde or glutaraldehyde. Cytobiologie 8:97–111

Yamazaki K, Nagata A, Karo Y, Imamoto F (1984) Isolation and characterization of nucleoid proteins from Escherichia coli. Mol Gen Genet 196:217–224

Detection and Identification of Proteins Cross-Linked in Vivo to Nucleic Acids

J.P. SCHOUTEN[1]

1 Introduction

Bacterial genomes cannot be purified as a stable, compact DNA-protein complex like chromatin. They can be isolated intact in a compact, supercoiled form from *Bacillus* species (Sloof et al. 1983), but the forces which maintain a compact conformation are easily disrupted, preventing extensive purification. Chromosomes from *E. coli* have only been isolated in an artificially compacted form. Compaction has been induced, for example, by addition of polyamines (Kornberg et al. 1974) or of large amounts of basic proteins like lysozyme (Guillen et al. 1978), or by using high salt concentrations (Stonington and Pettijohn 1971). These preparations of intact chromosomes have been too crude to identify the proteins which keep the bacterial chromosomes in vivo in a compact form. The well-documented negative superhelicity of the bacterial DNA, part of which appears not to be stabilised (Sinden et al. 1980), will be responsible for a large part of the compaction of the chromosome. There must be other factors, however, presumably mainly nucleic acid-protein interactions (Sloof et al. 1983), resulting in a further level of compaction.

Despite many attempts of various research groups, the early claims for the visualisation of nucleosomes in bacteria (Griffith 1976) have never been confirmed. It is clear now that no structure comparable to the stable eukaryotic nucleosomes exist in bacteria.

Some information about bacterial DNA-binding proteins has been obtained by analysis of highly purified preparations of plasmid-protein complexes (Busby et al. 1979; Wu et al. 1982) or of chromosomal DNA fragments complexed with protein (Varshavsky et al. 1977).

However, many DNA-binding proteins may have been lost during the long isolation procedure, and because of the use of detergents to open the cells. Further artefacts will have been caused by the use of very low ionic strength solutions. Under the conditions often applied (1–100 mM salt), most proteins released from degraded RNA as well as the basic lysozyme used to open the cells bind electrostatically to the DNA.

Cross-linking of proteins to nucleic acids before cell disruption makes it possible to avoid these artefacts and also to study labile nucleic acid-protein complexes. The

1 Department of Electron Microscopy and Molecular Cytology, University of Amsterdam, Amsterdam, The Netherlands

Bacterial Chromatin
(ed. by C.O. Gualerzi and C.L. Pon)

covalent nature of the nucleic acid-protein bond will also allow the use of hybridisation-selection procedures to purify the proteins cross-linked to a specific nucleic acid sequence.

I have recently developed some techniques that can be used to detect and eventually identify the proteins interacting in vivo with specific nucleic acid sequences, as well as a method to study contacts between different nucleic acid-bound polypeptides (Schouten 1985a,b). These methods make use of nucleic acid-protein cross-linking by means of ultraviolet light irradiation or formaldehyde treatment of intact cells.

These methods may complement the nuclease and ultraviolet light foot-printing techniques that have been developed to probe the nucleic acid bases that interact with protein. These foot-printing techniques are most valuable when the identity of the protein that interacts with a particular nucleic acid sequence is already known, e.g. from molecular genetic experiments.

Recently, a protocol has been described for the analysis of the DNA fragments that interact with a specific protein (Gilmour and Lis 1984). This method may permit the identification of the DNA sequences to which a particular protein, against which antibodies have been raised, binds in vivo.

In contrast to the methods of Gilmour and Lis, and the various foot-printing methods, the techniques I have developed were designed to characterise and eventually identify all proteins bound in vivo to a specific mRNA or DNA sequence.

2 Nucleic Acid-Protein Cross-Linking

Nucleic acid-protein cross-linking in vivo can be accomplished by treating intact cells with physical agents, like UV- and γ-irradiation (Wagenmakers et al. 1980; Smith 1976), or with chemicals such as formaldehyde (Jackson and Chalkley 1981) and dichloro-diammine-platinum compounds (Ward et al. 1984). I have restricted myself to the use of UV irradiation and formaldehyde treatment of intact bacterial cells.

2.1 Ultraviolet Light

The three main advantages of UV-induced cross-linking are: (1) Only polypeptides in direct contact with nucleic acid are cross-linked. (2) Cross-linking can be accomplished before the intracellular environment is drastically changed by inactivation of certain membrane proteins. Cross-linking can even be induced in frozen cells (Smith 1976). (3) No large protein aggregates are formed as a result of extensive protein-protein cross-linking. The most serious limitation of UV-induced cross-linking is the dependence of the efficiency of cross-linking on the steric arrangement of reactive amino acids that are in the close neighbourhood of the nucleic acid. Thus, although all polypeptides detected must have been bound directly to nucleic acids in vivo, some nucleic acid-binding proteins might be overlooked. The reactivity of the DNA upon irradiation can be enhanced by the substitution of thymine residues by 5-bromo deoxyuridine residues. It has been reported that the *E. coli lac* repressor either is not, or else is very inefficiently cross-linked to the *lac* operator by UV-irradiation in vitro (Lin and Riggs

1974). The *lac* repressor could, however, be efficiently cross-linked to BrUrd-substituted DNA.

2.2 Formaldehyde

Formaldehyde, H_2CO, is a bifunctional chemical reagent, and has been used for a variety of purposes in both protein and nucleic acid research. Like UV-irradiation, formaldehyde fixation results in the cross-linking of proteins to RNA, as well as to single-stranded (ss) and double-stranded (ds) DNA (Jackson 1978; Möller et al. 1977). In contrast to UV-irradiation, however, the much larger amount of protein-protein cross-links formed may result, even after a limited reaction period, in the formation of a substantial number of complexes in which proteins are cross-linked indirectly to nucleic acids by virtue of their cross-linking to nucleic acid-bound proteins. These indirectly cross-linked proteins will be difficult to distinguish from proteins directly cross-linked to nucleic acids. Thus, although all polypeptides cross-linked to nucleic acids by UV-irradiation are almost certainly directly interacting with that nucleic acid (Schouten 1985a), this cannot be assumed for the proteins cross-linked by formaldehyde treatment.

Most, if not all, protein-protein and nucleic acid-protein cross-links formed during a limited formaldehyde treatment are destroyed upon heating for 48 h at 37°C (Jackson 1978). This is fortunate in that it permits the use of gel electrophoresis for the separation and detection of all polypeptide components of a purified formaldehyde cross-linked complex. However, the lability of the cross-links precludes a detailed analysis at the molecular level of the cross-links formed.

3 Detection of Cross-Linked Proteins

3.1 UV-Induced Cross-Links

An outline of the cross-linking and detection method used, is shown in Fig. 1. Exponentially growing cultures of *E. coli* were rapidly cooled by addition of crushed ice, and were irradiated with germicidal lamps, emitting mainly at 254 nm. The low UV absorbance of the minimal growth medium used made washing of the cells superfluous.

Irradiation was for 2–5 min, and was started within 30 s after removal of the culture from a 37°C waterbath. From these cells, a crude lysate was prepared by sonication in the presence of large amounts of detergents. Specific nucleic acid sequences and the proteins covalently bound to them are purified by hybridisation selection, and proteins present in these preparations are labeled to high specific activity with [^{125}I]. An RNAse-digestion step can be included at this stage to remove all mRNA-bound proteins. The [^{125}I]-labeled protein-nucleic acid sequences are further purified by a second cycle of hybridisation selection. Finally, nucleic acids are digested by nucleases, and the freed labeled proteins are separated by gel electrophoresis and detected by autoradiography. We did not succeed in obtaining reproducibly, samples containing only mRNA-bound proteins. The use of some hybridisation probes, e.g. pBR322 DNA-

Mid-log bacterial cells, rapidly cooled to 0°C

↓

UV-Irradiation for 2–5 min

↓

Centrifugation

↓

Sonication of 1 g cell paste in 6 ml detergent-solution

↓

Hybridisation selection of nucleic acid-protein complexes with cellulose-bound ss-DNA (3 h 65°C)

↓

[^{125}I]-Labeling of the protein moieties of the recovered nucleic acid-protein complexes

↓

RNAse Digestion when only DNA-bound proteins are to be detected

↓

Second cycle of hybridisation selection

↓

Nuclease digestion, gel electrophoresis of proteins

↓

Visualisation of cross-linked proteins by autoradiography

Fig. 1. Outline of the cross-linking and detection procedures

cellulose, yields preparations containing at least five times more mRNA- than DNA bound proteins. In Fig. 2, the pattern of proteins cross-linked to pBR322 mRNAs + DNA (lane 1), or to pBR322 DNA only (lane 2) is shown. The patterns of proteins cross-linked to trp, ompA, and other sequences look very similar. Differences were mainly due to variations in the ratio of DNA-bound and mRNA-bound proteins (unpublished results).

Slightly different patterns of cross-linked proteins were obtained when the *E. coli* cells were grown for 1.5 generations in the presence of 5-bromodeoxyuridine before irradiation (Fig. 2, lane 3). Several parameters had a more or less pronounced effect on the patterns of proteins obtained. Higher irradiation doses increased the amount of cross-linked protein, but resulted also in some protein breakdown. Instead of the oxidative iodination method used in the standard procedure, which results in the formation of [^{125}I]-tyrosine residues, the Bolton and Hunter method of labeling could be used, resulting in the formation of [^{125}I]-lysine residues. For the detection of the tyrosine-lacking HU-proteins, the latter labeling procedure was essential, but otherwise no detectable differences were found (Schouten 1985a).

Finally, some UV-induced nucleic acid-protein cross-links were found to be heat-labile. These cross-links were destroyed during the two hybridisation selection procedures at 65°C (3 h incubation each). Lowering the hybridisation temperature to 28°C by the inclusion of 70% formamide resulted in the recovery of a larger number of cross-linked proteins (Schouten 1985a).

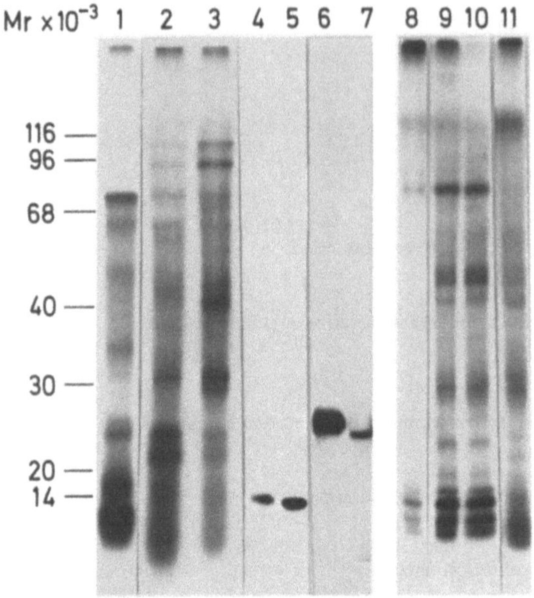

Fig. 2. Detection and identification of cross-linked proteins. *Lanes 1–3* Polypeptides cross-linked during UV-irradiation of intact *E. coli* cells to pBR322 mRNAs + DNA (*lane 1*) to pBR322 DNA (*lane 2*), or to BrUrd-substituted pBR322 DNA (*lane 3*). *Lane 4* HU proteins immunopurified from a sample as shown in *lane 3*, but labeled with [^{125}I]-Bolton and Hunter reagent. *Lane 5* Purified [^{125}I]-labeled HU protein. *Lane 6* SSB protein immunopurified from a sample as shown in *lane 2*. *Lane 7* Purified [^{125}I]-labeled SSB protein. *Lanes 8–11* Polypeptides cross-linked by formaldehyde treatment of intact *E. coli* cells to pBR322 mRNAs + DNA (*lanes 8–10*) or to pBR322 DNA only (*lane 11*). *Lane 8* Cross-links were not destroyed by heat treatment. *Lanes 9 + 11* Cross-links were reversed by heating 6 h at 65°C. *Lane 10* Cross-links were reversed by heating 10 h at 65°C. Molecular weight indication is valid for *lanes 1–3* only

3.2 Formaldehyde-Induced Cross-Links

Formaldehyde treatment of cells was performed by addition of formaldehyde to exponentially growing *E. coli* cultures to a concentration of 0.75% and further incubation at 37°C for 3 h.

Formaldehyde cross-links are readily destroyed upon heating. The two hybridisation selection steps were therefore performed at 28°C in the presence of 70% formamide. In contrast with the UV-induced cross-links, no nuclease digestion was required to free the polypeptides from their attached nucleic acid chains. Instead, samples were heated to eliminate all cross-links. In Fig. 2 lanes 8–10, the effect of heating on the gel patterns of the polypeptides cross-linked to pBR322 mRNAs and DNA by formaldehyde treatment of intact cells is shown. Similar to the results obtained with UV-irradiation, samples containing only DNA-bound proteins could also be prepared (lane 11).

The finding that even after a rather drastic formaldehyde treatment, mRNAs as well as DNA fragments with attached proteins could still be purified by hybridisation selection was a surprise. In this respect, the stabilisation of mRNA secondary structure, and thus the limitation of the formaldehyde reaction with mRNAs by the Mg^{2+} ions present in the fixation solution, might be important (Schouten 1985b).

4 Identification of Cross-Linked Proteins

Identification of cross-linked proteins can be accomplished in at least three ways. First, using specific antisera, polypeptides can be purified from a sample of cross-linked proteins, and visualised by a combination of gel electrophoresis and autoradiography. As shown in Fig. 2, the presence of the single-stranded binding protein (SSB) and one or both of the HU proteins among the proteins cross-linked to pBR322 DNA, could be demonstrated.

Second, differences in the pattern of cross-linked proteins between mutant and wild-type strains can sometimes be used for identification purposes. In this way, I have demonstrated the presence of the *E. coli* ribosomal protein S1 among the proteins cross-linked to mRNAs by UV-irradiation of intact cells (Schouten 1985a).

Finally, comparison of protease digestion products (Cleveland peptide mapping) between purified [^{125}I]-labeled polypeptides and cross-linked proteins can prove the identity of a specific cross-linked protein. The presence of the RNA polymerase σ subunit among the proteins cross-linked by UV-irradiation to DNA was demonstrated in this way. The use of specific antisera provides the most powerful method for identification of cross-linked polypeptides. Immunopurification also provides a method for the analysis of protein-protein contacts. Formaldehyde treatment of intact cells results in the formation of nucleic acid-protein as well as protein-protein cross-links. With the use of hybridisation selection, protein complexes cross-linked to specific DNA or RNA sequences can be purified. When these samples are used for immunopurification, not only the polypeptides recognised by the antiserum are purified, but also all polypeptides cross-linked to the recognised polypeptide. Following reversion of the formaldehyde cross-links by heating, all constituents of the immunopurified protein complexes can be visualised. Despite the changes in the intracellular environment during the formaldehyde treatment, highly specific cross-links are formed, resulting in the formation of discrete protein complexes, some of which are also bound to nucleic acids. Together with the results obtained with UV-induced cross-linking, very detailed information can be gathered. I was able to demonstrate, for instance, that the *E. coli* ribosomal proteins S1 and S21 are in direct contact with each other, while both of them are in direct contact with the mRNA in vivo. Furthermore, it could be shown that the three translational elongation factors from *E. coli* were cross-linked indirectly through a specific set of 30S ribosomal subunit proteins to mRNAs, while, for example, initiation factor 1 was cross-linked directly to mRNAs without having any contacts with other proteins. This latter method may have use in the analysis of contacts between various chromosomal proteins in both prokaryotes as well as eukaryotes.

5 Conclusions

In this chapter I have described briefly a method that can be used to detect polypeptides cross-linked in vivo to specific mRNAs or DNA sequences. Nucleic acid-protein cross-link methods that are compatible with the detection method include irradiation of intact cells with ultraviolet light and formaldehyde treatment of intact cells. It is likely that proteins cross-linked to nucleic acids by other means, like γ-irradiation or treatment of cells with dichlorodiammineplatinum compounds, can also be detected with this method.

So far I have successfully used the method with gram-positive and gram-negative bacteria, as well as with yeast. Detection of proteins cross-linked to nucleic acids in higher eukaryotes should also be possible. Approximately 3×10^{11} copies of the mRNA or DNA sequence studied need to be present in a few grams of cells treated with a suitable cross-linking agent. For most bacteria, this corresponds to single-copy sequences.

Protein present in low amounts will often be obscured on the autoradiographs by the major cross-linked proteins. Comparison of the proteins cross-linked to different nucleic acid sequences will be much easier when the major cross-linked proteins (e.g. 30S proteins in the case of *E. coli* mRNAs) can be removed before gel electrophoresis. In the case of formaldehyde cross-linking, 2-D gel electrophoretic techniques might be of use. The presence of a variable number of nucleotides attached to the cross-linked proteins precludes this approach when other cross-linking inducing agents are used.

The results obtained so far with the cross-linking and detection techniques described include the following.

1. At least some of the *E. coli* HU proteins interact with DNA in vivo. The small amount of cross-linked HU detected may be due to inefficient cross-linking of this small protein. The possibility that only a small part of the HU protein is interacting in vivo with ss- or ds-DNA cannot be excluded. It has to be kept in mind that a large part of the cellular HU content interacts with ribosomes (Suryanarayana and Subramanian 1978; Dijk et al. 1983).
2. The *E. coli* single-strand binding protein (SSB) is one of the two major proteins cross-linked to *E. coli* DNA upon UV-irradiation of intact cells. Since SSB is not an abundant protein in the cell, these results suggest that the efficiency of nucleic acid-protein cross-linking by UV-irradiation depends strongly on the type of contact between the protein and the nucleic acid.
3. The major proteins cross-linked to *E. coli* DNA by formaldehyde treatment, as well as to BrUrd-substituted DNA by UV-irradiation, have molecular weights of 29,000, 31,000, 40,000 and 45,000. In addition a Mr 9000 protein containing tyrosine (in contrast to HU) constituted a substantial part of the formaldehyde cross-linked proteins.
4. Most, but not all, proteins are cross-linked more efficiently by UV-irradiation to BrUrd-substituted DNA than to normal DNA. The SSB protein is among the proteins cross-linked to both types of DNA at the same rate.
5. Some UV-induced nucleic acid-protein cross-links are heat-labile.
6. The use of formaldehyde as a cross-linking agent provides the opportunity to study interactions between nucleic acid-bound polypeptides.

The results obtained thus far have shed little light on the question of the nature of the bacterial "histonelike" proteins. Only part of the HU proteins are undoubtedly in direct contact with the *E. coli* DNA in vivo. No major protein in the molecular weight range of protein H1 (Mr 15,000; Spassky et al. 1984), or the *firA* gene product (Mr 17,000; Lathe et al. 1980) was detected among the proteins cross-linked to DNA. Two cross-linked proteins, however, had molecular weights close to that of protein H (Mr 28,000; Hübscher et al. 1980). These proteins, of molecular weight 29,000 and 31,000; respectively were cross-linked to BrUrd-substituted DNA by UV-irradiation as well as to normal DNA by formaldehyde treatment. They were not always resolved as separate bands on gel, and appeared to be almost identical as deduced from comparison of their protease digestion products. No similarity between their protease digestion products and those of purified protein H was, however, observed.

These results indicate that the designation "histonelike" proteins should be reserved for structural proteins that are known to complex with DNA in vivo. Proteins that are acid-soluble, heat-stable and that lack affinity for hydroxyapatite do not need to have any functional resemblance to true histones. Until now, the HU proteins and the neutral H1 protein are the only well-studied abundant DNA-binding *E. coli* proteins for which no enzymatic function is known. Whether they bear functional resemblance to eukaryotic histones is doubtful. Functional similarity to some of the eukaryotic non-histone proteins, like the high mobility group of proteins should be subject of further study. First, however, it has to be determined which of the abundant *E. coli* proteins is bound in vivo to RNA, ss-DNA or to ds-DNA. Cell fractionation and reconstitution experiments have not provided unequivocal results in the past 10 years. The use of in vivo localisation techniques as described in this chapter may provide a more clear picture.

References

Busby S, Kolb A, Buc H (1979) Isolation of plasmid-protein complexes from Escherichia coli. Eur J Biochem 99:105–111

Dijk J, White SW, Wilson KS, Appelt K (1983) On the DNA binding protein 2 from Bacillus stearothemophilus. J Biol Chem 258:4003–4006

Gilmour DS, Lis JT (1984) Detecting protein-DNA interactions in vivo: Distribution of RNA-polymerase on specific bacterial genes. Proc Natl Acad Sci USA 81:4275–4279

Griffith JD (1976) Visualization of prokaryotic DNA in a regularly condensed chromatin-like fiber. Proc Natl Acad Sci USA 73:563–567

Guillen N, Le Hegarat F, Fleury AM, Hirschbein L (1978) Folded chromosomes of vegetative Bacillus subtilis: composition and properties. Nucleic Acids Res 5:475–489

Hübscher U, Lutz H, Kornberg A (1980) Novel histone H2A-like protein of Escherichia coli. Proc Natl Acad Sci USA 77:5097–5101

Jackson V (1978) Studies on histone organization in the nucleosome using formaldehyde as a reversible cross-linking agent. Cell 15:945–954

Jackson V, Chalkley R (1981) Use of whole-cell fixation to visualize replicating and maturing Simian Virus 40: Identification of a new viral gene product. Proc Natl Acad Sci USA 78:6081–6085

Kornberg T, Lockwood A, Worcel A (1974) Replication of the Escherichia coli chromosome with a soluble enzyme system. Proc Natl Acad Sci USA 71:3189–3193

Lathe R, Buc H, Lecocq JP, Bautz EKF (1980) Procaryotic histone-like protein interacting with RNA-polymerase. Proc Natl Acad Sci USA 77:3548–3552

Lin SY, Riggs AD (1974) Photochemical attachment of lac-repressor to bromodeoxyuridine-substituted lac operator by ultraviolet irradiation. Proc Natl Acad Sci USA 71:947–951

Möller K, Rinke J, Ross A, Buddle G, Brimacombe R (1977) The use of formaldehyde in RNA-protein cross-linking studies with ribosomal subunits from Escherichia coli. Eur J Biochem 76: 175–187

Schouten JP (1985a) Hybridisation-selection of nucleic acid-protein complexes 1. J Biol Chem 260:9916–9928

Schouten JP (1985b) Hybridisation-selection of nucleic acid-protein complexes 2. J Biol Chem 260:9929–9935

Sinden RR, Carlson JO, Pettijohn DE (1980) Torsional tension in the DNA double helix measured with trimethylpsoralen. Cell 21:773–783

Sloof P, Maagdelijn A, Boswinkel E (1983) Folding of prokaryotic DNA; isolation and characterisation of nucleoids from Bacillus licheniformis. J Mol Biol 163:277–297

Smith KC (1976) In: Wang SY (ed) Photochemistry and photobiology of nucleic acids, vol 2. Academic Press, New York, pp 187–218

Spassky A, Rimsky S, Garreau H, Buc H (1984) H1a, an E. coli DNA-binding protein which accumulates in stationary phase, strongly compacts DNA in vitro. Nucleic Acids Res 12:5321–5340

Stonington OG, Pettijohn DE (1971) The folded genome of Escherichia coli isolated in a protein-DNA-RNA complex. Proc Natl Acad Sci USA 68:6–9

Suryanarayana T, Subramanian AR (1978) Specific association of two homologous DNA-binding proteins to the native 30-S ribosomal subunits of Escherichia coli. Biochem Biophys Acta 520: 342–357

Varshavsky AJ, Nedospasov SA, Bakayef VV, Georgiev GP (1977) Histone-like proteins in the purified Escherichia coli deoxyribonucleoprotein. Nucleic Acids Res 4:2725–2745

Wagenmakers AJM, Reinders RJ, Van Venrooy WJ (1980) Cross-linking of mRNA to proteins by irradiation of intact cells with ultraviolet light. Eur J Biochem 112:323–330

Ward WS, Schmidt WN, Schmidt CA, Hnilica LS (1984) Association of cytokeratin p39 with DNA in intact Novikoff hepatoma cells. Proc Natl Acad Sci USA 81:419–423

Wu FYH, Kolb A, Buc H (1982) A transcriptionally active plasmid-protein complex isolated from Escherichia coli. Biochem Biophys Acta 696:231–238

Proteins Associated with Chromatin from *Escherichia coli*

I. LOSSIUS, A. HOLCK, R. AASLAND, L. HAARR, and K. KLEPPE[1]

Introduction

The chromosome from prokaryotic organisms can be isolated in a packaged form termed the nucleoid (Stonington and Pettijohn 1971). The nucleoid is obtained either as membrane-free or membrane-associated depending on the type of detergent and ionic strength used when lysing the cells (Worcel and Burgi 1972; Pettijohn 1976). Studies with in vitro high-salt nucleoids isolated from *E. coli* have shown that the DNA in these particles is separated into a number of superhelical domains and that they contain an RNA-protein core. The protein components of the high-salt nucleoid include the $\beta\beta'$ and α subunits of RNA-polymerase, the major outer membrane proteins of *E. coli* and a 56 kD protein (Portalier and Worcel 1976). Two acid extractable polypeptides with molecular weights of 10 and 17 kD are enriched in the membrane-associated compared to the membrane-free nucleoid (Varshavsky et al. 1977).

Some authors have proposed that the DNA loops are further condensed in vivo in nucleosomelike particles (Rouviere-Yaniv et al. 1979). This assumption has been based on the appearance of beaded structures extruding from lysed *E. coli* cells (Griffith 1976), limited digestion of membrane-associated nucleoids (Varshavsky et al. 1977; Georgiev et al. 1981) and the presence of histonelike proteins in bacteria (Cukier-Kahn et al. 1972; Varshavsky et al. 1977; Hübscher et al. 1980).

We have previously suggested that the membrane-free nucleoid is partially an artificial structure; the high-salt concentrations and detergents present during lysis could induce nucleic acid aggregation (Kleppe et al. 1979). Recent results of Sinden and Pettijohn (1981) and Sloof et al. (1983) clearly show that protein and not RNA is the major component determining superhelical domains in the nucleoid in vivo.

The membrane-associated nucleoids which have been used for isolation of histone-like proteins (Varshavsky et al. 1977; Rouviere-Yaniv 1978) also contain large amounts of RNA (Worcel and Burgi 1974; Pettijohn 1976). Some of the histonelike proteins derived from these structures may therefore be associated with RNA.

1 Department of Biochemistry, University of Bergen, Bergen, Norway

Bacterial Chromatin
(ed. by C.O. Gualerzi and C.L. Pon)

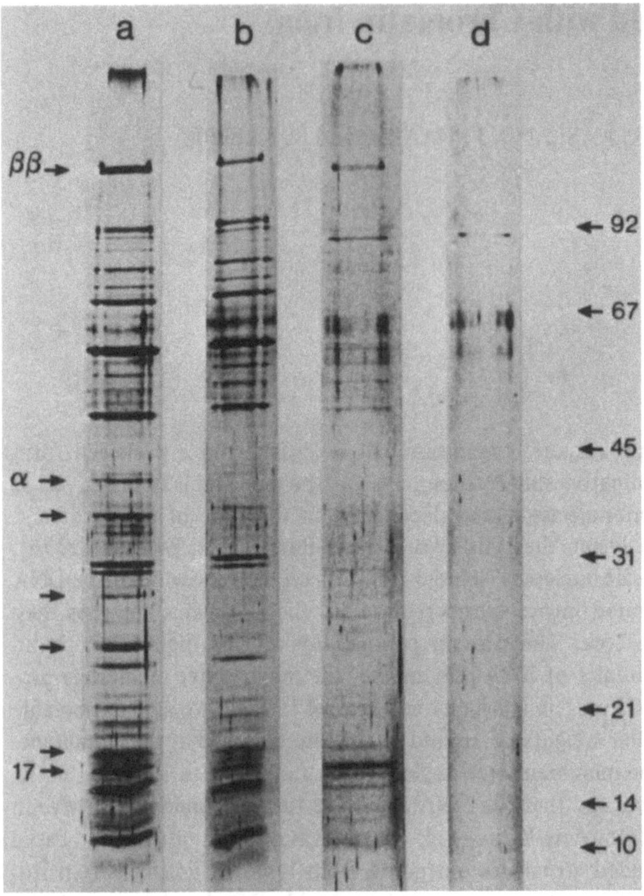

Fig. 1. Effect of DNase I and RNase treatment on the polypeptide pattern of the DNA- and RNA-protein complexes. The pattern of the 20 S DNA fraction is shown. An aliquot of 3 ml of *E. coli* chromatin was diluted to 5 ml in buffer, final concentration 60 mM NaCl, 2 mM EDTA, 10 mM Tris-HCl, pH 8.2. The sample was given three brief sonic treatments of 0.5 s on ice. Undissolved material was pelleted by centrifugation at 12,000 × g for 30 min. The samples were applied to sucrose gradients and centrifuged for 5.5 h at 38,000 rpm and 4°C in a SW 41 Ti-rotor. The sucrose gradients contained 5% to 28.8% sucrose, 60 mM NaCl and 1 mM EDTA, pH 7.5. *a* Polypeptides co-sedimenting with the DNA-containing nucleoprotein complexes. *b* The sample was incubated with 30 units DNase I ml^{-1} for 40 min at 37°C. *c* Incubated for 10–20 h on ice to facilitate degradation of the RNA by endogenous RNases prior to sucrose gradient centrifugation. *d* Incubation with DNase as well as RNase prior to centrifugation

Results and Discussion

Bacterial Chromatin

We have recently developed a new method for the isolation of the *E. coli* chromosome (Sjåstad et al. 1982). The resulting viscous material contains equal amounts of DNA and RNA, a large number of DNA- and RNA-binding proteins as well as loosely associated proteins and outer membrane fragments (Lossius et al. 1984). The chromatin can easily be solubilized by brief sonication or micrococcal nuclease treatment and the nucleoprotein complexes separated from free proteins by sucrose or metrizamide gradient centrifugation (Lossius et al. 1985a; Aasland et al. 1985).

DNA complexes with an average S-value of 20 S were obtained when the chromatin was sonicated briefly on ice and subjected to sucrose gradient centrifugation. Approximately 30 major polypeptides cosedimented with the DNA in the presence of 60 mM NaCl (Fig. 1, lane a). The effects of RNase treatment on the nucleoprotein complexes are shown in Fig. 1. Most of the nucleic acid-associated polypeptides were found to be bound to RNA, whereas seven of the polypeptides were significantly affected by the DNase treatment (Fig. 1, lane b). The polypeptides of the DNA-containing nucleoprotein complexes have been analyzed by two-dimensional gel electrophoresis (Lossius et al. 1984; Aasland et al. 1985). Three low molecular weight polypeptides with molecular weights 16, 17 and 18 kD and approximate isoelectric points of pH 7.0, 9.5 and 9.0 were associated with the DNA. All three had an affinity for double-stranded DNA-Sepharose (Aasland et al. 1985). The 16 kD polypeptide may be identical to the H-NS protein (Cuckier-Kahn et al. 1972), whereas the 18 kD polypeptide has been tentatively identified as an outer membrane protein (Lossius et al. 1984). The 17 kD polypeptide is probably either the *fir A* gene product (Lathe et al. 1980) or the BH I protein (Varshavsky et al. 1977). It has been purified to homogeneity and the amino acid sequence has been partially determined (Holck, unpublished).

The results were slightly different when the chromatin was treated with micrococcal nuclease at 37°C instead of sonication to yield high-molecular weight DNA-protein complexes which were separated from free protein by gel filtration (Fig. 2). The additional proteins, which are associated with DNA from micrococcal nuclease-treated chromatin, may originate from native DNA-membrane complexes, which are affected by the residual amount of NP-40 detergent present in the chromatin during nuclease treatment. They have been tentatively identified as membrane proteins and some of them have been shown to have affinity for double-stranded DNA-Sepharose (Aasland et al. 1985). The 56 and 84 kD polypeptides are of special interest, since cross-linking experiments have shown that polypeptides of similar molecular weights are closely associated with DNA in the membrane-associated nucleoids (Portalier and Worcel 1976). The results are summarized in Table 1.

Protein HU

When the total double-stranded DNA-cellulose-binding proteins of *E. coli* are analyzed by two-dimensional gel electrophoresis approximately 350 different polypeptides can be detected. In this case, the HU protein is one of the major DNA-binding proteins of *E. coli* (Sjåstad et al. 1983).

The bacterial chromatin contains, however, only a small amount of the HU protein (Lossius et al. 1984), and the present results suggest that the HU protein in this case is bound to RNA rather than to DNA.

Affinity studies with the HU protein (equimolar mixtures of NS 1 and NS 2) show that this protein has a higher affinity for RNA and single-stranded DNA than for

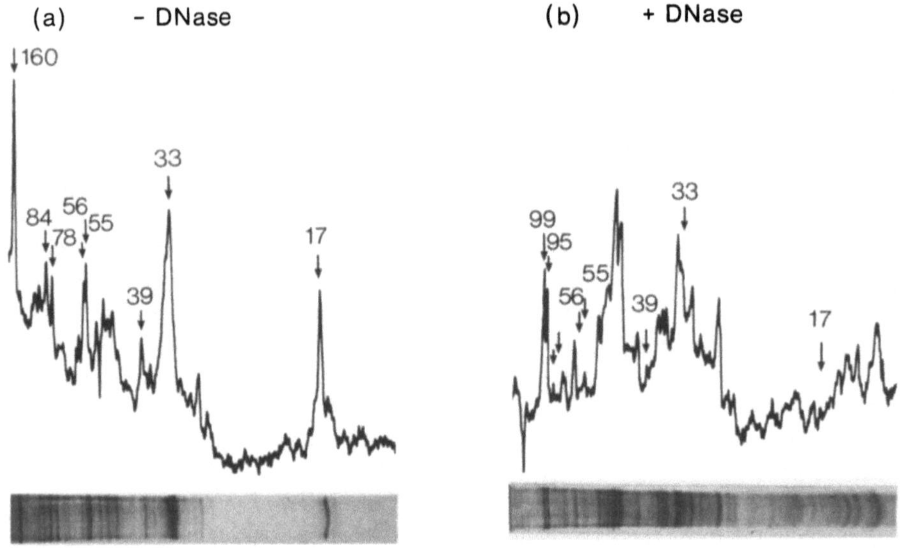

Fig. 2a,b. Effect of DNase I treatment on the polypeptide pattern of the DNA-protein complexes from micrococcal nuclease-treated chromatin. Autoradiographs of [^{35}S]-protein-labelled samples as well as soft-laser tracings of these are shown. The chromatin was treated with 0.5 units of micrococcal nuclease for 15 min at 37°C, undissolved material removed by centrifugation at 12,000 × g for 10 min and the supernatant subjected to sucrose gradient centrifugation as described in the legend to Fig. 5, but omitting NaCl. The 10–20 S DNA-protein complexes were collected and separated from free protein by Sepharose CL-6B gel filtration. The running buffer contained 1 mM EDTA, 10 mM Tris-HCl, pH 8.0, with 10 mM $NaHSO_3$ as a protease inhibitor. The DNA-protein complexes eluted in the void volume were lyophilized and analyzed by SDS polyacrylamide gel electrophoresis. **a** Protein pattern of the void fraction (24 h exposure). **b** Protein pattern of the void fraction when the sample was treated with 50 units DNase I ml^{-1} for 60 min at 37°C prior to gel filtration (72 h exposure)

Table 1. Major polypeptides liberated by DNase I treatment of nucleoprotein complexes from *E. coli* chromatin

Molecular weight (kD)	Sonicated chromatin	Micrococcal nuclease-treated chromatin	Present in membrane-free nucleoid
160 ($\beta\beta'$)	+	+	+
84		+	
78		+	
56		+	+
55		+	
39 (α)	+	+	+
35	+		+
33		+	
28	+		+
25	+		+
18	+		
17	+	+	

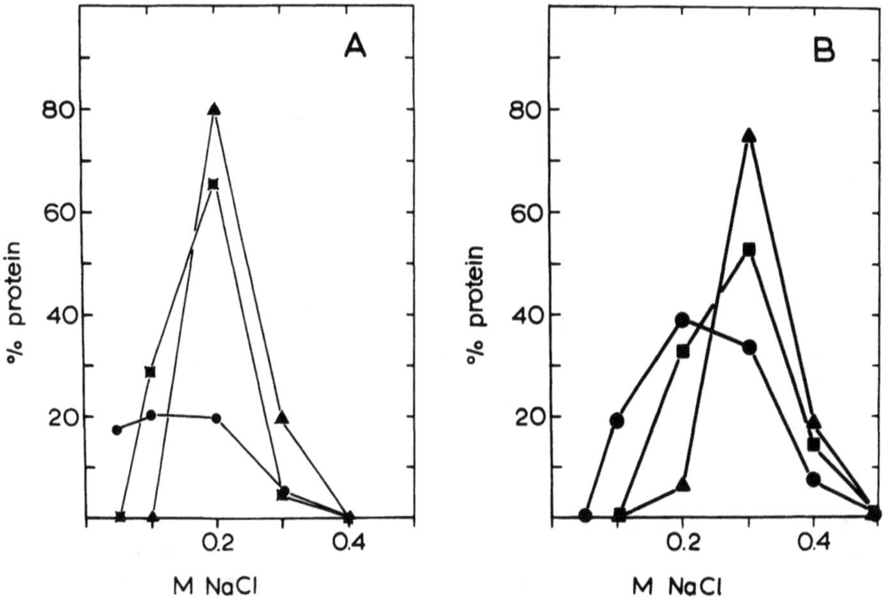

Fig. 3A,B. Effect of spermidine on the binding of the HU protein to double-stranded (ds) as well as single-stranded (ss) DNA-Sepharose. The HU protein was bound to dsDNA-Sepharose (A) or to ssDNA-Sepharose (B) and eluted stepwise by increasing the ionic strength. The concentration of spermidine was: –▲– 0 mM, –■– 1 mM, –●– 5 mM. The amount of protein eluted was assayed as described previously. The figure is from Holck and Kleppe (1985) and reproduced by permission from Elsevier Biomedical Press

double-stranded DNA and ribosomal RNA of another species (Holck and Kleppe 1985). The binding properties are affected by the presence of polyvalent cations such as spermidine (Fig. 3). It also binds much more strongly to linearized than to supercoiled ϕX174 RFI DNA (Fig. 4).

The HU protein form nucleosomelike structures with circular DNA in vitro in the presence of the nicking-closing enzyme and low concentration of NaCl (Rouviere-Yaniv et al. 1979). This could indicate a similar function for this protein in vivo. Equivalent structures are formed, however, also with other DNA-binding proteins and apparently must reflect the general physicochemical properties of DNA and the proteins employed (Chrysogelos and Griffith 1982). Our results also show that the HU protein only has a low affinity for double-stranded DNA and this is affected by the presence of physiological concentrations of polyvalent cations such as spermidine. The HU protein has been suggested to be involved in transcription and/or replication (Rouviere-Yaniv and Gros 1975; Suryanarayana and Subramanian 1978; Dixon and Kornberg 1984). Recent results by Bjornsti et al. (1985) show that the HU protein is located mainly on the nucleoid-cytoplasm border with a similar spatial distribution as RNA and single-stranded DNA in the cell.

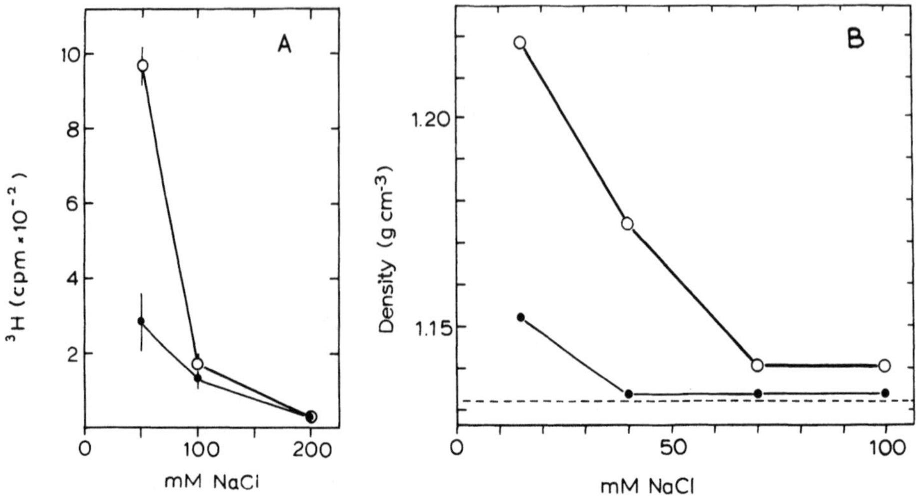

Fig. 4A,B. Affinity of the HU protein for supercoiled ϕX174 RF I DNA and for ϕX174 RF I DNA linearized by treatment with the restriction enzyme Ava I. The affinity was measured either by the retention of the HU-DNA complexes on nitrocellulose filters (**A**) or by estimating the density of the complexes by metrizamide gradient centrifugation (**B**). The affinity was measured in the presence of different concentrations of KCl (**A**) or NaCl (**B**). •–• Supercoiled DNA; o–o linearized DNA. From Holck and Kleppe (1985) reproduced by permission from Elsevier Biomedical Press

Micrococcal Nuclease Digestion of Bacterial Chromatin

When bacterial chromatin was digested with micrococcal nuclease and the DNA-containing complexes analyzed by sucrose gradient centrifugation, weak restrictions towards the nucleolytic activity were detected. DNA-protein complexes with sedimen-

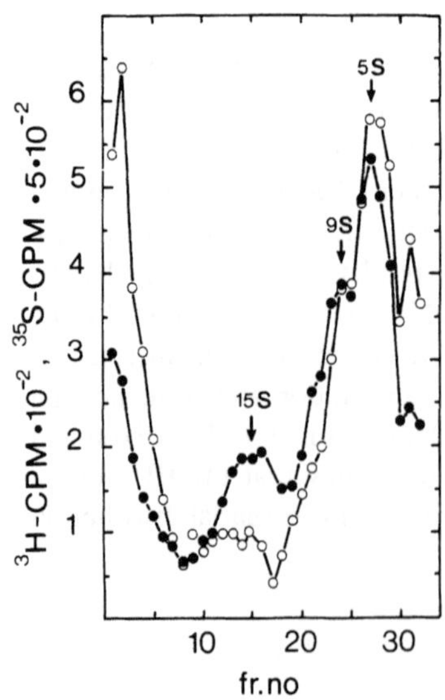

Fig. 5. Sedimentation pattern of the DNA-containing nucleoprotein complexes liberated from *E. coli* chromatin by micrococcal nuclease treatment. An aliquot of 300 μl of [^{3}H]-thymidine and [^{35}S]-labelled chromatin was diluted with buffer to final concentrations of 0.5 mM $CaCl_2$ and 10 mM Tris-HCl, pH 8.0, in a final volume of 1 ml. The sample was treated with 5 units of micrococcal nuclease for 15 min at 37°C. Undigested material, which was enriched in outer membrane fragments, was removed by centrifugation at 12,000 × g for 10 min at 4°C. Aliquots of 500 μl were applied to 5% to 28.8% sucrose gradients, which contained 150 mM NaCl and 1 mM EDTA, pH 7.5 and centrifuged for 21 h at 28,000 rpm and 4°C in a SW 41 Ti-rotor. •–• [^{3}H]-labelled DNA; o–o [^{35}S]-labelled protein

tation coefficients 5 S, 9 S, 15 S and >25 S were released by the nuclease treatment (Fig. 5). Nucleoprotein complexes with similar sedimentation coefficients, but enriched in the faster sedimenting species, were obtained when the amount of enzyme was reduced. Polyacrylamide gel electrophoresis showed that these complexes contained corresponding DNA fragments of average lengths 115, 190 and 680 base pairs as well as high molecular weight DNA (Lossius et al. 1985a). Similar results have been obtained with membrane-associated nucleoids from *E. coli* (Varshavsky et al. 1977) and *B. subtilis* (Le Hegarat et al. 1985) for the size of the limit digest. Metrizamide gradient analysis of the digestion products have shown that the 9 S and 15 S complexes and the complexes containing high molecular weight DNA have a protein to DNA ratio of 1:1 (Lossius et al. 1985a), which is slightly higher than the corresponding value for sonicated chromatin (Kleppe et al. 1984). The 5 S DNA complex was shown to have a buoyant density of 1.22 g ml^{-1}, indicating a protein to DNA ratio of 1.7:1. This heavy complex was detected only when the chromatin was solubilized by micrococcal nuclease treatment.

Polypeptides Present in the DNA Complexes Liberated by Micrococcal Nuclease Treatment

Purification of the 5 S DNA-protein complex showed that the major polypeptides possessed molecular weights of 10, 17, 32 and 56 kD. These were associated with an average sized DNA of 115 base pairs (Lossius et al. 1985a). The 10 and 56 kD polypeptides have been tentatively identified as membrane proteins (Lossius et al. 1984). The 17 kD polypeptide has a pI of approximately 9.5 and may be the BH I protein or the *fir A* gene product. It is solible in perchloric acid (Kleppe et al. 1984). The major proteins co-purifying with the 9 S complex had molecular weights of 18, 78 and 84 kD, respectively. The 15 S DNA complex contained mainly $\beta\beta'$ and α subunits of RNA polymerase (Lossius et al. 1985a). Previous studies by Portalier and Worcel (1976) have shown that polypeptides with molecular weights of 31, 56 and 80 kD are tightly associated with DNA in the membrane-associated nucleoid.

When the membrane-free nucleoid is isolated in the presence of 1 M NaCl, two different DNA-containing complexes are obtained; a 1600 S DNA-RNA-protein complex as well as a DNA fraction enriched in membrane components with a sedimentation coefficient >3200 S. In the case of the membrane-free nucleoid, a 56 kD polypeptide was shown to be a major component, whereas two polypeptides with molecular weights of 10 and 18 kD, both possessing acidic isoelectric points (results not shown), were among the major components in the outer membrane-enriched DNA fraction (Fig. 6).

The Three-Dimensional Structure of the in Vivo Nucleoid

It has been shown by Sinden and Pettijohn (1981) that the *E. coli* chromosome is separated into superhelical domains in vivo. The number of superhelical regions appears to be significantly lower when estimated in vivo than for the isolated nucleoid in vitro. This was the case when the superhelicity was probed both by psoralen binding (Sinden and Pettijohn 1981), as well as by estimation of the number of topoisomerase II

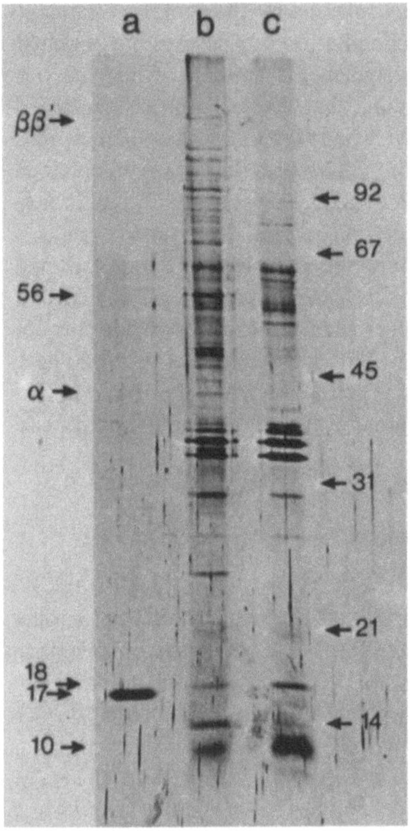

Fig. 6. Polypeptide pattern of membrane-free nucleoids from *E. coli.* The membrane-free nucleoids were isolated as previously described (Lossius et al. 1983). The 10–20% polyacrylamide gel was stained with silver essentially according to Marshall and Latner (1981) with modifications (Lossius et al. 1985a). The same amount of DNA was present in each of the nucleoid samples. *Lane a* Purified 17 kD polypeptide from *E. coli* chromatin. *Lane b* Membrane-free nucleoid. *Lane c* Membrane-enriched (>3200 S) DNA fraction from the high-salt nucleoid preparation. The positions of the proteins of the Biorad low molecular weight standard are indicated

binding sites (Snyder and Drlica 1979). Numbers of 43 and 45 regions per *E. coli* genome were found in these cases. When the number of supercoiled loops was estimated by repair enzyme-mediated DNA strand-breakage, a number of 45 loops per *E. coli* genome was found, assuming a Poisson distribution for the introduction of breaks in the separate superhelical domains (Lossius et al. 1981). These values are similar to the number of DNA-membrane attachment sites found by several authors (Dworsky and Schaechter 1973; Portalier and Worcel 1976; Abe et al. 1977; Craine and Rubert 1979).

Studies on the nucleolytic breakdown of *E. coli* chromatin do not provide evidence for the presence of a structural subunit analogous to the eukaryotic nucleosome. Our results with sonicated as well as micrococcal-nuclease-digested *E. coli* chromatin suggest that the DNA-protein particles are derived from replication and transcription complexes as well as from membrane-associated DNA. The latter assumption is supported by the fact that additional DNA-protein complexes containing membrane protein were obtained by micrococcal nuclease digestion of bacterial chromatin in the presence of low concentrations of non-ionic detergent. The major polypeptide of these complexes had a molecular weight of 56 kD. The complexes liberated by sonication contained only few proteins which were affected by DNase I treatment. The major ones were the $\beta\beta'$ and α subunits of RNA polymerase and two basic polypeptides with molecular weights of 17 and 18 kD. The HU protein and a 16 kD polypeptide, which may be the H-NS protein were in both cases mainly associated with RNA.

Further studies are needed to clarify the roles of the 16, 17 and 18 kD polypeptides. Recent results suggest that the 56 kD protein, which is one of the major DNA-binding proteins in the *E. coli* cell, may be involved in the cell-division process (Lossius et al. 1985b).

Acknowledgements. This work was partly supported by grants from the Norwegian Research Council for Science and Humanities.

References

Abe M, Brown C, Hendrickson WG, Boyd DH, Clifford P, Cote RH, Schaechter M (1977) Release of *Escherichia coli* DNA from membrane complexes by single-strand endonucleases. Proc Natl Acad Sci USA 74:2756–2760

Aasland R, Holck A, Lossius I, Haarr L, Kleppe K (1985) Proteins associated with the *Escherichia coli* chromosome (to be published)

Bjornsti MA, Hobot JA, Kelus AS, Villiger W, Kellenberger E (1985) Immunocytochemical localization of protein and DNA compounds of the bacterial nucleoid. Symp Chromatin Struct Funct. Camerino 1985, pp 161–162

Chrysogelos S, Griffith J (1982) *Escherichia coli* single-strand binding protein organizes single-stranded DNA in nucleosome-like units. Proc Natl Acad Sci USA 79:5803–5807

Craine BL, Rupert CL (1979) Deoxyribonucleic acid-membrane interactions near the origin of replication and initiation of deoxyribonucleic acid synthesis in *Escherichia coli.* J Bacteriol 137:740–745

Cukier-Kahn R, Lacquet M, Gros F (1972) Two heat-resistant, low molecular weight proteins from *Escherichia coli* that stimulate DNA-directed RNA synthesis. Proc Natl Acad Sci (USA) 69: 3643–3647

Dixon NE, Kornberg A (1984) Protein HU in the enzymatic replication of the chromosomal origin of *Escherichia coli.* Proc Natl Acad Sci USA 81:424–428

Dworsky P, Schaechter M (1973) Effect of rifampin on the structure and membrane-attachment of the nucleoid of *Escherichia coli.* J Bacteriol 116:1365–1374

Georgiev GP, Bakayev VV, Nedospasov SA, Razin SV, Matieva VL (1981) Studies on structure and function of chromatin. Mol Cell Biochem 40:29–48

Griffith J (1976) Visualization of procaryotic DNA in regularly condensed chromatin-like fiber. Proc Natl Acad Sci USA 63:563–567

Holck A, Kleppe K (1985) Affinity of protein HU for different nucleic acids. FEBS Lett 185:121–124

Hübscher U, Lutz H, Kornberg A (1980) Novel histone H2A-like protein of *Escherichia coli.* Proc Natl Acad Sci USA 77:5097–5101

Kleppe K, Øvrebø S, Lossius I (1979) The bacterial nucleoid. J Gen Microbiol 112:1–13

Kleppe K, Lossius I, Aasland R, Sjåstad K, Holck A, Haarr L (1984) The structure of the bacterial nucleoid. In: Hübscher U, Spadari S (eds) Proteins involved in DNA replication. Plenum, New York, pp 457–466

Lathe R, Buc H, Lecocq JP, Bautz EKF (1980) Prokaryotic histone-like protein interacting with RNA-polymerase. Proc Natl Acad Sci USA 77:3548–3552

Le Hegarat F, Salti V, Hirschbein L (1985) Symp Chromatin Struct Funct. Camerino 1985, p 163

Lossius I, Krüger PG, Kleppe K (1981) Effect of methyl methanesulphonate on the nucleoid structure of *Escherichia coli.* J Gen Microbiol 124:159–171

Lossius I, Krüger PG, Male R, Kleppe K (1983) Mitomycin-C induced changes in the nucleoid of *Escherichia coli* K12. Mutat Res 109:13–20

Lossius I, Sjåstad K, Haarr L, Kleppe K (1984) Two dimensional gel-electrophoretic separation of the proteins present in chromatin of *Escherichia coli*. J Gen Microbiol 130:3153–3157

Lossius I, Aasland R, Holck A, Sjåstad K, Kleppe K (1985a) Analysis of micrococcal nuclease digestion products of *Escherichia coli* chromatin by isokinetic and isopycnic gradient centrifugation (Submitted for publication)

Lossius I, Sjåstad K, Haarr L, Kleppe K (1985b) Effect of mitomycin-C on the proteins associated with *Escherichia coli* chromatin (in preparation)

Marshall T, Latner AL (1981) Incorporation of methylamine in an ultrasensitive silver stain for detecting protein in thick polyacrylamide gels. Electrophoresis 2:228–235

Pettijohn DE (1976) Prokaryotic DNA in nucleoid structure. CRC Crit Rev Biochem 4:175–202

Portalier R, Worcel A (1976) Association of the folded chromosome with the cell envelope of *Escherichia coli:* Characterization of the proteins at the DNA-attachment site. Cell 8:245–255

Rouviere-Yaniv J (1978) Localization of the HU protein on the *Escherichia coli* nucleoid. Cold Spring Harbour Symp Quant Biol 42:439–447

Rouviere-Yaniv J, Gros F (1975) Characterization of a novel low-molecular-weight DNA-binding protein from *Escherichia coli*. Proc Natl Acad Sci USA 72:3428–3432

Rouviere-Yaniv J, Yaniv M, Germond J-E (1979) *Escherichia coli* DNA binding protein HU form a nucleosome-like structure with circular double-stranded DNA. Cell 17:265–274

Sinden RR, Pettijohn DE (1981) Chromosomes in living *Escherichia coli* cells are segregated into domains of supercoiling. Proc Natl Acad Sci USA 78:224–228

Sjåstad K, Fadnes P, Krüger PG, Lossius I, Kleppe K (1982) Isolation, properties and nucleolytic degradation of chromatin from *Escherichia coli*. J Gen Microbiol 128:3037–3050

Sjåstad K, Haarr L, Kleppe K (1983) Characterization of the DNA-cellulose-binding proteins from *Escherichia coli*. Biochem Biophys Acta 739:8–16

Snyder M, Drlica K (1979) DNA gyrase on the bacterial chromosome: DNA cleavage induced by oxolinic acid. J Mol Biol 131:287–302

Sloof P, Magdelijn A, Boswinkel E (1983) Folding of prokaryotic DNA. Isolation and characterization of nucleoids from Bacillus licheniformis. J Mol Biol 163:277–297

Stonington OG, Pettijohn DE (1971) The folded genome of *Escherichia coli* isolated in a protein DNA-RNA complex. Proc Natl Acad Sci USA 68:6–9

Suryanarayana R, Subramanian A-R (1978) Specific association of two homologous DNA-binding proteins to the native 30-S ribosomal subunits of *Escherichia coli*. Biochem Biophys Acta 520: 342–357

Varshavsky AJ, Nedospasov A, Bakayev VV, Bakayeva TG, Georgiev GP (1977) Histone like proteins in the purified *Escherichia coli* deoxyribonucleoprotein. Nucleic Acids Res 4:2725–2745

Worcel A, Burgi E (1972) On the structure of the folded chromosome of *Escherichia coli*. J Mol Biol 71:127–147

Worcel A, Burgi E (1974) Properties of a membrane-attached form of the folded chromosome of *Escherichia coli*. J Mol Biol 82:91–105

Proteins from the Prokaryotic Nucleoid. Structural and Functional Characterization of the *Escherichia coli* DNA-Binding Proteins NS (HU) and H-NS

C.O. GUALERZI[1,2], M.A. LOSSO[3], M. LAMMI[3], K. FRIEDRICH[2], R.T. PAWLIK[2], M.A. CANONACO[1,2], G. GIANFRANCESCHI[1], A. PINGOUD[4], and C.L. PON[2]

1 Introduction

In spite of the commonplace notion that prokaryotes do not contain histones and that their DNA is not organized in chromatin, evidence is accumulating that bacterial DNA is indeed organized in nucleosomelike structures by histonelike proteins. Since the problem of the physical packaging and confinement of the genetic material can be regarded as a fundamental problem in all biological systems and in light of the underlying architectural unity of all living cells, it appears unlikely that completely different strategies have evolved to meet the same basic need. If the general strategies are the same, or at least, very similar, then it is possible that the differences existing between the pro- and eukaryotic systems may merely reflect differences in the tactics that each system must have perfected in order to harmonize its DNA-packaging mechanism with its specific physiological requirements.

In eukaryotes, large portions of the genome are not transcriptionally active. Transcriptionally active and inactive DNA is organized in "dynamic" and "static" chromatin, respectively (Luchnik et al. 1982; Ryoji and Worcel 1984; Glikin et al. 1984; Ryoji and Worcel 1985). Only the latter chromatin behaves like conventional nucleosomal chromatin. The resemblance between eukaryotic static chromatin and bacterial chromatin is, of course, rather remote. The dynamic chromatin, on the other hand, even though it certainly contains histones, probably lacks a canonical histone octamer and may possess additional protein factors. Unlike the transcriptionally inactive static chromatin, where the histones restrain the supercoiling of DNA even in the presence of a swivel, the dynamic chromatin is torsionally strained and becomes fully relaxed upon treatment with DNase I, topoisomerase I or novobiocin (to inhibit gyrase activity). Furthermore, dynamic chromatin is sensitive to DNase I and micrococcal nuclease and the size-limit digest with the latter nuclease corresponds to half-nucleosome and lacks the typical nucleosomal repeat (Ryoji and Worcel 1985).

In prokaryotes, a large proportion of the genome must be expressed and the entire genome must be replicated within a generally short generation time. Furthermore, the

1 Department of Cell Biology, Laboratory of Genetics, University of Camerino, 62032 Camerino (MC), Italy
2 Max-Planck-Institut für Molekulare Genetik, D–1000 Berlin 33
3 Department of Cell Biology, University of Calabria, 87036 Rende (CS), Italy
4 Zentrum Biochemie, Medizinische Hochschule Hannover, 3000 Hannover, FRG

Bacterial Chromatin
(ed. by C.O. Gualerzi and C.L. Pon)

system must be very flexible, in order to respond rapidly to changed environmental conditions by modulating its gene activity. Thus, it is interesting, but not entirely surprising, that the bacterial "chromatin" has many characteristics in common with the eukaryotic dynamic chromatin (see below).

Even in actively growing bacteria, the chromosome is not evenly distributed within the cell, but is confined (Kellenberger et al., this volume) in a skeinlike structure named the nucleoid. Up to four nucleoids can be found within a single cell. Each nucleoid is believed to be anchored to the cellular membrane at one or more points (Ogden and Schaechter, this volume).

Early experiments on the structure of the nucleoid had led to the conclusion that RNA molecules might form a central core and that RNA-DNA interactions might be responsible for the stabilization of the nucleoid structure and for the separation of the chromosome into a number of topologically-independent domains (Worcel and Burgi 1972; Kavenow and Bowen 1976). More recent studies have led to the general conclusion, however, that RNA does not play any important role in the organization of the nucleoid and that the presence of RNA in high-salt prepared nucleoids should be regarded as a preparation artefact (Sloof et al. 1983; Kleppe et al. 1984). Within each nucleoid, the DNA is segregated into negatively supercoiled, topologically independent domains (Worcel and Burgi 1972; Sinden et al. 1980; Sinden and Pettijohn 1981; Pettijohn 1982; Sloof et al. 1983). The supercoiling of the chromosomal DNA is under the antagonistic control of topoisomerases I and II (Menzel and Gellert 1983; Drlica et al., this volume).

When *E. coli* cells were treated with gamma rays to induce single-stranded breaks in the DNA and subsequently allowed to repair the nicks under conditions where DNA gyrase was inhibited, it was found that only half of the supercoils were maintained. This finding clearly indicated that in vivo some interaction is able to restrain part of the negative supercoils, but that a large portion of them is not restrained (Pettijohn and Pfenninger 1980). Similar conclusions were reached when the in vivo torsional tension of *E. coli* DNA was measured using trimethylpsoralen as a photoaffinity probe (Sinden et al. 1980). This situation would indicate a fundamental difference in the organization of the DNA supercoils in the bacterial and eukaryotic chromatin. If one considers, however, that only a small part of the latter is transcriptionally active, then the reason for the difference is obvious. Indeed, the situation of the bacterial chromatin is clearly reminiscent of that described for a roughly 50:50 mixture of dynamic and static eukaryotic chromatin produced in transcriptionally active cells (Ryoji and Worcel 1985).

Even though the nature of the interactions responsible for restraining the negative supercoils of prokaryotic DNA is not known, it should be mentioned that previous studies in vitro have shown that the histonelike protein HU (NS) can introduce negative superhelical turns into relaxed DNA in the presence of pure nicking-closing enzyme (Rouviere-Yaniv et al. 1979).

Gentle lysis methods and special sample preparation techniques allowed the visualization, under the electron microscope, of particularly unstable chromatinlike fibers attached to the cell envelope, having a diameter of 120 Å and a repeated nucleosomelike granular structure. It was estimated that each granule (∿130 Å) contained approximately 250 bp of DNA, corresponding to a condensation factor of about

6.5-fold (Griffith 1976). Even though it is not known what kind of interactions are responsible for the formation of the beaded structures observed by Griffith, it is nevertheless tempting to speculate that these short-lived granules are held together by the same histonelike proteins found in association with DNA by various biochemical approaches. In fact, after isolation of the *E. coli* chromosomes under mild conditions, Varshavsky et al. (1977a,b) and Bakayev (1981) obtained deoxyribonucleoprotein particles in which the DNA was associated with two major proteins, B1 and B2, which have molecular weights of 17 kD and 9 kD, respectively. It was estimated that B1 and B2 were present in a ratio of approx. 1:2 molecules every 150–200 bp of DNA. Protein B2 was identified as being equal to a protein described, with different names, in other laboratories: HU (Rouviere-Yaniv and Gros 1975), HD (Berthold and Geider 1976) and NS (Suryanarayana and Subramanian 1978). While the first two groups must be credited for the functional characterization of this protein as an important constituent of the "bacterial chromatin", the third group provided the first reliable chemical characterization, showing that it consists of two subunits (NS1 and NS2), and also determined their primary structures (Mende et al. 1978). Concerning the relevance of NS (HU, HD, B2) in the condensation of the prokaryotic chromosome, Rouviere-Yaniv (1977) showed, by electron microscopy, that this protein is localized in the *E. coli* nucleoids, provided that these are prepared at low (0.1 M NaCl) and not at high (1 M NaCl) salt concentration. Furthermore, NS protein was found to condense circular SV40 DNA molecules into beaded structures which, upon fixation with glutaraldehyde, resembled eukaryotic nucleosomes (Rouviere-Yaniv et al. 1979).

NS proteins have been described in several types of eubacteria as well as in blue-green algae (Haselkorn and Rouviere-Yaniv 1976) and in chloroplasts and mitochondria (Rouviere-Yaniv, personal communication). It is remarkable that the gene for an NS-like protein has also been found in a phage genome; the structural gene for the transcription factor 1 (TF1) of *B. subtilis* phage SP01 has recently been sequenced and found to code for a 99-amino-acid-long protein which is approximately 50% homologous to *B. stearothermophilus* NS. The TF1 protein inhibits transcription and binds selectively to the phage DNA, which is characterized by having 5-hydroxymethyluracil in place of thymine. It is believed that the function of TF1 is to organize SP01 DNA into a chromatinlike structure (Greene et al. 1984).

The identity of the second chromatin-associated protein (B1) described by Varshavsky et al. (1977a,b), with respect to other proteins of similar size described in other laboratories, is less sure. Several DNA-binding proteins larger than NS (HU), for which an important role in the condensation of bacterial DNA has been proposed, have been described: k protein (Schäfer and Zillig 1973); B1 (Varshavsky et al. 1977a; Bakayev 1981); H1 (Spassky and Buc 1977); HLP I (Lathe et al. 1980); H-NS (Lammi et al. 1984b); H1a (Spassky et al. 1984). The relationship between these proteins is rather confusing but they all share some important characteristics. They are the most stable, are relatively abundant (estimated at approximately 20 000 copies per cell), bind DNA rather tightly, have molecular weights in the 15–17 kD range and affect, one way or the other,transcription (Schäfer and Zillig 1973; Lathe et al. 1980; Spassky et al. 1984).

In addition to the above proteins, two other abundant histonelike proteins have been described in *Escherichia coli* (Hübscher et al. 1980; Kishi et al. 1982).

The first of these two proteins (H), with a monomer molecular weight of 28 kD, was estimated to be present in about 30 000 copies per cell and was found to behave as a dimer. Its amino acid composition, which is different from that of NS and H-NS, is characterized by the lack of methionine, cysteine and tryptophan, and bears a striking resemblance to eukaryotic histone H2A with which it also shows immunological cross-reactivity. The H protein was found to be heat- and acid-stable and was found to bind to both ss- and ds-DNA, if the salt (NaCl) concentration did not exceed 100 mM. Upon binding to DNA, this protein inhibits DNA duplication by several types of DNA polymerases, transcription, topoisomerase activity, as well as the ATPase activity of several DNA-dependent ATPases involved in DNA replication (rep, dnaB, n′ proteins). Furthermore, the H protein was found to favor the reannealing of DNA and to protect DNA against micrococcal nuclease digestion. For most of these activities, a stoichiometry of one H protein dimer per approximately 70 nucleotides of DNA was calculated (Hübscher et al. 1980).

The second protein (22 kD protein) was estimated to be present in approximately 10 000 copies per cell and was characterized by its strong preference for binding to superhelical and ssDNA and, because of this property, it was suggested that it may play a role in maintaining the superhelical conformation of the nucleoid loops. Unlike other histonelike proteins, however, the 22 kD protein behaves as a monomer, is heat-labile, is retained by hydroxylapatite and does not affect either DNA synthesis or DNA transcription (Kishi et al. 1982). In light of these properties, the grouping of the 22 kD protein with the other histonelike proteins does not appear to be entirely justified even though "histonelike" is still a rather ill-defined term.

Here, we shall describe some properties of the histonelike proteins NS (and of its subunits NS1 and NS2) and H-NS.

2 The NS Proteins

2.1 General Characteristics

In *Escherichia coli,* NS (HU) is composed of two polypeptide chains, NS1 and NS2, whose primary structures (Mende et al. 1978; Laine et al. 1980) display 69% homology, a clear indication that they originated from duplication of a common gene. Proteins homologous to *E. coli* NS have been isolated from a large number of prokaryotes. All these proteins display a very high degree of evolutionary conservation (Dijk et al., this volume) and it is noteworthy that *E. coli* is the only case known so far where two types of NS subunits exist. Although any speculation concerning the functional significance of two separate subunits is premature, their existence offers some advantages for the analyses of their structure-function relationships (Paci et al., this volume).

In spite of their large structural homology, NS1 and NS2 are characterized by a substantial difference in their Lys:Arg ratios (1.8 in NS1 and 3.7 in NS2) and by the presence of a histidine residue in NS2. The two subunits can be separated by ion-exchange chromatography under denaturing conditions; owing to the presence of the histidine, NS2 is eluted after NS1 at pH 5.2, while it is eluted before NS1 at pH 7.1. The

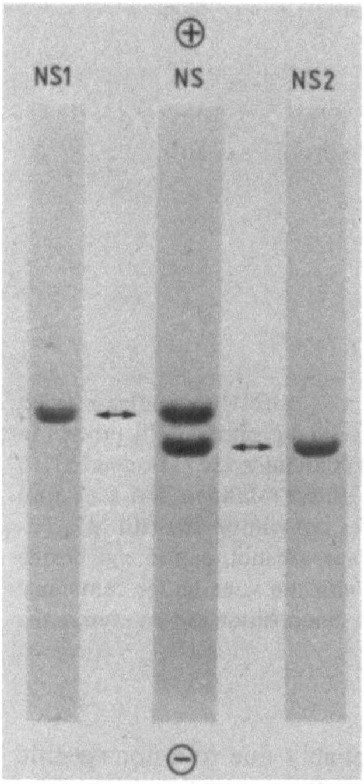

Fig. 1. Electrophoretic separation of NS1 and NS2 at pH 4.5 in Triton X100-urea polyacrylamide gels (From Paci et al. 1984)

two proteins can be separated electrophoretically at pH 4.5 on polyacrylamide gels containing Triton X100 and urea (Fig. 1).

2.2 Interaction of NS with Nucleic Acids

NS1, NS2 and NS can retain radioactive DNA (Lammi et al. 1984b) or RNA (Berthold and Geider 1976) on nitrocellulose filters. NS is slightly more efficient in retaining DNA than either NS1 or NS2 alone. With all three proteins the dose-response curve appears sigmoidal, suggesting that the binding of these proteins to DNA, or at least the event responsible for the retention of DNA on the filter, is cooperative (Lammi et al. 1984b). This is in contrast with the case of H-NS which, in comparable binding experiments, displays a hyperbolic binding curve (see Fig. 21A). The binding of NS to DNA was also studied by competition between NS and proflavin for DNA binding. In this experiment, the proflavin absorbance at 420 nm is quenched by DNA and the competition between proflavin and NS for DNA binding results in a decrease of the quenching effect. The dose response curve is sigmoidal also for this effect (Lammi et al. 1984b). The interaction between NS and DNA, as measured by nitrocellulose filtration, is very sensitive to the salt concentration (Fig. 2). As seen in Fig. 2, the amount of radioactive DNA retained on nitrocellulose filters by a given amount of NS is maximum at very low ionic strength and abruptly decreases above 50 mM NH_4^+ to reach a minimum at 80 mM. Above 80 mM, the capacity of NS to retain DNA increases steadily and reaches a broad maximum between 180 and 300 mM. Since at low ionic strength the negative charges of DNA are only partially compensated by counterions, the high

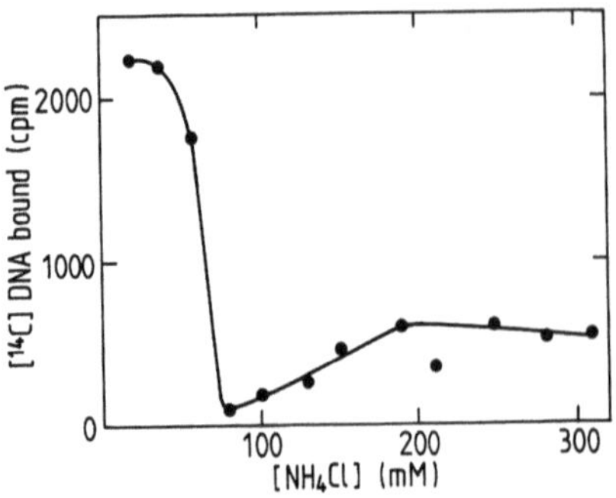

Fig. 2. Effect of salt concentration on the NS-DNA interaction. 3.5 μg [^{14}C]thymidine-labelled *E. coli* chromosomal DNA (650 cpm μg^{-1}) were incubated with 3.5 μg purified NS protein in 0.3 ml 0.1 × SSC buffer (Na citrate 1.5 mM, pH 7.0; NaCl 15 mM) containing the indicated NH_4Cl concentrations. The bound DNA was collected on nitrocellulose filters (Millipore HA 0.45 μm) pretreated by soaking for 60 min at room temperature in buffer containing Tris-HCl, pH 7.2, 20 mM; KCl, 100 mM; $MgCl_2$, 5 mM; NaEDTA, 1 mM; 2-mercaptoethanol, 5 mM and bovine serum albumin, 50 μg ml^{-1}. Before use, the filters were washed with the same buffer containing no bovine serum albumin and then with a buffer having the same composition as that used in the binding reaction

retention of DNA by NS at low salt concentration is probably due to a non-specific neutralization interaction. Indeed, other nucleic acid-binding proteins (e.g. some ribosomal proteins, translational initiator factor IF3), which certainly have functions distinct from that of histonelike proteins, can retain radioactive DNA on nitrocellulose filters at low ionic strength. In contrast, the retention of DNA on the filters at higher ionic strength indicates that NS1 is capable of interacting with DNA even at a physiological concentration of salt. From this point of view, the behavior of NS resembles somewhat that of H-NS, whose ability to bind DNA is also increased by increasing concentrations of monovalent and divalent cations (cf. Fig. 21C,D).

In addition to binding to ss-DNA and ds-DNA (Berthold and Geider 1976), NS was also found to bind to different types of RNAs. Indeed, Berthold and Geider (1976) found that NS (HD) binds to phage fr ssRNA (Berthold and Geider 1976) and, in our laboratory, NS was first isolated and identified as a tRNA-binding protein interfering (by producing high backgrounds) with the functional assay for translational initiation factor IF1 (the stimulation of the binding of fMet-tRNA to poly(AUG)-programmed 70S ribosomes); in addition, other groups had reported the association of NS with *E. coli* 30S ribosomal subunits (Suryanarayana and Subramanian 1978) or with 30S, 50S and 70S ribosomes (Dijk et al. 1983). To determine the relative affinity of NS for various nucleic acids, we made use of affinity chromatography. In Fig. 3 we show the elution of radioactive NS from four affinity columns, each containing a different type of nucleic acid. The conductivity measurements allowed the direct comparison of the four elution profiles.

As seen from Fig. 4, NS displays a marked preference in binding to DNA over RNA. Furthermore, NS binds better to tRNA than to rRNA and to ss-DNA than to ds-DNA. The latter finding, which agrees with that recently reported by Holck and Kleppe (1985), is somewhat puzzling in light of the fact that NS protects DNA against heat

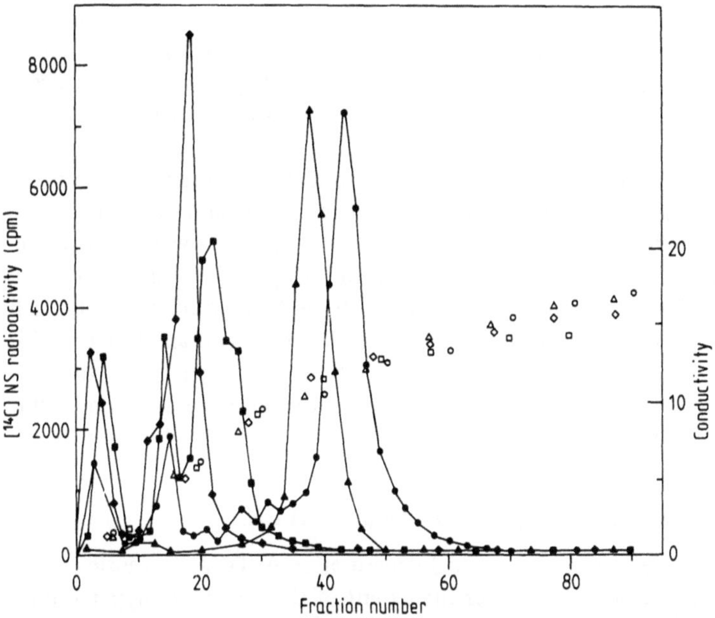

Fig. 3. Comparative affinity of NS for different nucleic acids. Radioactively-labelled NS was eluted from four different affinity chromatographic columns (0.5 × 5 cm) containing matrix-bound ds-DNA (▲), ss-DNA (●), tRNA (■) and rRNA (◆) with a 50 ml linear NaCl gradient (25–700 mM in 20 mM Tris-HCl, pH 7.2 containing 5 mM β-mercaptoethanol). The conductivity of the indicated fractions of each eluate was measured (indicated by the corresponding *open symbols*) in order to align the elution profiles. The in vitro labelling of NS was carried out by reductive methylation as previously described under particularly mild reaction conditions (low pH, low concentration of formaldehyde) (Gualerzi and Pon 1979). Control experiments showed that radioactively-labelled NS and non-labelled NS (as detected by electrophoresis) were eluted at the same ionic strength

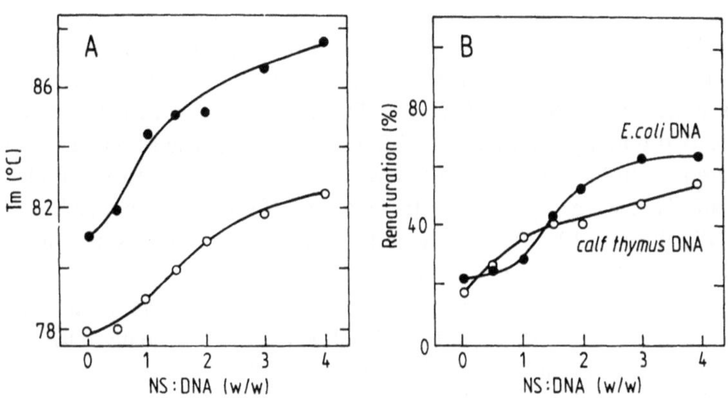

Fig. 4A,B. Effect of NS on the thermal stability of DNA as a function of the protein:DNA (w/w) ratio. Comparative effect of NS on: **A** the Tm of *E. coli* (●) and calf thymus (○) DNA; **B** the extent of renaturation of *E. coli* (●) and calf thymus (○) DNA. With renaturation we indicate the loss of hyperchromicity recorded upon slow cooling of thermally-denatured DNA (Modified from Miano et al. 1982)

denaturation (see Fig. 4). It should be noted, however, that NS was found to condense ss-DNA and ds-DNA in a very similar manner (Zentgraf et al. 1977). The very efficient condensation of ss-DNA by NS can also perhaps explain the large hypochromic effect induced by NS proteins on thermally-denatured DNA (Fig. 4B and Miano et al. 1982; Lammi et al. 1984b).

The preference of NS for binding to ss-DNA is also difficult to reconcile with the proposed role of this protein in the condensation of the bacterial chromosome. It is likely, however, that NS has no chance to interact with ss-DNA in vivo if the single-stranded regions of the chromosome are covered with SSB proteins. This seems to be substantiated by the finding that by virtue of their strong binding cooperativity, SSB proteins bind much more efficiently than NS to ss-DNA (Uhlmann and Geider 1977; Zentgraf et al. 1977). In light of the preference of NS for ss-DNA, however, the term "DNA-binding protein II", proposed for this protein (Geider and Hoffmann-Berling 1981), appears to be inappropriate.

2.3 Effect of NS on Thermal Stability and Transcription of DNA

As mentioned above, NS affects the thermal denaturation of DNA by causing an increase in the Tm without affecting either the extent of the maximum hyperchromic effect or cooperativity of the process (Miano et al. 1982). As seen in Fig. 4, with *E. coli* DNA, a Tm increase of over 6°C is observed at a protein:DNA ratio (w/w) of 4:1. Also in this case, the binding curve appears sigmoidal. From the qualitative point of view, *E. coli* NS affects, in a similar manner, all DNAs examined (*E. coli, Bacillus stearothermophilus,* calf thymus, synthetic polydeoxynucleotides). From the quantitative point of view, however, *E. coli* NS is somewhat more efficient in protecting *E. coli* than *B. stearothermophilus* DNA (Lammi et al. 1984b) and, most important, substantially more efficient with prokaryotic DNA than with eukaryotic DNA (Fig. 4A).

In addition to the effect on denaturation, NS also affects the renaturation of DNA as estimated from the extent of the hypochromic effect which accompanies the cooling of fully denatured DNA. As mentioned above, it is possible that a tight condensation of ss-DNA by NS might account for at least part of the large hypochromic effect observed. Whatever the nature of this phenomenon, however, it is noteworthy that also this effect is more pronounced with *E. coli* DNA compared with calf thymus DNA, and that the two binding curves appear different (Fig. 4B). The activities of NS and of the separated subunits NS1 and NS2 were found to be qualitatively similar, but isolated NS1 and NS2 were slightly less efficient than NS in increasing the Tm of the DNA, while NS2 displayed by far the greatest efficiency in producing the hypochromic effect (Miano et al. 1982; Lammi et al. 1984b).

Contrary to an earlier report (Rouviere-Yaniv and Gros 1975), we found that transcription of various template DNAs [λ, *E. coli,* calf thymus DNA and poly d(A,T)] by *E. coli* RNA polymerase is inhibited by NS (Losso et al. 1982). Compared to NS, isolated proteins NS1 and NS2 produced similar though slightly less efficient inhibitions. Interestingly, the inhibitory effect of a given amount of NS on transcription is about two-fold greater with *E. coli* than with calf thymus DNA (Fig. 5).

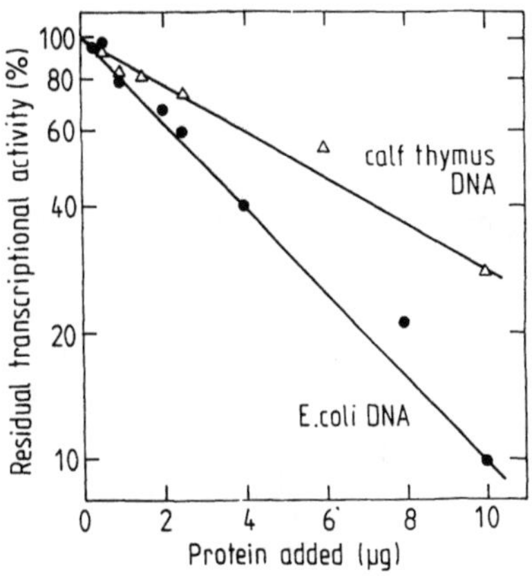

Fig. 5. Differential effect of NS on transcription of *E. coli* and calf thymus DNA. From Losso et al. (1982) where the experimental conditions are given

2.4 NS Can Inhibit Restriction Endonuclease Cleavage of DNA

Within the *E. coli* cell the DNA is presumably complexed with several natural ligands, among which the prokaryotic histonelike proteins are important and abundant representatives. Since both optimum rate and substrate specificity of restriction endonuclease cleavage may depend on the presence on DNA of these natural ligands, the effect of NS1 and NS2 on the restriction endonuclease cleavage of DNA was investigated (Pingoud et al. 1984; Ehbrecht et al. 1985). Thus, pEMBL DNA was incubated with EcoR I in the presence and absence of NS1 and NS2 and the rate and the extent of cleavage was determined by densitometric scanning of the bands separated on agarose gels. It was found that the rate of cleavage is increased at low NS:DNA (w/w) ratios and decreased when the NS:DNA ratio is increased. Almost complete protection was obtained at an NS1 (or NS2):DNA ratio of 2. This protection was found to be somewhat specific since, under the same conditions, comparable amounts of cytochrome c, a known DNA-binding protein, gave substantially less protection, while bovine serum albumin gave no protection at all. The protection by NS was also greater with supercoiled DNA than with relaxed DNA, suggesting that, as reported earlier (Rouviere-Yaniv et al. 1979), NS might preferentially bind to the former. These results can be explained assuming that, at non-saturating concentrations of NS, competition between NS and EcoRI for non-specific binding sites on the DNA would have the same effect (i.e. an increased rate of cleavage) as increasing the concentration of free EcoRI; conversely, competition between EcoRI and NS for the real restriction site, occurring at saturating NS concentrations, would cause the inhibition of cleavage. An additional (or alternative) explanation could be that NS binding causes a structural alteration of the DNA which affects the restriction endonuclease activity. The finding that NS alters somewhat the circular dichroism spectrum of supercoiled DNA, producing a decreased positive ellipticity accompanied by a slight red shift of its maximum wavelength would support this premise (Pingoud et al. 1984). Suppression of the positive ellipticity band of free DNA at 260–290 nm upon addition of histones has been observed by several authors (see, for instance, Read et al. 1985) and has been attributed to alteration of the secondary and/or tertiary structure of DNA, such as tilting

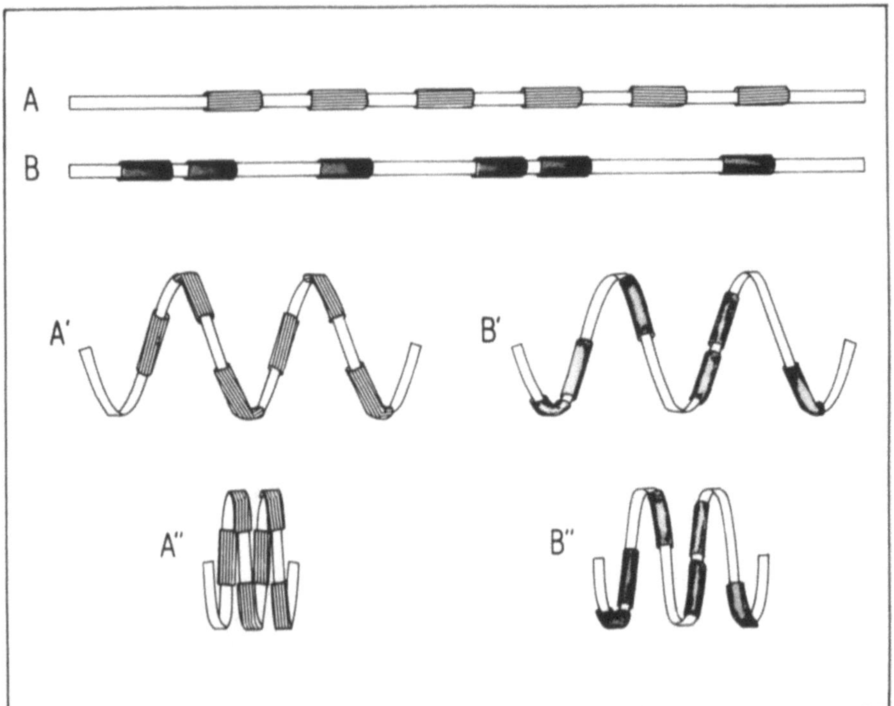

Fig. 6A,B. A hypothetical mechanism of species-specific phasing of NS on DNA. The same number of NS molecules is shown bound to hypothetical preferential binding sites regularly distributed along the homologous DNA (A) or irregularly distributed along the heterologous DNA (B). The maximum packaging efficiency is obtained in A′ and A″ making use of protein-protein interactions

of the base pairs with respect to the helical axis and decrease of the number of base pairs per turn. It is likely that the intensity of the positive CD band of DNA is reduced by NS for similar reasons.

Finally, it was found in separate experiments that under conditions where linear diffusion of the restriction endonuclease EcoRI on DNA was shown to accelerate the rate of DNA cleavage, NS is an obstacle to the linear diffusion of EcoRI along the DNA. In fact, it was found that while in the absence of NS the rate of cleavage can be a function of the length of the DNA (i.e. being faster with longer DNA fragments than with shorter ones), in the presence of NS, all the cleavage rates become equal (Ehbrecht et al. 1985).

2.5 Is There a Mechanism to "Phase" NS on the Chromosome?

As seen above, NS displays a low degree of "quantitative species specificity" in both denaturation-renaturation and transcription experiments. If the inhibition of transcription is due – like the increase in Tm – to a stabilization of the DNA double-helical structure, then both effects could simply reflect the same basic mechanism (i.e. the binding of NS to DNA). In this case, since the two types of DNA were prepared under rigorously the same conditions and displayed, as far as we could judge, identical properties (absence of segments of ss-DNA, and of contaminating RNA or proteins), we must conclude, at least tentatively, that the interaction of NS is more efficient with the homologous than with the heterologous DNA. An attractive hypothesis to explain this finding could be the existence, only in prokaryotic DNA, of a pattern of regularly repeated "signals" to which NS binds preferentially. These signals would produce a phasing of proteins on the DNA so as to favor the establishment of protein-protein interactions; in this way the maximum packaging effect would be obtained with the lowest amount of protein. This hypothesis can be visualized in the model presented in Fig. 6. More experiments are obviously needed to verify the validity of this hypothesis and, in particular, to identify the possible nature of the signals. Nevertheless, it seems relevant to the present discussion to recall that there is good evidence for the existence of a phasing mechanism, at least in the case of eukaryotic nucleosomes (Zachau and Igo-Kemenes 1981) and to mention (1) that some of the data suggest the existence of a low degree of cooperativity in the binding of NS to DNA; (2) that at high concentrations (i.e. $>5 \times 10^{-5}$ M), NS seems to undergo extensive aggregation to produce higher-order structures (see below); and (3) even though one would not expect a protein like NS to be able to recognize a specific DNA sequence, nevertheless some recent results raise the possibility that NS might have, after all, some degree of binding specificity. In fact, it was found that two *E. coli* proteins, IHFα and IHFβ required for the site-specific recombination of phage λ DNA are homologous in primary structure to NS proteins and recognize a specific DNA sequence (Craig and Nash 1984; Flamm and Weisberg 1985). Furthermore, it was found that a host factor identified to be NS (HU) (Mizuuchi 1983 and personal communication) is necessary, together with two Mu gene products (A and B proteins) for the transposition in vitro of phage Mu DNA. Since the transposition of Mu DNA, unlike the integration of λ DNA, occurs ubiquitously on the chromosome without being totally random, it is possible that this NS activity might be due to its preferential binding to a frequently occurring DNA sequence or structure.

2.6 Identification of the NS Region Responsible for the Interaction with DNA

Several lines of evidence indicate that protein-protein interactions in NS are primarily hydrophobic (Paci et al., this volume). Circumstantial evidence indicates, on the other hand, that ionic interactions play an important role in the interaction between NS and DNA.

The mechanism of the interaction of NS with DNA was studied by [^{1}H]-NMR spectroscopy and by protein chemical modification experiments (Lammi et al. 1984a). It was found that when increasing amounts of deuterated DNA were added to NS, the resolution of the NMR spectrum of the protein was progressively lost owing to the broadening of all resonance lines and that all resonances, including those due to tertiary and quaternary structure, broaden simultaneously. Thus, it was concluded from these experiments that NS binds to DNA in its aggregated form and without gross alteration of its tertiary structure (Lammi et al. 1984a; Paci et al., this volume). In the same experiments, we also obtained some indications that one or more Arg residues of NS are implicated in DNA binding. Direct evidence for this premise was sought by

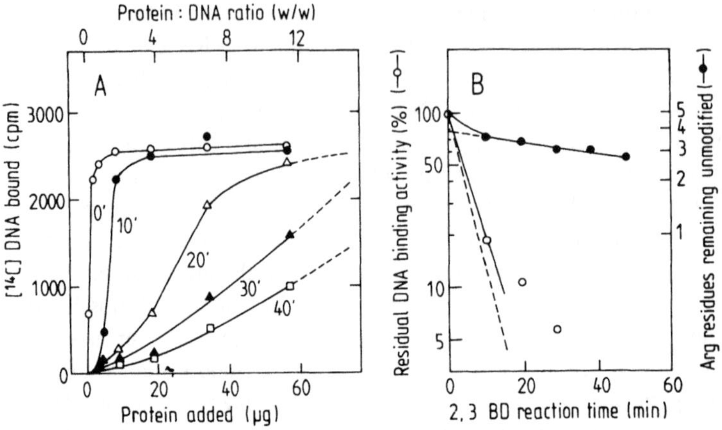

Fig. 7A,B. Inactivation of the DNA-binding site of *B. stearothermophilus* NS protein following chemical modification of the arginines with 2,3 butanedione. **A** DNA-binding activity of NS modified for the indicated times. **B** Kinetics of arginine modification and loss of DNA-binding activity (From Lammi et al. 1984a)

selective chemical modification. To simplify the identification of the reacted amino acid residues, we decided in this case to work with BS-NS, the protein homologous to NS extracted from *B. stearothermophilus* which possesses only one type of subunit.

After reacting with 2.3 butanedione for various lengths of time, the residual capacity of BS-NS to bind DNA was tested by nitrocellulose filtration experiments. As seen in Fig. 7, the DNA-binding capacity of BS-NS. is rapidly lost. The inactivation is accompanied by the loss of 1.5 (out of 5) Arg residues; the modification of the remaining arginines follows at a much slower rate. The fast-reacting arginine residue essential for DNA binding was identified by protein chemical analysis and found to be located in the peptic peptide P8 which contains a cluster of four Arg residues. The fifth Arg residue of the protein (Arg 37), located in a different region of the molecule, was found to be either poorly accessible or completely inaccessible to the reagent. The Arg residues in peptide P8 were also found to be protected from chemical modification by DNA. Thus, a positively charged, hydrophilic region of the protein molecule implicated in the interaction with DNA via at least one of its arginine residues was identified (Fig. 8). Our results did not allow us, however, to determine which of the four Arg residues of peptide P8 is responsible for the inactivation. Comparison of the primary structures of NS-related proteins from other bacteria revealed that Arg 61 has been conserved in all sequences determined up to now. Arg 55 and 58 also appear to be highly conserved; Arg 53 is the least conserved of these residues (it is missing in *E. coli* NS1 and NS2), but is usually replaced by either lysine or by hydrophilic residues. Therefore, all Arg residues of P8, with the only possible exception of Arg 53, are equally likely candidates to participate in DNA binding. Furthermore, it is probable that more than one arginine residue of NS is involved in the interaction, even though the modification of just one of them is sufficient to inactivate the binding site.

The implication of the hydrophilic domain (residues 51–70) in the interaction of NS with DNA is further substantiated by the finding that dye-sensitized photooxidation

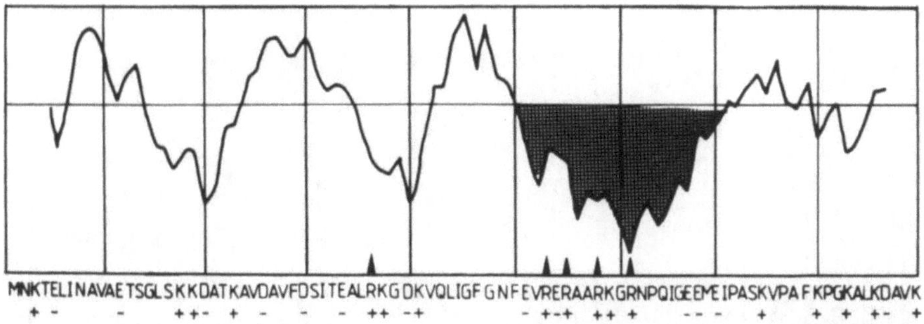

Fig. 8. Location of the Arg residues in the primary structure of *B. stearothermophilus* NS protein. The primary structure of *B. stearothermophilus* NS was analyzed according to the computer program of Kyte and Doolittle (1982) to display the hydrophobic (*above the horizontal line*) and the hydrophilic (*below the horizontal line*) regions of the protein. The primary sequence (Kimura and Wilson 1983) and the charged amino acids are indicated. The *arrows* indicate the positions of the five Arg residues in the primary structure. The *hatched area* represents the hydrophilic, positively-charged region implicated in DNA binding (From Lammi et al. 1984a)

of His 54 of *E. coli* NS2 was also found to prevent the interaction of this protein with DNA (Lammi, unpublished results). Finally, it should be mentioned that the X-ray diffraction studies of BS-NS crystals revealed that the 52-77 residue region extends as a structurally, poorly defined arm region which, by virtue of its positive charge and location in the molecule was proposed to be the DNA-binding site of the protein (Tanaka et al. 1984).

2.7 The Nature of the Quaternary Structure of NS

As mentioned above, NS binds to DNA in the aggregated form, but there is no agreement so far concerning the nature of its quaternary structure. Thus, NS consists of tetramers according to some authors (Berthold and Geider 1976; Suryanarayana and Subramanian 1978) and of dimers according to others (Rouviere-Yaniv and Kjeldgaard 1979). Hydrodynamic studies performed on *B. stearothermophilus* NS did not resolve the controversy since the results were consistent with either a tetrameric or with an elongated dimeric structure (Dijk et al. 1983). Since the three-dimensional structure obtained from X-ray crystallography of BS-NS protein revealed the presence of a dimeric structure and suggested the existence of two extended arms, however, it was concluded that the dimer is the fundamental unit of the protein. Furthermore, even though no dimer-dimer contacts suggestive of a possible higher-order structure were found, it was proposed that BS-NS forms a nucleosomelike object containing 8–10 dimers with a diameter of approximately 140 Å able to induce DNA supercoiling (Tanaka et al. 1984).

The problem of the quaternary structure of NS in solution was investigated by NMR spectroscopy (see Paci et al., this volume), as well as by protein-protein cross-linking reactions. The results of the latter experiments are presented here. After incubation with dimethylsuberimidate (DMS) at least six discrete products, detectable by both one- and two-dimensional PAGE analyses were obtained (Fig. 9). From their

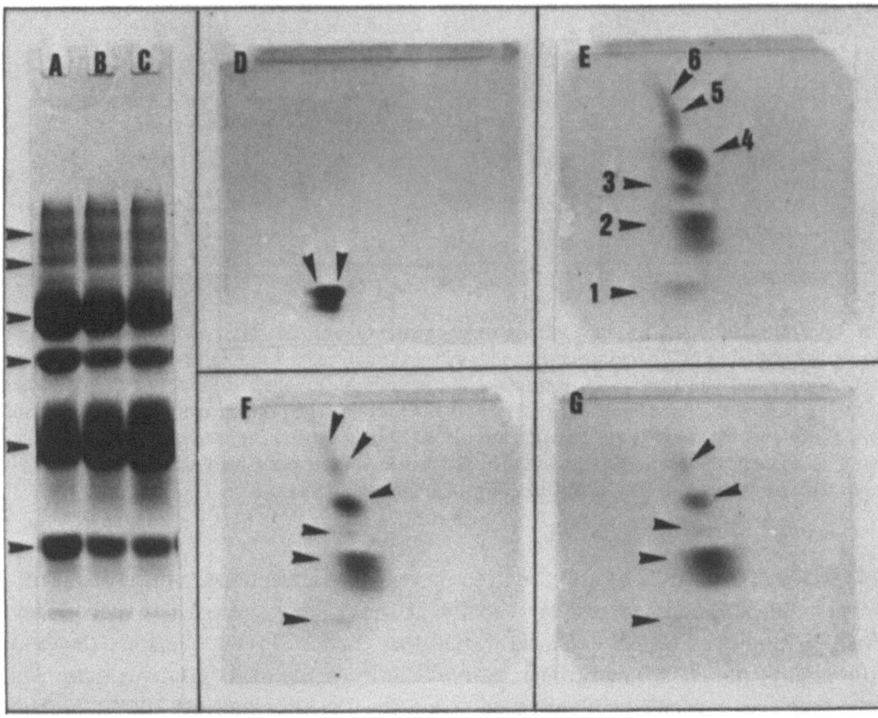

Fig. 9A–G. Electrophoretic separation of dimethylsuberimidate cross-linked products of NS. One-dimensional SDS-containing PAGE (15% acrylamide) of cross-linked NS1 (A), NS (B) and NS2 (C); two-dimensional polyacrylamide gel electrophoresis (Mets and Bogorad 1974) of non-reacted NS (D), cross-linked NS1 (E), NS (F) and NS2 (G). Further experimental details are given in Losso et al. (1986)

respective molecular weights, these products were identified as being monomers (i.e. monomeric proteins monofunctionally reacted with DMS), dimers, trimers, tetramers, pentamers and hexamers. With both NS1 and NS2 and with NS, dimers and tetramers generally represent the main reaction products, but the amount of each component, quantified by densitometric scanning of the one-dimensional stained gels, can vary depending upon some reaction parameters. With all combinations of proteins, the cross-linked dimers, as evidenced by both one- and two-dimensional electrophoreses (Figs. 9 and 10) migrate as heterogeneous bands in contrast to the other products which appear homogeneous. In addition, the cross-linked dimers obtained with NS1, NS2 and NS show different degrees of heterogeneity with the NS dimers displaying the largest extent (Fig. 10). Having also ruled out that the heterogeneity of the band of dimers was due to an overloading artefact (not shown), we concluded that the electrophoretic behavior of the dimers is due to a real heterogeneity of this cross-linked product (see below).

The yield of cross-linking with dimethyladipimidate (DMA), whose aliphatic arm is approximately 1.6 Å shorter than DMS, was found to be somewhat lower than with DMS. In particular, covalent aggregates larger than tetramers were virtually absent

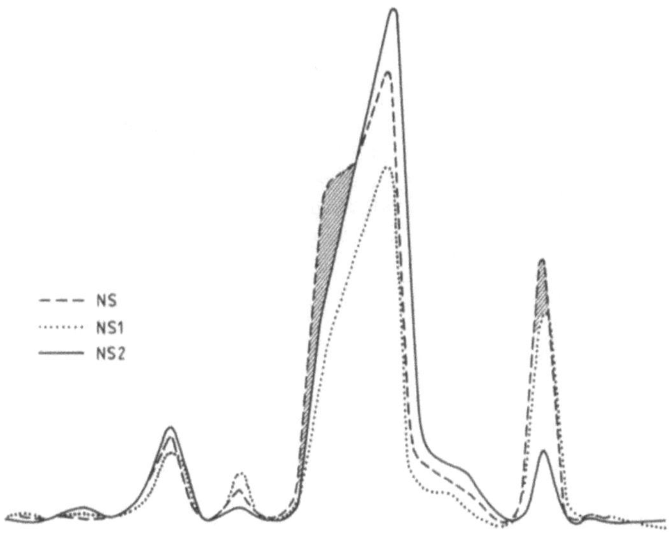

Fig. 10. Electrophoretic heterogeneity of DMS-cross-linked dimers of NS proteins. Densitometric scans of one-dimensional SDS-containing PAGE (15% acrylamide) separation of the cross-linking reaction products of NS1 (····), NS2 (——), and NS (----). Electrophoretic migration is from *left to right.* The peaks are (from *right to left*): monomers, dimers, trimers, tetramers

(see Fig. 13C,D). This finding indicates that the distance between some lysine residues implicated in the cross-linking reaction is critical and implies that aggregates larger than tetramers result from a cross-linking bond(s) distinct from those giving rise to dimers, trimers and tetramers. These considerations are particularly relevant for the interpretation of the cross-linking data (see below).

To investigate the nature of the quaternary structure of NS and of its subunits NS1 and NS2, we examined the protein concentration dependence of the yield of formation of individual cross-linked products. The reaction conditions were chosen so that the yield of each product would reflect the state of aggregation of the protein and the protein concentrations were varied over a range covering over four orders of magnitude. Following each reaction, the amount of each cross-linked product, analyzed and quantified electrophoretically, was expressed as percentage of the total amount of protein resolved and plotted as a function of the protein concentration in the reaction mixture. The results are presented in Fig. 11. Superficially, NS1, NS2 and NS seem to display different patterns of behavior. Three important features, however, are common to all reactions. These are that (1) sizable amounts of trimers and tetramers are obtained even at the lowest protein concentration; (2) below a critical protein concentration (approximately 10^{-6} M), the yield of trimers is higher than that of tetramers, while above this concentration, the yield of tetramers exceeds the yield of trimers; and (3) the yield of tetramers increases slightly and linearly with protein concentration up to a critical concentration above which a steep increase in the amount of tetramers and larger aggregates formed takes place; this is accompanied by a corresponding reduction in the yield of dimers and, to a lesser extent, of trimers. These results can be explained by the model presented in Fig. 12. The model is based on the behavior

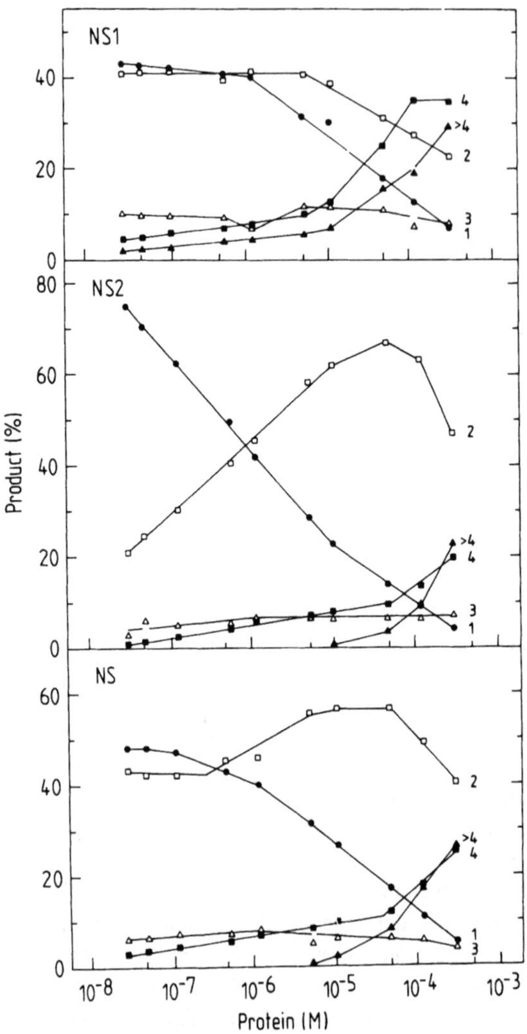

Fig. 11. Dependence of the yield of formation of cross-linked products on the NS1, NS2 and NS concentrations. The concentration of each protein (taken as monomers) in the DMS cross-linking reaction mixtures is indicated in the figure. *1* Monomers (●); *2* dimers (□); *3* trimers (△); *4* tetramers (■); *> 4* pentamers + hexamers (▲). Further details are given in Losso et al. (1986)

of NS1, but can also be extended to explain the results obtained with NS and, with slight modifications, those obtained with NS2. The large amount of cross-linked dimers produced at even the lowest protein concentration (3×10^{-8} M) and the finding that their yield is, by and large, insensitive to increases of protein concentrations up to at least 5×10^{-6} M, indicates that the yield of dimers depends only on the efficiency of the chemical reaction(s) leading to their formation and not on the position of the monomers ⇌ dimer equilibrium. Thus, even at the lowest protein concentration tested, this equilibrium must be completely shifted to the right. The substantial amount of trimers and tetramers obtained at the same protein concentration indicates – on the other hand – the existence of an equilibrium dimers ⇌ tetramers. The steady increase of tetramers, which takes place at the expense of monomers and trimers between 3×10^{-8} and 5×10^{-6} M, indicates that this equilibrium is progressively shifted from

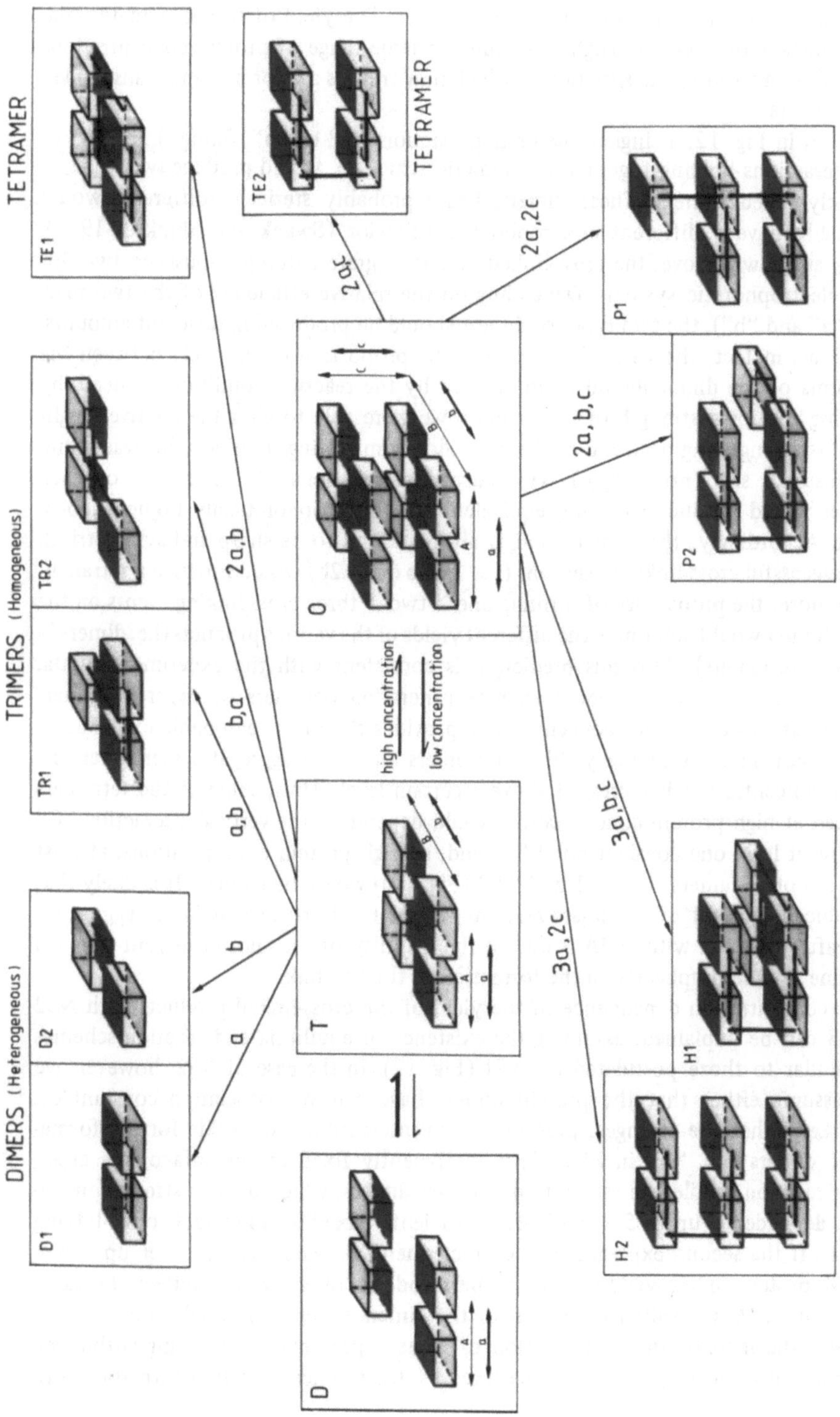

Fig. 12. Model of quaternary structure of NS illustrating the various patterns of DMS covalent bonding. Protein-protein interactions are indicated by the arrows *with capital letters; arrows with lower case letters* indicate the corresponding covalent bonds. Modified from Losso et al. (1986)

left to right as the protein concentration increases. The yield of trimers remains relatively constant or decreases slightly within the same range of protein concentrations as a result of a steady state situation in which new trimers are formed and transformed into tetramers.

As seen in Fig. 12, a single cross-linking reaction, "a" or "b", fixing either "A" or "B" interactions holding together a tetrameric structure, would produce two types of covalently-linked dimers. These dimers, being probably sterically different, would presumably have a different electrophoretic behavior (Swank and Munkres 1971). Indeed, as shown above, the cross-linked dimers migrate heterogeneously in two different electrophoretic systems. Depending on the relative efficiency of the two reactions ("a" and "b"), the two types of dimers should be produced in different amounts; and this is, in fact, the case. Furthermore, the proportion of the yield between the two forms of the dimer should be influenced by the reaction conditions; indeed, by increasing the ionic strength of the reaction we were able to alter the relative height of the faster migrating peak and of the more slowly migrating shoulder; this was found to increase in size (not shown). Two successful reactions ("a" and "b") on each tetramer would produce sterically equivalent (i.e. electrophoretically homogeneous) trimers. Accordingly, the trimer band is always found to be sharp and symmetrical. Three successful cross-linking reactions (i.e. 2a + b or a + 2b) would produce a tetramer. Furthermore, the probability of forming one > two > three cross-linking events on the same tetramer would determine the different yields of the various products (i.e. dimers > trimers > tetramers). Also this prediction is consistent with the experimental data. Finally, an interaction between tetramers generating octamers or larger multimers occurring at higher protein concentrations provides the most reasonable explanation for the sharp increase in the yield of tetramers and large aggregates seen when the protein concentration is increased above a certain level. Thus, some of the tetramers produced at high protein concentration would depend on the vertical interaction "C" fixed by at least one covalent bond "c" and, at high protein concentrations, at least two types of tetramers [i.e. TE1 and TE2 (Fig. 12)] would be formed. It is likely that the reduced amount of tetramers and the nearly total absence of larger aggregates found after reaction with DMA is due to the inability of the shorter reagent to reach the lysine residues implicated in the formation of the "c" link.

The concentration dependence of the yield of the cross-linked products with NS2 and NS can be explained, assuming the existence of equilibria and reaction schemes very similar to those postulated for NS1 (Fig. 12). In the case of NS2, however, we must assume either that the protein dimers have a lower association constant or, most likely, that the strongest protein-protein interaction responsible for the formation of dimers (i.e. "A" in Fig. 12) is inefficiently fixed by an unfavorable cross-linking reaction, while the association between dimers, which display strong concentration dependence up to 5×10^{-5} M, is efficiently fixed by a favorable cross-linking reaction. If the second explanation is correct, then the linear increase, seen up to 5×10^{-5} M protein in the yield of both dimers and tetramers, would reflect the same phenomenon, i.e. the shift to the right of the dimers $\rightleftharpoons$ tetramer equilibrium. Also in this case, the increase in the formation of trimers, presumably occurring within the same concentration range, must be obscured by their concurrent transformation into tetramers.

The results obtained with NS (NS1 + NS2) are very similar, overall, to those obtained with NS1. As in the case of NS2, however, a concentration-dependent dimerization of dimers (i.e. formation of tetramers) is probably responsible for the marked increase in the formation of covalently-linked dimers occurring between 5×10^{-7} and 10^{-5} M protein. Thus, we interpret the cross-linking results to indicate that even at 3×10^{-8} M, no monomeric form of the protein is present in the samples of NS1, NS (NS1 + NS2) and, very likely, of NS2; instead, we propose that a multiple equilibrium dimers $\rightleftharpoons$ tetramers $\rightleftharpoons$ larger aggregates exists. As the protein concentration is increased, the equilibria are shifted to the right and one can estimate the K_d to be $\leqslant 10^{-9}$ M for the dimers and between 10^{-6} and 10^{-7} M for the tetramers. The latter value is in good agreement with that determined for the quaternary interaction involving the stacking between phenylalanine residues monitored by [^{1}H]-NMR spectroscopy (Paci et al., this volume). Thus, if our interpretation of the data is correct, NS would be found in equilibrium between tetramers and larger aggregates at its presumed (10^{-4} M) in vivo concentration (Rouviere-Yaniv 1977).

The present data do not allow us to identify the nature of the larger aggregates of NS; the hexamer is the largest cross-linked product identified with certainty but, in spite of our inability to detect octamers, it is reasonable to assume that they exist if the tetramer is indeed the key element of the quaternary organization of NS, and the higher-order structures are produced by interactions between tetramers.

Finally, it should be noted that the model of the quaternary structure of NS derived from the present data is inconsistent with the previously mentioned model of NS structure in which 8–10 dimers combine to form a disclike object (Tanaka et al. 1984).

2.8 DNA Favors the Formation of Higher-Order Aggregates of NS

To determine whether DNA has any influence on the quaternary structure of NS, we investigated the effect of DNA on the yield of the individual cross-linked products of the protein. The results of these experiments, shown in Figs. 13 and 14, indicate that while the overall extent of cross-linking as measured from the amount of residual monomers is not, or only slightly, affected by the presence of DNA, the yield of the individual cross-linked products is substantially altered. In fact, when the cross-linking reaction with DMS is carried out in the presence of DNA, a larger proportion of trimers, tetramers and larger aggregates is produced at the expense of the dimers, whose yield is drastically reduced. These effects are much more pronounced after a short reaction time (2 min) than after a long one (30 min), suggesting that not only the extent, but also the rate of formation of the larger aggregates, is stimulated by the presence of DNA (Fig. 13A,B). That this is indeed the case can be seen from the time course of appearance of the various cross-linked products shown in Fig. 14A. In a parallel experiment, the effect of DNA on the yield of the cross-linking reaction with the shorter cross-linking reagent DMA was investigated. As seen from Figs. 13C,D and 14B, the effect of DNA in this case is much less pronounced than with DMS. The only difference worth mentioning is a slight increase in the amount of trimers and tetramers formed in the presence of DNA and a rate increase in the formation of trimers which can be explained by a somewhat tighter packing of a tetrameric structure in the presence of DNA.

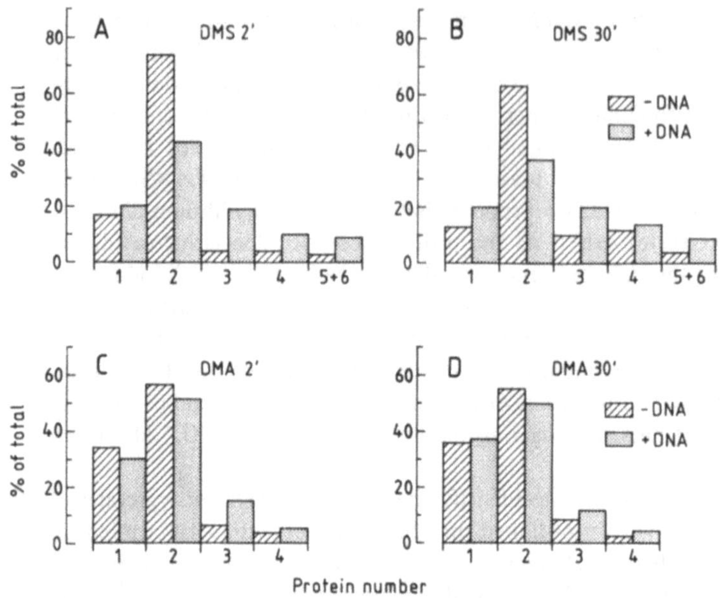

Fig. 13A–D. Effect of DNA on the yield of the various DMS and DMA cross-linked products of NS. The cross-linking reactions were carried out for the indicated times with either DMS or DMA; each reaction mixture also contained about 10^{-5} M NS and, where indicated, one-fourth by weight of *E. coli* chromosomal DNA sonicated to reduce its size to ∿500 bp (From Losso et al. 1986)

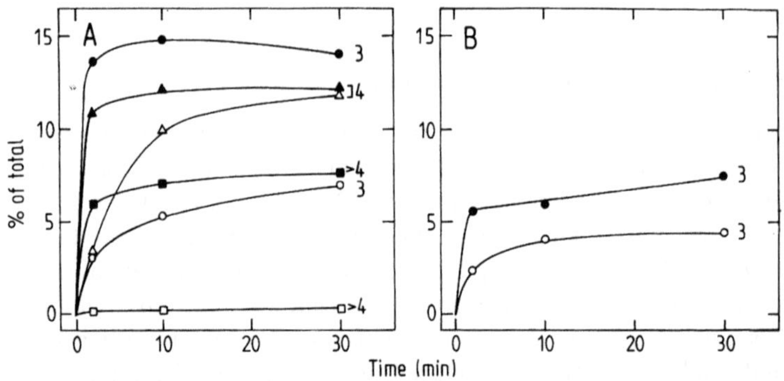

Fig. 14A,B. Effect of DNA on the time course of formation of various NS cross-linked products. **A** Time course of formation of DMS cross-linked trimers (●, ○), tetramers (▲, △), pentamers + hexamers (■, □) (From Losso et al. 1986). **B** Time course of formation of DMA cross-linked trimers (●, ○). *Open symbols* no DNA present; *closed symbols* DNA present. Other conditions as in Fig. 13

Overall, the finding that DNA stimulates the formation of cross-linked products larger than dimers when the reaction is carried out with DMS, but only has marginal effects when the reaction is carried out with the shorter reagent, can be explained assuming that DNA favors the formation of higher-order aggregates which henceforth are covalently linked by DMS, but which DMA, as mentioned above, fails to fix by covalent bonds.

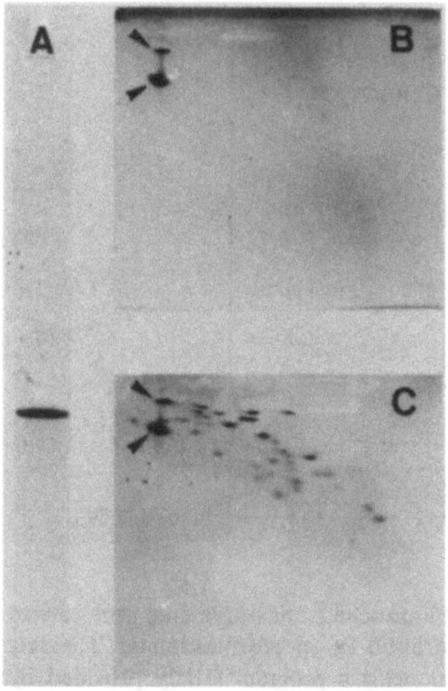

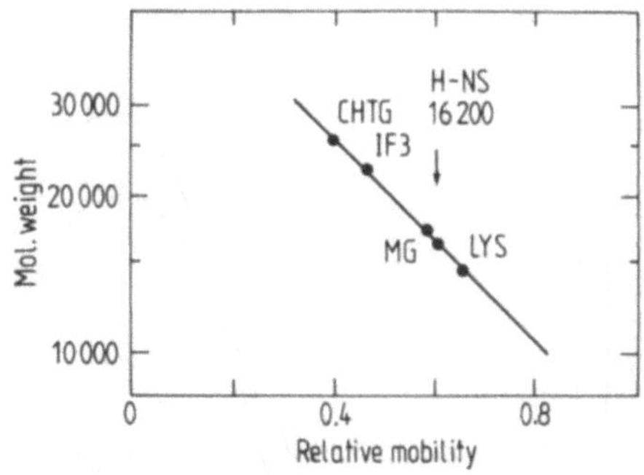

Fig. 16. Determination of the molecular weight of H-NS. Molecular weight standards were chymotrypsinogen (CHTG), translational initiation factor 3 (IF3), myoglobin (MG) and lysozyme (LYS)

Fig. 15A–C. One- and two-dimensional electrophoretic analysis of H-NS. **A** One-dimensional SDS-containing PAGE (15% acrylamide) of 10 μg of purified H-NS. **B** Two-dimensional PAGE (Geyl et al. 1981) of H-NS. **C** Same as in **B** in the presence of *E. coli* ribosomal proteins

3 The H-NS Protein

3.1 General Characteristics

In addition to NS, we have detected the presence of another abundant DNA-binding protein of higher molecular weight (H-NS) in *E. coli* extracts (Lammi et al. 1984b). As in the case of NS, also H-NS was originally purified from the high-salt ribosomal wash in our laboratory (Pawlik et al. 1981). The possible homology between this protein and other proteins with similar properties described by other groups is still unclear (see Sect. 3.4). The purification of H-NS from the post-ribosomal fraction of *E. coli* MRE600 cells grown in continuous culture yields approximately 60 mg of pure protein per kilogram of wet cell paste. This would correspond to well over 10 000 copies of H-NS monomers per cell assuming 60% recovery. The purified protein migrates as a single band in one-dimensional SDS containing PAGE (Fig. 15A).

From the electrophoretic mobility in this gel, the molecular weight of H-NS can be estimated to be about 16 000 (Fig. 16). This molecular weight agrees fairly well with the actual size of the protein determined from its amino acid composition. The electophoretic mobility of H-NS in two-dimensional PAGE alone and in relation to *E. coli* ribosomal proteins is shown in Fig. 15B,C. In the electrophoretic system used (Geyl et al. 1981), H-NS migrates as a double spot, the major one of which is found just below L9 and partially overlapping S6. The minor spot, which migrates more slowly in the second dimension, is due to an aggregate (dimer) of H-NS.

The elucidation of the primary sequence of H-NS has been nearly completed (M. Falconi and M.T. Gualtieri, unpublished results). No sequence homology has been

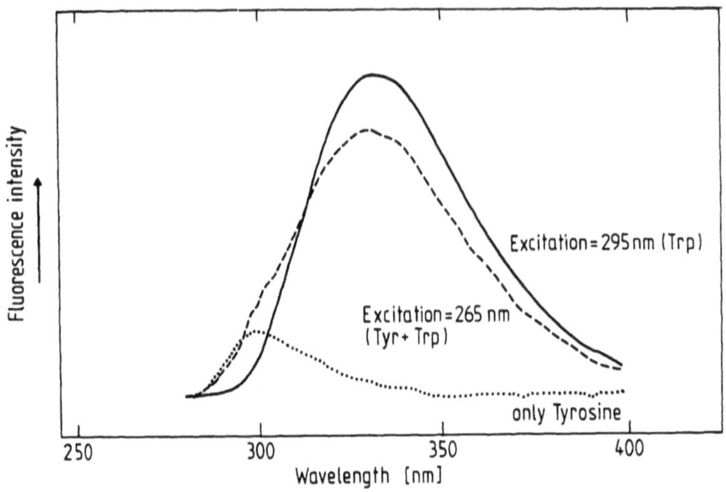

Fig. 17. Intrinsic fluorescence emission spectra of H-NS. Fluorescence measurements were carried out as described by Steinhäuser et al. (1982) with an SLM 8000 DS spectrofluorimeter. The data were processed with a Hewlett-Packard 9825A computer using a program kindly provided by Dr. P. Woolley. All measurements were done, unless otherwise specified, at 20° ± 2°C in 5 × 5 mm quartz cuvettes. Excitation was at either 265 or 295 nm as indicated, with a band width of 4 nm. The emission band width was 2 nm

found so far between H-NS and NS1, NS2 or any other *E. coli* protein, any eukaryotic histone or, for that matter, between H-NS and any of the proteins listed in the NBRF protein sequence data bank (as of May 1985). The sequence of the first 20 amino acids from the amino terminus is nearly the same as that reported for Hla (Laine et al. 1984); the overall amino acid composition of H-NS, however, differs somewhat from that reported for Hla.

H-NS contains one tryptophan and three tyrosine residues. The intrinsic fluorescence emission spectra of the protein are shown in Fig. 17. Normalization and subtraction of the emission spectrum obtained when only tryptophan is excited at 295 nm from that obtained when tyrosines and tryptophan are excited at 265 nm, yields the separate emission spectra of the tyrosines and of tryptophan (Fig. 17), provided there is no efficient Förster transfer between the tyrosine residues and the tryptophan residue.

From the emission maximum of the tryptophan (327 nm), which is considerably blue-shifted with respect to free tryptophan (348 nm), from the high value of the fluorescence anisotropy of tryptophan in native H-NS (Fig. 18C) and from the fact that tryptophan fluorescence is not affected by either cationic or anionic quenchers (Fig. 19A,B), it can be concluded that this residue is localized in an internal hydrophobic region of the molecule and that its mobility is strongly restricted. This conclusion is consistent with the spectral properties of the tryptophan ring proton resonances seen in NMR spectroscopy (Paci et al., this volume).

Upon increasing the temperature, the emission maximum of the tryptophan fluorescence undergoes a progressive red shift (Fig. 18A), while the fluorescence anisotropy diminishes (Fig. 18C), indicating that the tryptophan ring becomes progressively

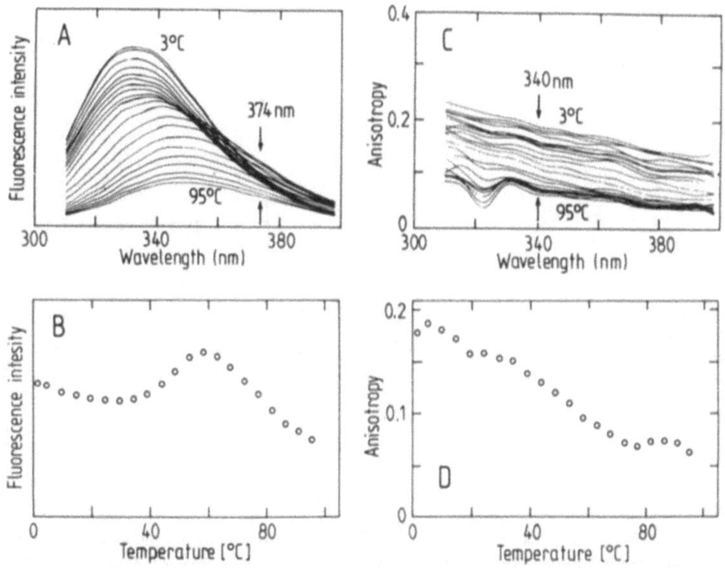

Fig. 18A–D. Thermal denaturation of H-NS monitored by changes of tryptophan fluorescence. **A** Emission spectra (λ_{ex} = 295 nm) of H-NS (10 μM in Tris-HCl, pH 7.2; KCl 500 mM; DTE, 0.5 mM) recorded between 3° and 95°C at approximately 5° intervals. **B** The fluorescence intensities at λ_{em} = 374 nm, derived from the spectra shown in **A**, are plotted against the respective temperatures. **C** Tryptophan fluorescence anisotropy between 3° and 95°C. **D** Plot of fluorescence anisotropy at λ_{em} = 340 nm vs temperature. The trough at ∿320 nm seen in C is due to a measurement artefact

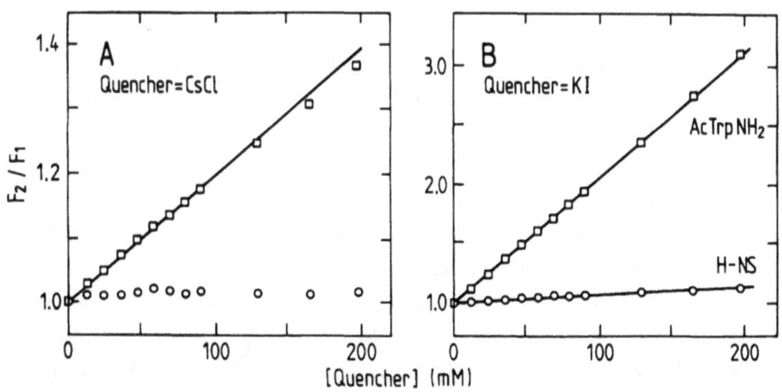

Fig. 19A,B. Inability of cationic and anionic fluorescence quenchers to affect the H-NS tryptophan fluorescence. Stern-Volmer plots for acetyltryptophanamide ($AcTrpNH_2$) (□) and H-NS (○) quenching by (A) CsCl and (B) KI

exposed to the solvent and increasingly mobile. Also, these findings are in good agreement with the NMR spectroscopic data. In Fig. 18, both red shift and change in anisotropy are plotted as a function of temperature; the tryptophan red shift appears as an increase in intensity of the fluorescence emission at 274 nm, superimposed on the overall temperature-dependent decrease of the fluorescence intensity (Fig. 18B). As

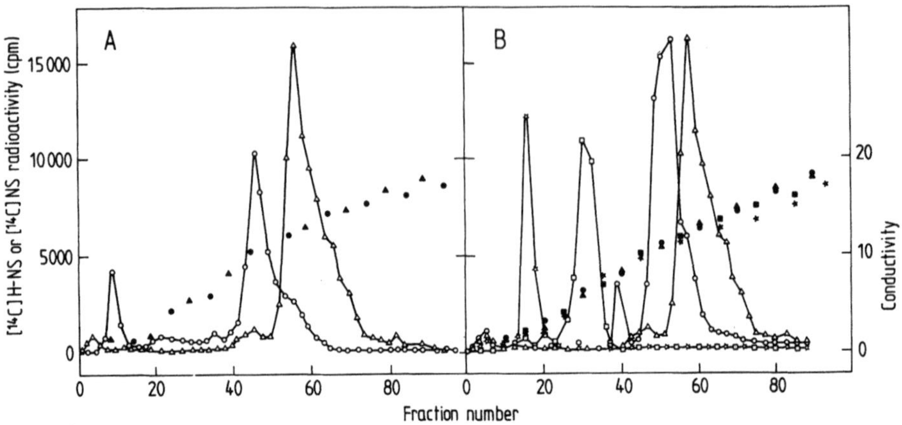

Fig. 20A,B. Comparative affinity of H-NS and NS for DNA and of H-NS for different nucleic acids. **A** Elution of radioactively-labelled NS (○) and H-NS (△) from affinity chromatographic columns containing matrix-bound ds-DNA. **B** Elution profiles of labelled H-NS from four different affinity chromatographic columns containing ds-DNA (△), ss-DNA (○), tRNA (□) and rRNA (☆). The experimental conditions are the same as described in Fig. 3. The conductivity measured for the indicated fractions of each eluate are indicated by the corresponding *closed symbols*

seen from Fig. 18, the red shift of the tryptophan begins around the physiological temperature (30°–35°C) and is virtually complete at 55°–60°C. The slight rise in anisotropy around 80°C was attributed to aggregation.

3.2 Interaction of H-NS with DNA

Compared to NS, H-NS is eluted at a higher ionic strength from a ds-DNA-cellulose affinity chromatography column (Fig. 20A). Furthermore, as seen from the elution profiles of H-NS from affinity chromatographies containing various types of nucleic acids, H-NS has a substantially greater affinity for DNA than for RNA and, unlike NS (cf. Fig. 3), is retained much more tightly by tRNA than by rRNA and by ds-DNA than by ss-DNA (Fig. 20B). Thus, H-NS not only has a higher affinity for DNA than NS, but also displays different binding selectivity for nucleic acids.

The interaction between H-NS and DNA can also be monitored, as in the case of NS, by the retention of radioactive DNA on nitrocellulose filters. In Fig. 21A, a binding curve is presented. Unlike with NS, the relationship between added protein and bound DNA is linear and at 150 mM NH_4Cl saturation is reached at a protein:DNA ratio of approximately 4 which corresponds to a stoichiometry of about one H-NS dimer per 12 bp. Monovalent (NH_4^+ or Na^+) and divalent (Mg^{2+}) cations (Fig. 21B–D) stimulate the retention of DNA on the filters by H-NS. Optimum binding is observed above 6 mM $MgCl_2$ and at about 300 mM NaCl; in the absence of Mg^{2+} the optimum concentration of NaCl for binding is shifted to higher concentrations (Fig. 21D).

The effect of the ions on the H-NS-DNA interaction is consistent with that expected for a physiological interaction; but it is difficult to assess whether this effect is primarily on DNA, on the protein or on both. In fact, although an effect of mono- and divalent cations on DNA structure can easily be anticipated, an effect on the protein structure

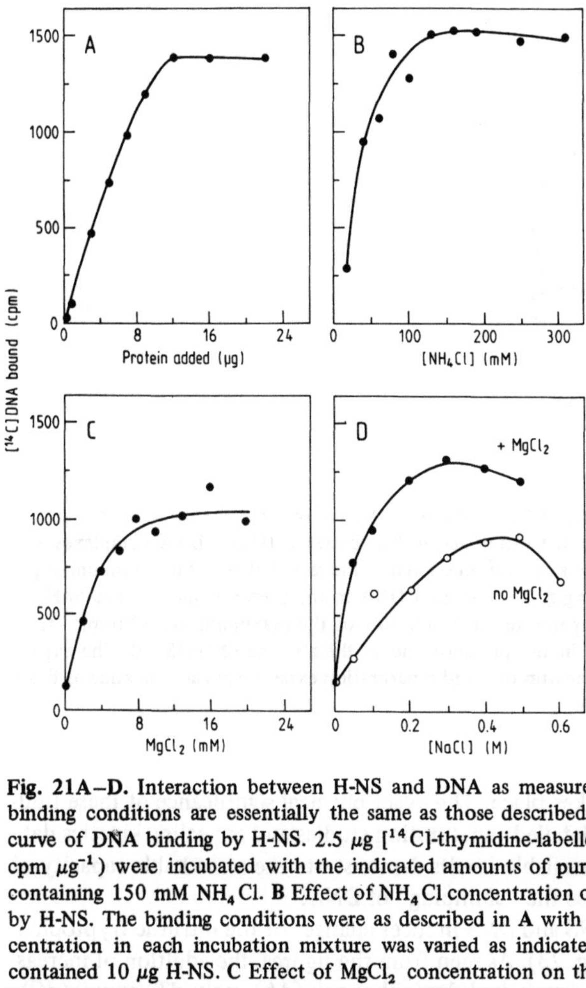

Fig. 21A–D. Interaction between H-NS and DNA as measured by nitrocellulose filtration. The binding conditions are essentially the same as those described for NS in Fig. 2. **A** Dose-response curve of DNA binding by H-NS. 2.5 μg [^{14}C]-thymidine-labelled *E. coli* chromosomal DNA (650 cpm μg^{-1}) were incubated with the indicated amounts of purified H-NS in 0.3 ml of 0.1 × SSC containing 150 mM NH_4Cl. **B** Effect of NH_4Cl concentration on the retention of radioactive DNA by H-NS. The binding conditions were as described in A with the exception that the NH_4Cl concentration in each incubation mixture was varied as indicated, and that each reaction mixture contained 10 μg H-NS. **C** Effect of $MgCl_2$ concentration on the retention of radioactive DNA by H-NS. The binding was performed in 0.1 × SSC as in A, but no NH_4Cl was present and 3 μg of H-NS were used. **D** Effect of NaCl and $MgCl_2$ on the retention of DNA by H-NS. The binding conditions were the same as in C with the exceptions that the reaction volume was 0.4 ml and, as indicated, 10 mM $MgCl_2$ was present and the NaCl concentration was varied

is also possible. It should be noted in this connection that the solubility of H-NS is markedly increased by increasing the ionic strength so that it is likely that its structure is also somehow affected.

As in the case of NS, H-NS affects DNA transcription (which is strongly inhibited), as well as the thermal stability of DNA (Fig. 22). In this case, however, the increase in Tm is rather modest (∿4°C) and, unlike the case of NS, we did not obtain any indication of a preferential effect of H-NS on the homologous DNA. Thus, the major effect of H-NS, as seen in these DNA thermal denaturation experiments, is the induction of a strong hypochromic effect on the denatured DNA accompanied by an increase of the temperature (by over 10°C) at which 50% of the maximum hypochromic effect

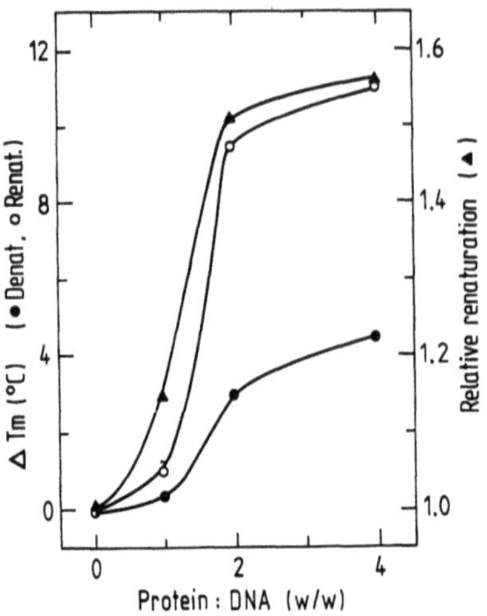

Fig. 22. Effect of H-NS on the thermal stability of *E. coli* DNA. Plotted in the figure are the effects of increasing amounts of H-NS on: the difference in Tm between (H-NS)-DNA complexes and DNA alone (●); the relative DNA renaturation measured as the ratio between the maximum hypochromic effect observed upon cooling fully denatured DNA in the presence and absence of H-NS (▲); the difference between the temperatures at which 50% of the maximum hypochromic effect (renaturation) of DNA takes place in the presence and in the absence of H-NS (○). The experimental conditions for the thermal denaturation and renaturation experiments are the same as those described for NS (Miano et al. 1982)

(referred to as renaturation) takes place. The exact physical significance of these findings will remain unclear until detailed renaturation studies and annealing kinetics data are obtained; nevertheless, the available results demonstrate the remarkable capacity of H-NS to compact and/or to cause the "clumping" of DNA.

The interaction between H-NS and DNA induces changes of the intrinsic tryptophan fluorescence of the protein (Fig. 23). As seen from the figures, the addition of increasing amounts of synthetic polydeoxynucleotides [either poly(dA) · poly(dT) or poly(dC) · poly(dG)] causes a decrease of the fluorescence intensity without shifting the emission maximum. In both cases, the quenching effect levels off when the fluorescence emission was reduced to 40%–60% of its original intensity. The precise molecular significance of this leveling off is not clear. The reduced fluorescence intensity probably reflects a change in the electronic environment of the tryptophan ring in the H-NS-DNA complex. However, in view of the dimeric or tetrameric nature of the H-NS protein (see below), we cannot exclude that the leveling off of the signal at about 50% of its original intensity might be due to an asymmetry of the complex between an H-NS dimer (or tetramer) and the nucleic acid with the tryptophan fluorescence in one H-NS molecule being quenched completely. Finally, the possibility that only a portion of H-NS molecules is active in DNA binding, though unlikely, cannot be completely ruled out at the present time. Assuming the last explanation to be incorrect, we calculated the apparent binding parameters (K_{diss}, n) for the H-NS-polydeoxynucleotide complex from the fluorescence titration curves of Fig. 23. K_{diss} was found to be

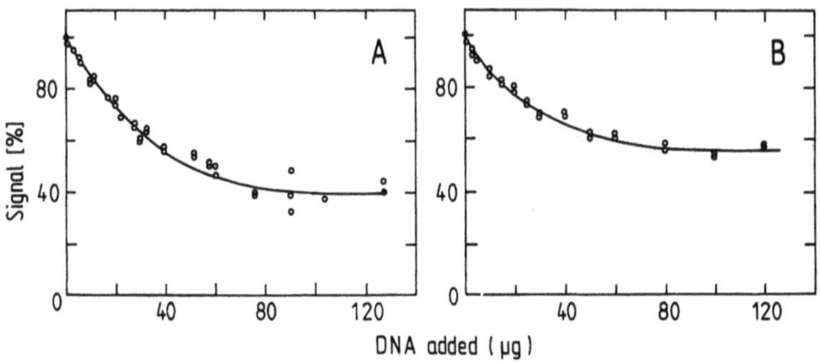

Fig. 23 A,B. Quenching the intrinsic fluorescence of H-NS by interaction with synthetic polydeoxynucleotides. The fluorescence intensity of a 10.3 μM solution of H-NS (monomer) in 0.32 ml 1.5 × SSC was recorded before (100% signal) and after addition of the indicated amounts of (A) poly(dA) · poly(dT) and (B) poly(dC) · poly(dG). To minimize the internal filter effect due to the polydeoxynucleotide absorption, the tryptophan was excited at 306 nm with a band width of 8 nm. The emission signal (λ_{em} = 327 nm, band width = 8 nm) was corrected for the dilution and absorption at 306 nm

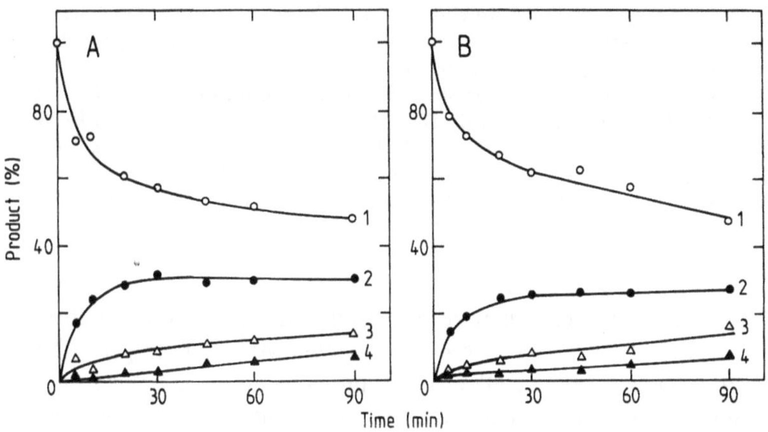

Fig. 24A,B. Time course of formation of cross-linked products of H-NS. The cross-linking reaction was carried out, stopped at the indicated times and analyzed electrophoretically essentially as described by Losso et al. (1986). The H-NS concentration in the reaction mixture was 1.5 × 10^{-4} M. **A** Reaction with DMS. **B** Reaction with DMA. Monomer (○), dimer (●), trimer (△), tetramer (▲)

1.1 × 10^4 M^{-1} per bp for both poly(dA) · poly(dT) and poly(dC) · poly(dG) and n to be approximately 40 bp per H-NS dimer.

3.3 The Nature of the Quaternary Structure of H-NS

The quaternary structure of H-NS in solution was investigated by means of protein-protein cross-linking reactions. With both DMS and DMA, H-NS yields dimers and, depending on the protein concentration (see below), small amounts of trimers and tetramers. Unlike the case of NS, DMA and DMS yield approximately the same amount

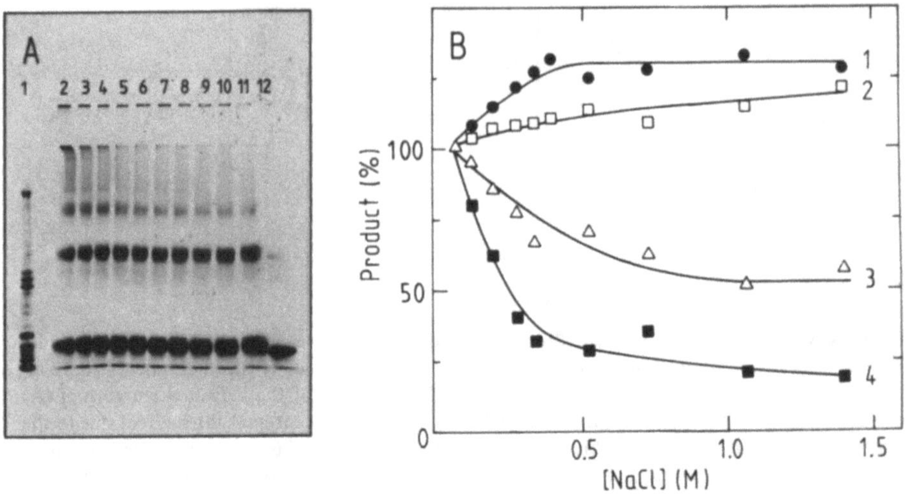

Fig. 25A,B. Effect of the ionic strength on the yield of DMS-cross-linked products of H-NS. The cross-linking reaction with DMS was carried out at 1.2×10^{-4} M H-NS in the presence of the indicated NaCl concentrations and analyzed by electrophoresis on SDS-containing PAGE (15% acrylamide). **A** *Lane 1* total 30S ribosomal proteins as molecular weight standards. The largest ribosomal protein S1 with mobility slightly more than the tetramer has a molecular weight of 61 000; *lanes 2–11* cross-linked products after 30 min reaction in the presence of the following NaCl concentrations; 66, 133, 199, 266, 333, 400, 532, 733, 1066, 1400 mM; *lane 12* non-reacted H-NS control. **B** Variation of the percentage of each cross-linked product as a function of the NaCl concentration in the reaction taking the yield obtained at 66 mM as 100%. The actual values were 51% for the monomer (•), 32% for the dimer (□), 13% for the trimer (△), and 5% for the tetramer (■)

of cross-linked products, but the efficiency of the reaction is much lower and covalent aggregates larger than tetramers have never been detected with either reagent. The time course of the reaction with either DMS or DMA, presented in Fig. 24, shows that the formation of dimers is virtually complete after 30 min, while a very slow accumulation of trimers and tetramers continues for at least 90 min.

The protein-protein interactions must have a very strong hydrophobic character since an increase in NaCl concentration up to 1.4 M does not affect the formation of cross-linked dimers and also the yield of tetramers and, even more so of trimers, remains relatively high after an initial drop caused by raising the NaCl concentration to 0.5 M (Fig. 25A,B).

The dependence of the yield of individual cross-linked products on the protein concentration between 1.65×10^{-8} and 2.2×10^{-4} M during the reaction is shown in Fig. 26. The yield of monomers and covalently-linked dimers, approximately 80% and 20% of the total protein, respectively, remains approximately constant between H-NS concentrations of 1.65×10^{-8} and 10^{-5} M. Within the same protein concentration range, the amount of trimers formed is negligible (<5%) and tetramers are not detectable. Above 10^{-5} M, the yield of dimers and trimers begins to increase; this increase becomes very steep above 5×10^{-5} M. At H-NS concentrations of 10^{-4} M and higher, also tetramers appear in increasing amounts among the reaction products. The steep

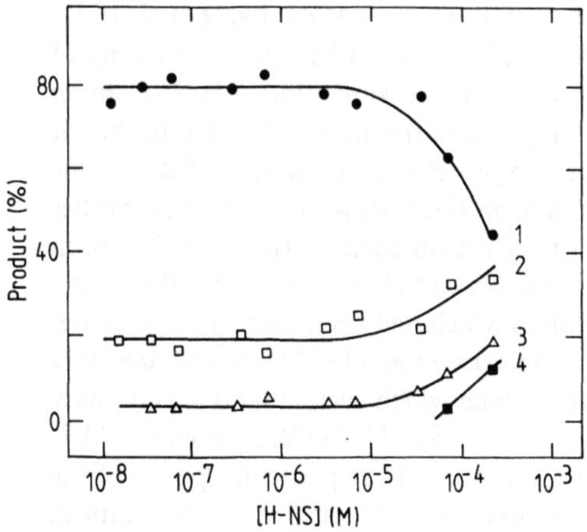

Fig. 26. Dependence of the yield of individual cross-linked products on the H-NS concentration between 1.65×10^{-8} and 2.2×10^{-4} M. The monomer concentrations of H-NS in the DMS cross-linking reactions are indicated in the figure. The yield (percent of total electrophoretically resolved protein) is plotted as a function of the protein concentration in each reaction mixture. Monomers (●), dimers (□), trimers (△), tetramers (■)

increase in cross-linked aggregates occurring above 10^{-5} M H-NS is accompanied by an equivalent decrease in the amount of residual monomers, which are reduced by half when the protein concentration increases by slightly more than one order of magnitude.

These results can be explained by the existence of a very strong dimeric interaction (K_d of dimer $\leq 10^{-9}$ M) and a much weaker tetrameric interaction (K_d for the equilibrium dimers $\rightleftharpoons$ tetramers $>10^{-5}$ M). Thus, our results indicate that the predominant quaternary structure of H-NS is the dimer. It is possible, however, that the equilibrium dimer $\rightleftharpoons$ tetramer might have a physiological relevance at the presumed intracellular concentration of H-NS ($\geq 10^{-5}$ M). It should be noted, however, that unlike the case of NS, where DNA favors the formation of larger protein aggregates, we could not detect any effect of DNA on the yield of the individual cross-linked products (not shown). Furthermore, the attempt to demonstrate, by cross-linking, the possible existence of an interaction in vitro, with or without DNA, between H-NS and NS was also unsuccessful.

As to the orientation of the H-NS monomers within the dimers, we have isolated dimers cross-linked with radioactive DMS and attempted to identify the points of contact between two monomers by comparing their peptide maps with those of monofunctionally modified monomers. These experiments did not reveal, however, any major point of cross-linking between the proteins (M. Falconi and F. Scarcello, unpublished). It is possible that our failure to detect a cross-linked peptide under a variety of conditions may stem from the fact that multiple points of contact, each accounting for a small fraction of the cross-linking reaction are formed, but it is also possible that the two monomers are arranged in a strictly parallel fashion so that each monomer contributes the same lysine residue and the same peptide to the cross-linking bond.

3.4 Relationship Between H-NS and Other Proteins

As mentioned above, several proteins (i.e. k, H1, B1, HLP1, H1a) with properties similar to those of H-NS have been reported (Schäfer and Zillig 1973; Spassky and Buc

1977; Varshavsky et al. 1977a,b; Lathe et al. 1980; Bakayev 1981; Laine et al. 1984; Spassky et al. 1984). Judging from its properties, H-NS could be equivalent to any of these proteins. As previously mentioned, the amino terminal peptide of H-NS (Lammi et al. 1984b; M. Falconi and M.T. Gualtieri, unpublished) is nearly identical to that of H1a and the amino acid compositions of the two proteins are similar, in spite of some important discrepancies (e.g. the lack of histidine in H1a). However, several properties of H-NS (e.g. its chromatographic behavior, its effect on transcription, its behavior in cross-linking reactions) are clearly different from those of H1a, if we accept the notion (Spassky et al. 1984) that H1a is the same protein which had been characterized earlier (Spassky and Buc 1977). On the other hand, Spassky et al. (1984) reported that H1a is one of three variants of the same polypeptide chain which accumulates in stationary phase cultures of *E. coli.* Indeed, the primary sequence of H-NS (M. Falconi and M.T. Gualtieri, unpublished results) displays amino acid variants at particular points of the molecule, indicating that we are dealing with a mixture of at least two nearly identical, but distinct proteins (two heterologous subunits of a dimer?) coded by different genes. However, since we purify H-NS from exponentially growing cells, H-NS should more likely correspond to H1b and/or H1c than to H1a. A further puzzle is that Laine et al. (1984) and Spassky et al. (1984) reported that H1a is also identical to protein H1 (Cukier-Kahn et al. 1972) and, in turn, to protein B1 (Bakayev 1981). Bakayev (1981), however, identified H1 (Cukier-Kahn et al. 1972) with his B2 protein and with HU (NS) (Rouviere-Yaniv and Gros 1975); indeed, like HU, H1 (Cukier-Kahn et al. 1972) was claimed to stimulate transcription and was found to have a molecular weight of less than 10 000; also the amino acid composition of H1 bears no resemblance whatsoever to that of H-NS. Thus, we suggest that H1 (Cukier-Kahn et al. 1972; Spassky and Buc 1977) is different from both B1 and H1a as well as from H-NS. In conclusion, we are tempted to suggest that H-NS is identical to B1 (Varshavsky et al. 1977; Bakayev 1981), to one or two of the H1 variants, most likely H1b and H1c (Spassky et al. 1984) and possibly to HLP1 (Lathe et al. 1980). Since there is no specific functional assay for these proteins, however, and since they also have very similar electrophoretic and chromatographic properties, only the elucidation of their primary structures will solve the problem of their identity and of the nature of the variants. If these proteins ultimately prove to be identical, a uniform nomenclature will have to be agreed upon.

4 Conclusion

In this paper, we have described the properties of the two *E. coli* histonelike proteins NS and H-NS. Even though NS and H-NS were found to have a different binding affinity and selectivity for nucleic acids, they were shown to have a number of properties in common; they are heat-stable, very slightly basic, are not retained by hydroxylapatite columns, inhibit DNA transcription, increase the Tm of DNA, produce large hypochromic effects on denatured DNA, form multimeric structures (heterologous tetramers for NS, dimers for H-NS), and there are indications that these proteins may form larger aggregates at their in vivo concentrations or, at least in the case of NS, at lower concentrations, but in the presence of DNA. Thus, it is conceivable that under physiological conditions, the weak forces responsible for these interactions may play an important role in the reversible and dynamic packaging of at least portions of the bacterial chromatin.

We have also presented some experimental data indicating a possible preference of NS protein for binding to prokaryotic vs eukaryotic DNA and suggested the possibility, also in light of recent findings, that the NS proteins may have a low degree of DNA-binding selectivity, which allows them, like eukaryotic histones, to be "phased" on the DNA.

The data reviewed here and in other chapters of this volume strongly suggest, in our opinion, a major involvement of NS and H-NS proteins in the organization of bacterial chromatin. It should be mentioned, however, that some data, including the failure to detect NS cross-linked to DNA in vivo or in vitro and the preference of NS for ssDNA as opposed to dsDNA, also described here, led some investigators to cast doubts on this putative role, at least as far as NS is concerned (Kellenberger et al.; Lossius et al.; Schouten, this volume). It is possible, indeed, that the NS proteins (as well as H-NS) might have more than one function in the cell. Still obscure is, for instance, the significance of the association of NS with ribosomes (Suryanarayana and Subramanian 1978; Dijk et al. 1983) and it is noteworthy that NS (HU) stimulates the in vitro initiation of DNA duplication at *oriC* (Dixon and Kornberg 1984), but that it strongly inhibits initiation of phage λ DNA synthesis (R. McMacken, personal communication). To what extent functional properties attributed to some preparations of NS (HU) protein, but not confirmed with others, might be due to different preparation methods or to preparation impurities cannot be assessed at present. Furthermore, even though it is possible that many of the functional properties described for NS and H-NS may be traced back to their fundamental property of binding DNA, thereby shielding it from other interactions and altering its topology, it cannot be excluded that, as in the case of some eukaryotic histones (Isenberg 1979; McGhee and Felsenfeld 1980), sequence variants of the bacterial histonelike proteins, each endowed with specific functions, might exist. In any event, it is rather surprising that after several decades of very active genetic and biochemical research on *E. coli* as well as other bacteria, so much is still unknown concerning the structure and function of these proteins which appear to be among the most abundant in the cell. This is perhaps the result of the long-standing opinion that histones and organized chromatin do not exist in prokaryotes. It is hoped that the at least partial elimination of this bias and the evidence that histonelike proteins may have a profound effect on cellular functions such as DNA duplication, transcription, restriction and recombination will stimulate more investigations in this field and that, through the biochemical and genetic characterization of these proteins, a better understanding of the organization of the bacterial chromatin will eventually emerge.

Acknowledgements. The help of M. Falconi, M.T. Gualtieri, A. Miano, and F. Scarcello in parts of this work, and the stimulating discussions of P. Woolley are gratefully acknowledged. We would also like to express our gratitude to Prof. H.G. Wittmann for his interest and support, to K. Ashman for assistance with the computer programs, to E. Philippi for her enthusiastic help in the preparation of the drawings and to J Belart for the careful and patient typing of the manuscript. This work was supported in small part by funds of the Italian Ministry of Public Education and of the Italian National Research Council.

References

Bakayev VV (1981) Structure of the chromosomal deoxyribonucleoproteins XI. Organization of deoxyribonucleoprotein complex in the bacterial cell. Mol Biol (Engl transl) 15:1045–1056

Berthold V, Geider K (1976) Interaction of DNA with DNA-binding proteins. The characterization of protein HD from *Escherichia coli* and its nucleic acid complexes. Eur J Biochem 71:443–449

Craig NL, Nash HA (1984) *E. coli* integration host factor binds to specific sites in DNA. Cell 39: 707–716

Cukier-Kahn R, Jacquet M, Gros F (1972) Two heat resistant, low molecular weight proteins from *Escherichia coli* that stimulate DNA-directed RNA synthesis. Proc Natl Acad Sci USA 69: 3643–3647

Dijk J, White SW, Wilson KS, Appelt K (1983) On the DNA binding protein II from *Bacillus stearothemophilus.* J Biol Chem 258:4003–4006

Dixon NE, Kornberg A (1984) Protein HU in the enzymatic replication of the chromosomal origin of *Escherichia coli.* Proc Natl Acad Sci USA 81:424–428

Ehbrecht HJ, Pingoud A, Urbanke C, Maass G, Gualerzi C (1985) Linear diffusion of restriction endonucleases on DNA. J Biol Chem 260:6160–6166

Flamm EL, Weisberg RA (1985) Primary structure of the hip gene of *Escherichia coli* and of its product, the β subunit of integration host factor. J Mol Biol 183:117–128

Geider K, Hoffmann-Berling H (1981) Proteins controlling the helical structure of DNA. Annu Rev Biochem 50:233–260

Geyl D, Böck A, Isono K (1981) An improved method for two-dimensional gel electrophoresis: analysis of mutationally altered ribosomal proteins of *Escherichia coli.* Mol Gen Genet 181: 309–312

Glikin GC, Ruberti I, Worcel A (1984) Chromatin assembly in Xenopus oocytes: in vitro studies. Cell 37:33–41

Greene JR, Brennan SM, Andrew DJ, Thompson CC, Richards SH, Heinrikson RL, Geiduschek EP (1984) Sequence of the bacteriophage SP01 gene coding for transcription factor 1, a viral homologue of the bacterial type II DNA-binding proteins. Proc Natl Acad Sci USA 81:7031–7035

Griffith JD (1976) Visualization of prokaryotic DNA in a regularly condensed chromatin-like fiber. Proc Natl Acad Sci USa 73:563–567

Gualerzi C, Pon CL (1979) Radioactive chemical labeling of ribosomal proteins and translational factors *in vitro.* Methods Enzymol 59:782–795

Haselkorn R, Rouviere-Yaniv J (1976) Cynaobacterial DNA-binding protein related to *Escherichia coli* HU. Proc Natl Acad Sci USA 73:1917–1920

Holck A, Kleppe K (1985) Affinity of protein HU for different nucleic acids. FEBS Lett 185: 121–124

Hübscher U, Lutz H, Kornberg A (1980) Novel histone H2A-like protein of *Escherichia coli.* Proc Natl Acad Sci USA 77:5097–5101

Isenberg I (1979) Histones. Annu Rev Biochem 48:159–191

Kavenoff R, Bowen BC (1976) Electron microscopy of membrane-free folded chromosomes from *Escherichia coli.* Chromosoma (Berl) 59:89–101

Kimura M, Wilson KS (1983) On the DNA binding protein II from *Bacillus stearothermophilus* II. The amino acid sequence and its relation to those of homologous proteins from other prokaryotes. J Biol Chem 258:4007–4011

Kishi F, Ebina Y, Miki T, Nakazawa T, Nakazawa A (1982) Purification and characterization of a protein from *Escherichia coli* which forms complexes with superhelical and single stranded-DNAs. J Biochem (Tokyo) 92:1059–1068

Kleppe K. Lossius I, Aasland R, Sjastad K, Holk A, Haarr L (1984) The structure of the bacterial nucleoid. In: Hübscher U, Spadari S (eds) Proteins involved in DNA replication. Plenum, New York, pp 457–466

Kyte J, Doolittle RF (1982) A simple method for displaying the hydropathic character of a protein. J Mol Biol 157:105–132

Laine B, Kmiecik D, Sautiere P, Biserte G, Cohen-Solal M (1980) Complete amino-acid sequences of DNA-binding proteins HU-1 and HU-2 from *Escherichia coli*. Eur J Biochem 103:447–461

Laine B, Sautiere P, Spassky A, Rimsky S (1984) A DNA-binding protein from *E. coli*. Isolation, characterization and its relationship with proteins H1 and B1. Biochem Biophys Res Commun 119:1147–1153

Lammi M, Paci M, Gualerzi CO (1984a) Proteins from the prokaryotic nucleoid. The interaction between protein NS and DNA involves the oligomeric form of the protein and at least one Arg residue. FEBS Lett 170:99–104

Lammi M, Paci M, Pon CL, Losso MA, Miano A, Pawlik RT, Gianfranceschi GL, Gualerzi CO (1984b) Proteins from the prokaryotic nucleoid. Biochemical and ^{1}H NMR studies on three bacterial histone-like proteins. In: Hübscher U, Spadari S (eds) Proteins involved in DNA replication. Plenum, New York, pp 467–477

Lathe R, Buc H, Lecocq JP, Bautz EK (1980) Prokaryotic histone-like protein interacting with RNA polymerase. Proc Natl Acad Sci USA 77:3548–3552

Losso MA, Miano A, Gianfranceschi GL, Gualerzi CO (1982) Proteins from the prokaryotic nucleoid II. Inhibition of DNA transcription by NS1 and NS2 (HU proteins). Biochem Int 5: 423–427

Losso MA, Pawlik RT, Canonaco MA, Gualerzi CO (1986) Proteins from the prokaryotic nucleoid. A protein-protein crosslinking study on the quaternary structure of *Escherichia coli* DNA-binding proteins NS (HU). Eur J Biochem 155:27–32

Luchnik AN, Bakayev VV, Zbarsky IB, Georgiev GP (1982) Elastic torsional strain in DNA within a fraction of SV40 minichromosomes: relation to transcriptionally active chromatin. EMBO J 1:1353–1358

McGhee JD, Felsenfeld G (1980) Nucleosome structure. Annu Rev Biochem 49:1115–1156

Mende L, Timm B, Subramanian AR (1978) Primary structures of two homologous ribosome-associated DNA-binding proteins of *Escherichia coli*. FEBS Lett 96:395–398

Menzel R, Gellert M (1983) Regulation of the genes for *E. coli* DNA gyrase: homeostatic control of DNA supercoiling. Cell 34:105–113

Mets LJ, Bogorad L (1974) Two-dimensional polyacrylamide gel electrophoresis: an improved method for ribosomal proteins. Anal Biochem 57:200–210

Miano A. Losso MA, Gianfranceschi GL, Gualerzi CO (1982) Proteins from the prokaryotic nucleoid I. Effect of NS1 and NS2 (HU) proteins on the thermal stability of DNA. Biochem Int 5:415–422

Mizuuchi K (1983) *In vitro* transposition of bacteriophage Mu: A biochemical approach to a novel replication reaction. Cell 35:785–794

Paci M, Pon CL, Losso MA, Gualerzi C (1984) High resolution ^{1}H NMR spectroscopic study of *Escherichia coli* DNA-binding proteins NS1 and NS2. Eur J Biochem 138:193–200

Pawlik RT, Littlechild J, Pon C, Gualerzi C (1981) Purification and properties of *Escherichia coli* initiation factors. Biochem Int 2:421–428

Pettijohn DE (1982) Structure and properties of the bacterial nucleoid. Cell 30:667–669

Pettijohn D, Pfenninger O (1980) Supercoils in prokaryotic DNA restrained *in vivo*. Proc Natl Acad Sci USA 77:1331–1335

Pingoud A, Urbanke C, Alves J, Ehbrecht HJ, Zabeau M, Gualerzi C (1984) Effect of polyamines and basic proteins on cleavage of DNA by restriction endonucleases. Biochemistry 23:5697–5703

Read CM, Baldwin JP, Crane-Robinson C (1985) Structure of subnucleosomal particles. Tetrameric $(H3/H4)_2$ 146 base pair DNA and hexameric $(H3/H4)_2$ (H2A/H2B), 146 base pair DNA complexes. Biochemistry 24:4435–4450

Rouviere-Yaniv J (1977) Localization of the HU protein in the *E. coli* nucleoid. Cold Spring Harbor Symp Quant Biol 42:439–447

Rouviere-Yaniv J, Gros F (1975) Characterization of a novel, low-molecular-weight DNA-binding protein from *Escherichia coli*. Proc Natl Acad Sci USA 72: 3428–3432

Rouviere-Yaniv J, Kjeldgaard NO (1979) Native *Escherichia coli* HU protein is a heterotypic dimer. FEBS Lett 106:297–300

Rouviere-Yaniv J, Yaniv M, Germond J-E (1979) *E. coli* DNA binding protein HU forms nucleosome-like structure with circular double-stranded DNA. Cell 17:265–274

Ryoji M, Worcel A (1984) Chromatin assembly in Xenopus oocytes: *in vivo* studies. Cell 37:21–32

Ryoji M, Worcel A (1985) Structure of the two distinct types of minichromosomes that are assembled on DNA injected in Xenopus oocytes. Cell 40:923–932

Schäfer R, Zillig W (1973) κ (kappa), a novel factor for the arrest of transcription *in vitro* by DNA-dependent RNA polymerase from *Escherichia coli* at specific sites of natural templates. Eur J Biochem 33:201–206

Sinden RR, Carlson JO, Pettijohn DE (1980) Torsional tension in the DNA double helix measured with trimethylpsoralen in living *E. coli* cells: analogous measurements in insect and human cells. Cell 21:773–783

Sinden RR, Pettijohn DE (1981) Chromosomes in living *Escherichia coli* are segregated into domains of supercoiling. Proc Natl Acad Sci USA 78:224–228

Sloof P, Maagdelijn A, Boswinkel E (1983) Folding of prokaryotic DNA. Isolation and characterization of nucleoids from *Bacillus licheniformis.* J Mol Biol 163:277–297

Spassky A, Buc HC (1977) Physico-chemical properties of a DNA binding protein: *Escherichia coli* factor H1. Eur J Biochem 81:79–90

Spassky A, Rimsky S, Garreau H, Buc H (1984) H1a, an *E. coli* DNA binding protein which accumulates in stationary phase, strongly compacts DNA *in vitro.* Nucleic Acids Res 12:5321–5340

Steinhäuser KG, Woolley P, Epe B, Dijk J (1982) Structure of ribosomal protein L6 from *Escherichia coli.* A fluorescence study. Eur J Biochem 127:587–595

Suryanarayana T, Subramanian AR (1978) Specific association of two homologous DNA-binding proteins to the native 30S ribosomal subunits of *Escherichia coli.* Biochim Biophys Acta 520: 342–357

Swank RT, Munkres KD (1971) Molecular weight analysis of oligopeptides by electrophoresis in polyacrylamide gel with sodium dodecyl sulfate. Anal Biochem 39:462–477

Tanaka I, Appelt K, Dijk J, White SW, Wilson KS (1984) 3-A resolution structure of a protein with histone-like properties in prokaryotes. Nature 310:376–381

Ullmann A, Geider K (1977) Interaction of DNA with DNA binding proteins III. Infectivity of protein-complexed phage fd DNA in *Escherichia coli* spheroplasts. Biochim Biophys Acta 474: 639–645

Varshavsky AJ, Bakayev VV, Nedospasov SA, Georgiev GP (1977a) On the structure of eukaryotic, prokaryotic and viral chromatin. Cold Spring Harbor Symp Quant Biol 42:457–473

Varshavsky AJ, Nedospasov SA, Bakayev VV, Bakayeva TG, Georgiev GP (1977b) Histone-like proteins in the purified *Escherichia coli* deoxyribonucleoprotein. Nucleic Acids Res 4:2725–2745

Worcel A, Burgi E (1972) On the structure of the folded chromosome of *Escherichia coli.* J Mol Biol 71:127–147

Zachau HG, Igo-Kemenes T (1981) Face to phase with nucleosomes. Cell 24:597–598

Zentgraf H, Berthold V, Geider K (1977) Interaction of DNA with DNA binding proteins II. Displacement of *Escherichia coli* DNA unwinding protein and the condensed structure of DNA complexed with protein HD. Biochim Biophys Acta 474:629–638

Proteins from the Prokaryotic Nucleoid. [^{1}H]-NMR Studies on *Escherichia coli* Histone-like Proteins NS and H-NS and Their Interaction with DNA

M. PACI[1], C.L. PON[2], M.A. CANONACO[2,3], and C.O. GUALERZI[2,3]

1 Introduction

Protein-protein and protein-DNA interactions are of fundamental importance in the physical packaging of DNA in all biological systems. In contrast to eukaryotes, where detailed information concerning the basic structural unit of chromatin (i.e. the nucleosome) is already available, studies on the mechanism responsible for the packaging of DNA in prokaryotes are still at an early stage. It is already clear, however, that new strategies must be devised when dealing with "bacterial chromatin" since no stable nucleosomelike structures can be obtained from bacterial cells. Furthermore, experiments with "intact" nucleoids are often thwarted by the difficulties intrinsic in work with an object of very large size and elusive chemical composition and physical properties (Hirshbein and Guillen 1982). Finally, the biological organization of the bacterial cell probably requires very plastic and flexible structures for the packaging of DNA, rather than structures endowed with the structural rigidity of eukaryotic chromatin.

For the above reasons, we are convinced that only detailed biochemical and physico-chemical information concerning individual interactions between the molecular components of the "bacterial chromatin" will lead to a better understanding of the mechanism by which the bacterial chromosome is physically organized. Thus, the aim of our research is to achieve a description, as complete as possible, of the mechanism governing the protein-protein and DNA-protein interactions of the components of the bacterial nucleoid at the structural, kinetic and thermodynamic level.

Among the proteins which may play a role in the physical packaging of bacterial chromosome, NS (HU) and H-NS (probably homologous to H1) are the most abundant and the best characterized (Gualerzi et al.; Lossius et al.; Rimsky and Spassky, this volume). In *E. coli,* NS consists of two 9000-dalton polypeptide chains, NS1 and NS2, displaying 69% sequence homology (Mende et al. 1978), while H-NS is a 15 kD protein whose primary structure has been nearly completely elucidated (Falconi and Gualtieri,

1 Department of Chemical Science and Technology, University of Rome, Tor Vergata, 00173 Rome, Italy
2 Max-Planck-Institut für Molekulare Genetik, 1000 Berlin 33, FRG
3 Department of Cell Biology, Laboratory of Genetics, University of Camerino, 62032 Camerino, Italy

Bacterial Chromatin
(ed. by C.O. Gualerzi and C.L. Pon)

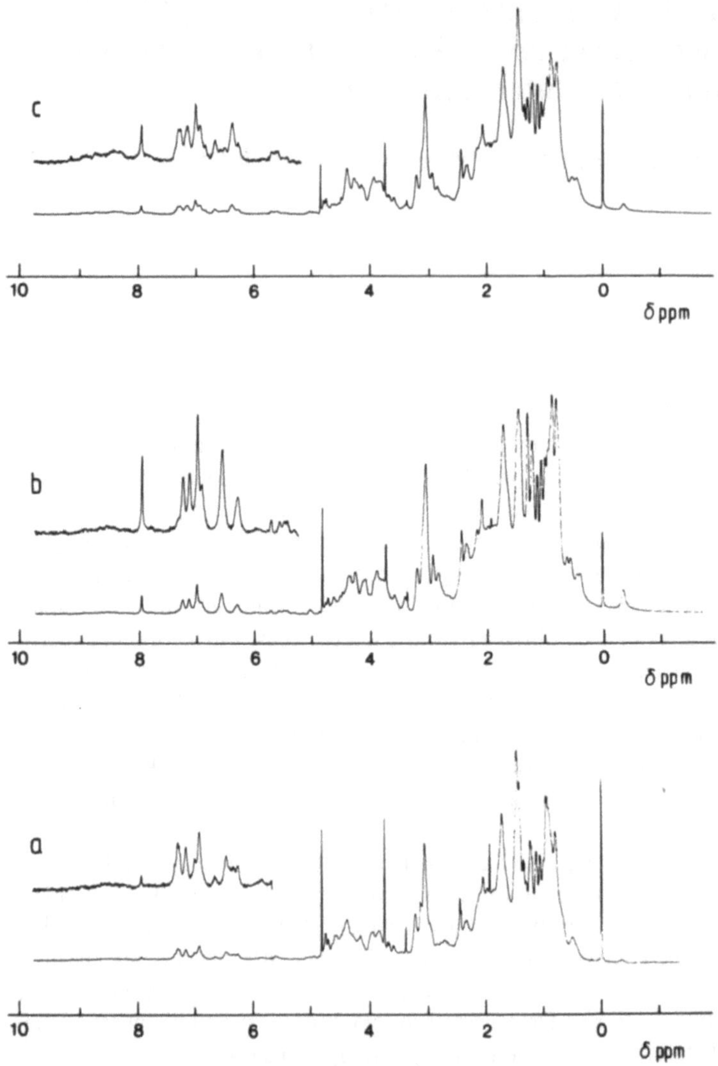

Fig. 1a–c. 400 MHz [^{1}H]-NMR spectra of NS1, NS2 and NS. The samples for NMR spectroscopy were concentrated by batch elution from small phosphocellulose columns to yield a protein concentration of 4–8 mg ml^{-1} (about 0.6 mM). The samples were dialyzed exhaustively against 10 mM potassium phosphate buffer, pH 7.5 (meter reading) containing 100 mM KCl in 99% deuterium oxide. [^{1}H]-NMR spectra were recorded on a Bruker WM-400 spectrometer operating at 400 MHz with 4 μs (∿30°) pulse, 6000 Hz spectral width and a relaxation delay of 2.0 s. Magnetization decays obtained in the quadrature detection mode were accumulated on 16 K of memory and a sensitivity enhancement of 0.5 Hz was applied before the FT transformation. Samples were observed in standard 5 mm tubes. HDO was partially suppressed using an inverse gate pulsed irradiation technique for a duration of 1.5 s. The spectra are (a) NS1; (b) NS2; (c) NS (NS1 + NS2). The figure also shows the aromatic region (10–5 ppm) amplified four times

unpublished results). Although the primary sequences of these proteins do not bear any resemblance to those of eukaryotic histones, they are commonly referred to as "histonelike" proteins since, like histones, they inhibit transcription (Hawkins and Wootton 1981; Losso et al. 1982), protect DNA from nuclease digestion (Varshavsky et al. 1977) and from thermal denaturation (Miano et al. 1982; Gualerzi et al., this volume) and probably have a physiological function related to that of eukaryotic histones. The extent to which the analogy between prokaryotic histonelike proteins and histones can be drawn from both structural and functional points of view, however, remains an open question.

In order to gain some insight into the tertiary and quaternary structure of NS and H-NS and to compare their properties with those of eukaryotic histones, we have examined these proteins by [^{1}H]-NMR spectroscopy under various experimental conditions. In the present paper, we outline these results.

2 Results

2.1 General Properties of the NS1, NS2 and NS Spectra

The 400 MHz [^{1}H]-NMR spectra of NS1, NS2 and of their native complex NS are presented in Fig. 1 (spectra a, b and c). The aromatic region of the spectra appears much more complex than expected from the amino acid composition of these proteins (Mende et al. 1978). This complexity is due to the presence of several high-field perturbed Phe resonances and, especially in NS, of several NH protons (9.5–7.5 ppm) rather inaccessible to the solvent.

The extensive high-field perturbations of the Phe ring proton resonances arise from hydrophobic interactions involving strong magnetic shielding effects, such as those expected from the stacking of phenylalanine rings. As will be shown below, these perturbations indeed result from stacking interactions which arise from the quaternary structure of these proteins. This premise is also supported by the finding that these residues are located in a position suitable for stacking interactions at the contact surface between subunits in the X-ray crystal structure of a *Bacillus stearothermophilus* DNA-binding protein homologous to NS (Tanaka et al. 1984).

The hard-to-exchange NH protons, on the other hand, indicate the existence of rigid, non-fluctuating structures possibly engaged in hydrogen bonding and most probably buried in a hydrophobic domain. The much greater abundance of these resonances in the spectrum of NS, compared to NS1 and NS2, indicates clearly that the homologous complexes are more labile than the heterologous one.

Owing to the presence of a His residue at position 54 in the primary sequence of NS2, the aromatic spectrum of this protein (as well as that of NS) appears even more complex than that of NS1. Histidine 54 gives rise to the resonances at approximately 8 and 7 ppm assigned to its C-2 and C-4 protons, respectively (see Fig. 1b,c).

The location of the three conserved Phe residues of NS1 and NS2 within their primary structures is indicated by the arrows in Fig. 2, which also displays a hydrophobicity analysis (Kyte and Doolittle 1982) of these proteins. As seen in the figure, each protein contains two major hydrophobic (A and C) and two major hydrophilic

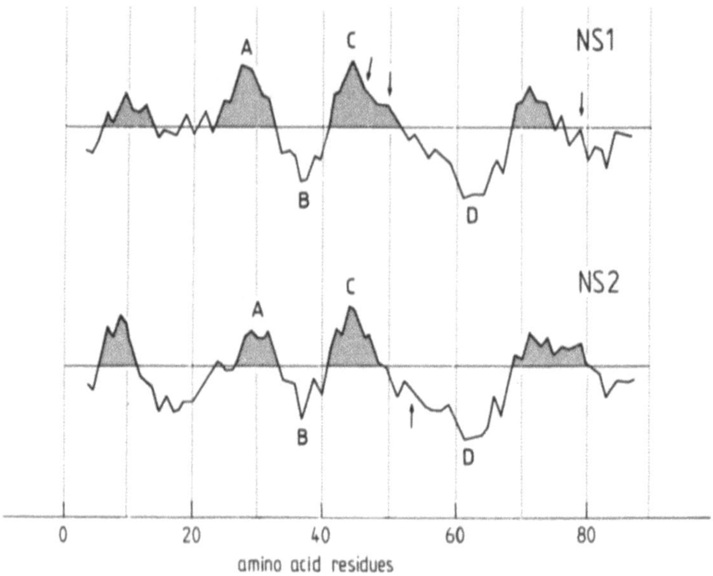

Fig. 2. Location of the phenylalanine and histidine residues in the primary structure of NS proteins. The primary structures of NS1 and NS2 were analyzed according to the computer program of Kyte and Doolittle (1982) to display the hydrophobic (*hatched areas above the horizontal line*) and the hydrophilic (*below the horizontal line*) regions. The *numbers* indicate the position of the residues. The *letters* (*A, B, C, D*) indicate the major hydrophobic or hydrophilic regions of the proteins. The three *arrows in NS1* indicate the position of the three Phe residues, which are also present in NS2 at the same positions. The *arrow in NS2* indicates the position of the His residue present exclusively in this polypeptide

(B and D) regions. Since two Phe residues are located in the hydrophobic region "C" and, as will be shown below, two Phe residues of each polypeptide chain are engaged in the quaternary structure, it can be concluded that contacts between the hydrophobic region "C" are involved in these quaternary interactions. The His residue of NS2, on the other hand, is located at the edge of the hydrophilic positively charged region "D", which was found to be implicated in the interaction with DNA (Lammi et al. 1984). We will show below that this residue is also very sensitive to quaternary interactions.

In the aliphatic region, the bands corresponding to the Arg δCH_2 and Lys ϵCH_2 (3.2 and 3.0 ppm, respectively) and a large number of CH_3 (1.5–0.8 ppm) as well as several ring current-shifted methyl (RCSM) resonances are easily recognizable.

2.2 Spectral Differences Between NS1, NS2, and NS

The spectra of NS1 and NS2 are similar. This is not surprising in view of the extensive sequence homology between these proteins, and it indicates that their overall tertiary and quaternary structures are also similar. The resemblance of the NS spectrum to that of a composite NS1-NS2 spectrum indicates, likewise, that the tertiary and quaternary structures of these proteins are not grossly altered in the NS complex. In

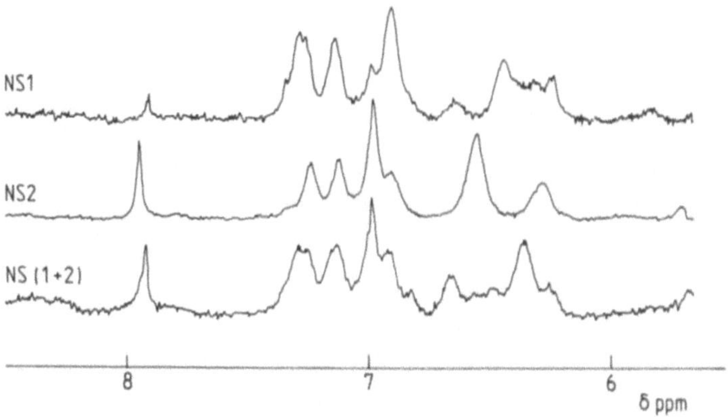

Fig. 3. Comparison of the enlarged aromatic region of the spectra of NS1, NS2 and NS. The 8.5–5.5 ppm spectra of NS1, NS2 and NS (NS1 + NS2) are juxtaposed so as to emphasize the differences. The peaks at 7.94 and 7.0 ppm are assigned to the His C-2 and C-4 proton resonances. The bands down-field the latter are attributed to freely rotating Phe rings, the up-field ones to Phe ring protons perturbed from their original position by magnetic shielding effects induced by stacking interactions

spite of the above mentioned similarities, however, important differences do exist among the spectra of the three proteins. The differences in the aromatic region are illustrated in Fig. 3; NS1, NS2 and NS differ in the position and intensity of the up-field perturbed Phe resonances between 6.95 and 6.1 ppm (Fig. 3) and in the chemical shift of the NS2 His 54 C-2 proton, which is not identical in the sample of NS2 alone (Fig. 3) and in that in which a small amount of NS2 (a preparation impurity) is present with a large excess of NS 1 (i.e. under conditions where NS2 is completely bound to NS1). In the spectrum of NS, which consists of NS1 and NS2 in a 1:1 stoichiometric ratio, the His C-2 proton resonance is composed of two bands, a major one with the same chemical shift as NS2 when bound to NS1 and a shoulder with the same chemical shift as NS2 alone (i.e. under conditions of self-aggregation). This finding shows that the His C-2 proton resonance is sensitive to the quaternary interactions and that the homologous (NS2-NS2) and the heterologous (NS2-NS1) interactions result in different chemical environments for the His C-2 proton.

In the aliphatic region, the main differences between the NS1, NS2 and NS consist in the presence, in NS2 and NS, of a multiplicity of the Lys ϵCH_2 resonances visible as a shoulder high-field of the main band. Also, the RCSM resonances are different in appearance and number. Particularly noteworthy is the presence in NS2 of a highly shielded resonance (–0.35 ppm), whose shift corresponds to a distance of 2–4 Å from the plane of an aromatic ring. From the inspection of the X-ray crystal structure of the *B. stearothermophilus* DNA-binding protein homologous to NS (Tanaka et al. 1984), this resonance can tentatively be assigned to the methyl group of Val 80. This resonance was found unchanged in the NS spectrum. The smaller number of RCSM resonances in NS1 indicates differences in the tertiary structure; the more homogeneous Lys resonance peak indicates that these residues are less involved in or affected by tertiary and/or quaternary interactions.

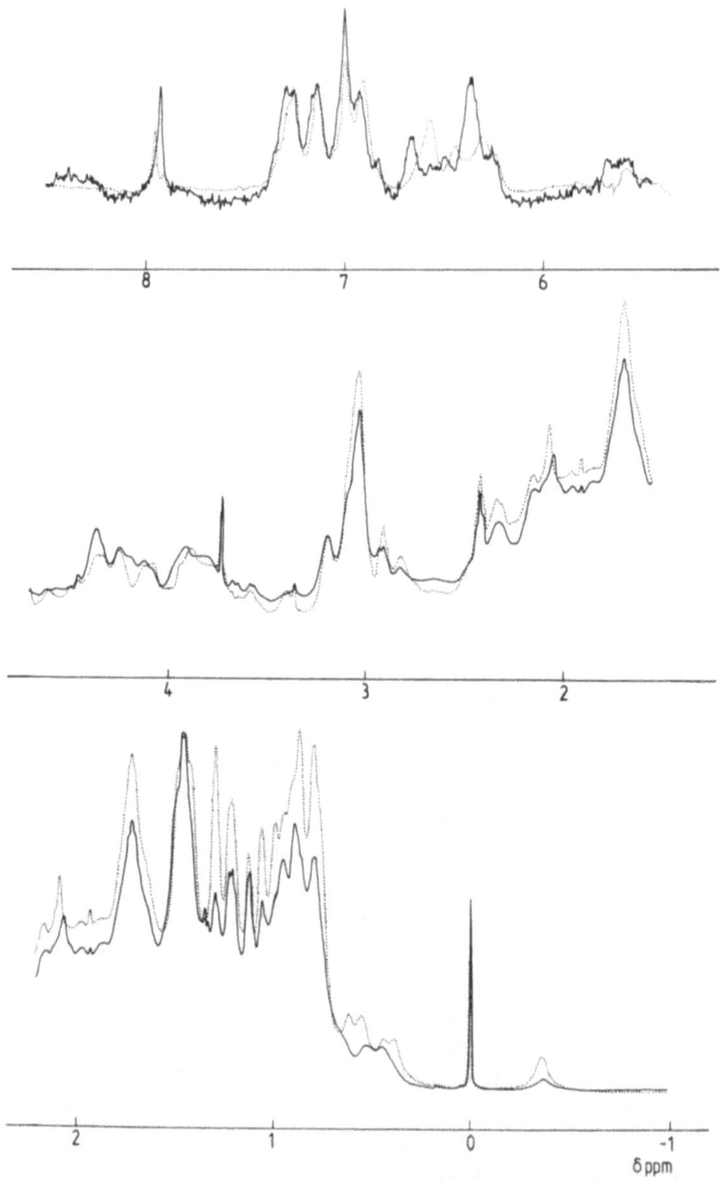

Fig. 4. The difference between the NS spectrum and a composite NS1-NS2 spectrum. Different regions of the spectra of NS (*solid line*) are compared with the corresponding regions of a spectrum generated by computer addition of the spectra of NS1 and NS2 of normalized intensity (*dotted line*)

A spectrum obtained by computer addition of NS1 and NS2 spectra of normalized intensity (dotted line) is superimposed on the spectrum of NS (solid line) (Fig. 4). As seen from this comparison, the spectrum of NS is somewhat different from the composite NS1-NS2 spectrum.

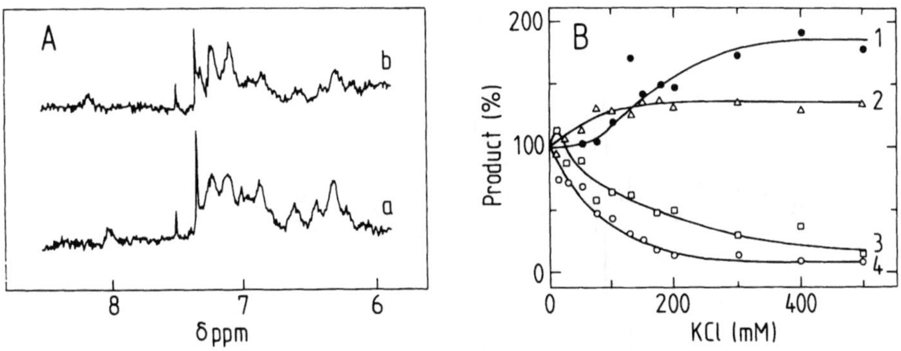

Fig. 5A,B. Effect of KCl concentration on the aromatic region of the NS spectrum and on the yield of protein-protein cross-linking reaction. **A** Comparison of the spectra of NS at 10 mM KCl (*spectrum a*) and at 1.0 M KCl (*spectrum b*). **B** Yield of monomers, dimers, trimers and tetramers of NS following cross-linking reaction for 20 min at 20°C in 50 mM triethanolamine-HCl, pH 8.5, containing 3 mg NS ml^{-1}, 4.5 mg dimethylsuberimidate-HCl ml^{-1} and the indicated KCl concentrations. The reaction products were quantified electrophoretically as described (Paci et al. 1984) and expressed as percentage of the amount of each product formed after reaction in the absence of KCl. Further details are found in Paci et al. (1984)

All the differences mentioned above clearly indicate that NS1 and NS2 self-aggregate in different ways and that, when present together, they form preferentially a heterologous aggregate (NS) endowed with a specific structure.

2.3 Effect of the Ionic Strength on the NS Spectra

It has been reported that eukaryotic histones exist primarily as random coils at low salt concentrations and upon increase in ionic strength, undergo drastic structural changes, acquiring secondary, tertiary and quaternary structures characterized by extensive hydrophobic interactions (Crane-Robinson 1979). These changes can be monitored by the selective broadening of some spectral lines, while the resonances of the segments left as random coil for the interaction with DNA remain relatively narrow. With the exception of a slight down-field shift of the His C-2 proton resonance and of a slight up-field shift of the "perturbed" Phe resonances and of the 3.1 ppm Lys ϵCH_2 peak, the spectrum of NS remains virtually unmodified through large variations of ionic strength (Fig. 5A). These results indicate that, unlike that of eukaryotic histones, the tertiary and quaternary structures of NS are unaffected by variations of ionic strength.

Protein-protein cross-linking reactions of NS with dimethylsuberimidate yield a large number of dimers, trimers and tetramers as well as a small amount of pentamers and hexamers. The change in yield of each cross-linked product as a function of ionic strength is depicted in Fig. 5B. The yield of dimers is not affected by increases in the KCl concentration up to 0.5 M (Fig. 5B) or higher (not shown). The production of trimers and tetramers, on the other hand, is reduced drastically, but this reduction might be due to a reduction in the efficiency of the cross-linking reaction rather than

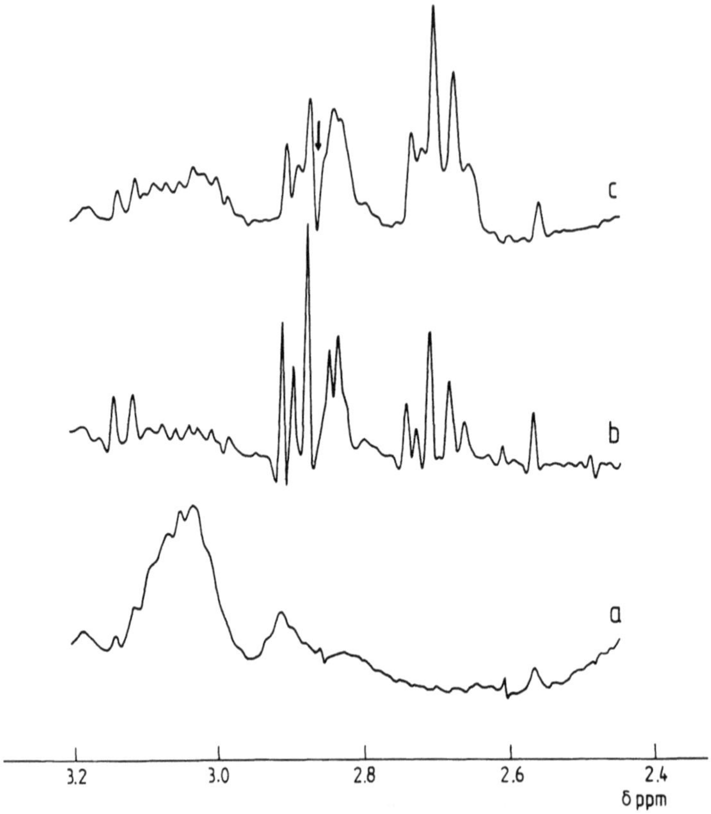

Fig. 6a–c. Spectra of native and reductively methylated NS2. NS proteins were reductively methylated using non-radioactive formaldehyde following the published procedure (Gualerzi and Pon 1979). Resolution enhanced spectra of (**a**) native NS2; (**b**) NS2 reductively methylated in the native form; (**c**) NS2 reductively methylated in the presence of 8 M urea. The *arrow* indicates the presence of an additional N-CH_3 dimethyllysine resonance appearing when the chemical modification is carried out in the presence of urea

to a dissociation of the larger aggregates. Regardless of the reason for the reduction in the yield of trimers and tetramers, these results show that at least the interaction responsible for dimer formation has a strong hydrophobic component and is preserved under conditions in which no alterations of the NMR spectra were observed.

2.4 Modification of Lysine Residues can Influence the Quaternary Interactions of NS Proteins

In order to gain information concerning the environment and the possible role of the Lys residues in the tertiary and quaternary structure of the NS proteins, purified NS1 and NS2 were subjected to reductive methylation (Gualerzi and Pon 1979) and their [^{1}H]-NMR spectra were compared with each other as well as with those obtained after

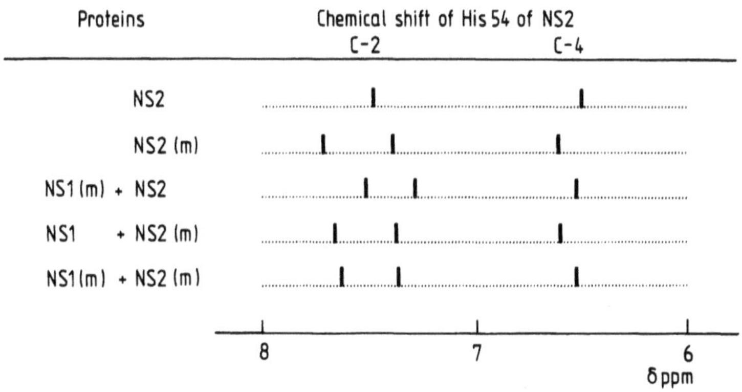

Fig. 7. Influence of lysine modification on the chemical shift of the C-2 and C-4 protons of His 54 of NS2. NS1 and NS2 were reductively methylated under native conditions as described in the legend of Fig. 6. Different combinations of proteins were prepared as indicated by mixing stoichiometric equivalent amounts of NS1 and NS2 either native or methylated (*m*). The spectral positions of the His C-2 and C-4 resonances are indicated

mixing modified NS1 and NS2 with stoichiometrically equivalent amounts of non-modified NS2 or NS1, respectively.

A representative example of this type of experiment is shown in Fig. 6, which presents the 3.2–2.4 ppm region of the spectra of native NS2 (spectrum a) and of NS2 reductively methylated under non-denaturing (b) or denaturing (i.e. in the presence of 8 M urea) (c) conditions. Regardless of the conditions under which the chemical modification was carried out, the distribution of the monomethyl- and dimethyllysine resonances, visible as bands centered around 2.5 and 2.9 ppm, respectively, is overall very similar in NS2 (Fig. 6) and in NS1 and NS (not shown). In both NS1 and NS2, however, one additional $N\text{-}CH_3$ band appears when the protein is methylated under denaturing conditions, suggesting that under native conditions at least one lysine is not amenable to chemical modification because of its involvement in the tertiary and/or quaternary structure. In addition to the bands due to the monomethyl- and dimethyllysine resonances, the spectra of methylated NS2 and NS1 show characteristic changes in the aromatic and aliphatic regions of the spectra. The His C2 proton resonance is shifted to lower field and splits into two bands of approximately equal intensity and different line widths ($\sim$12 and 24 Hz). Furthermore, the non-perturbed Phe ring resonances are shifted to lower field and the perturbed ones are reduced in intensity. Finally, the appearance of the RCSM peaks is modified. Particularly interesting is the finding that at least some of these effects (e.g. the splitting of His C-2 proton resonance) result from perturbations of the quaternary structure. In fact, splitting of the His C-2 proton resonance is observed not only when protein NS2 is chemically modified, but also when native NS2 is mixed with reductively methylated NS1. The results of these experiments are summarized schematically in Fig. 7, which shows the position of the C-2 and C-4 proton resonances of His 54 of NS2 under various experimental conditions. As seen in the figure, each combination of proteins is characterized by a different pattern of His resonances; the C-2 resonance is split whenever one pro-

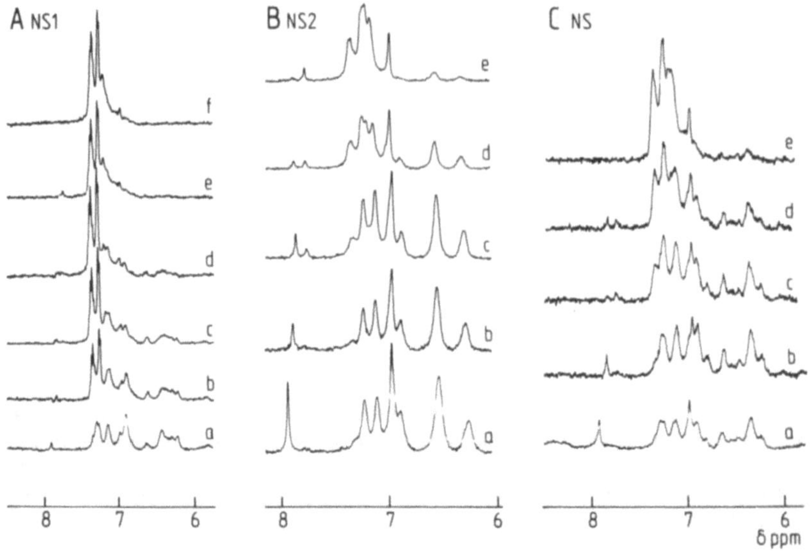

Fig. 8. Effect of temperature on the aromatic region of the spectra of NS1, NS2 and NS. The temperature of the samples was increased by 2°–5°C increments and kept at each temperature for approximately 60 min (an equilibration time of 20 min and an accumulation of 40 min). Only some representative spectra are reported. The temperatures at which the spectra were recorded are: **A** NS1: *a* 19°; *b* 29°; *c* 31°; *d* 35°; *e* 45°; *f* 50°C. **B** NS2: *a* 23°; *b* 34°; *c* 41°; *d* 45°; *e* 55°C. **C** NS: *a* 23°; *b* 40°; *c* 48°; *d* 53°; *e* 58°C (From Paci and Gualerzi 1986)

tein is methylated, while the C-4 resonance changes position in the spectrum, but never splits.

A very similar situation concerns the N-CH_3 resonances, which are also noticeably different in the homologous and heterologous interactions. Thus, when native NS2 is mixed with methylated NS1, two N-CH_3 resonances of the latter are shifted up-field and two down-field. Conversely, three N-CH_3 resonances of methylated NS2 are shifted up-field upon mixing this protein with native NS1.

Overall, the results of these experiments indicate that reductive methylation of the lysine residues causes perturbations of the tertiary and, even more so, quaternary structure of NS. Furthermore, regardless of whether the effects of the lysine methylation on the His resonances are direct or indirect, these results provide additional evidence that the His residue of NS2 is very sensitive to changes in the quaternary interactions of this protein.

Finally, the finding that at least some lysine residues of both NS1 and NS2 are directly or indirectly implicated in the quaternary interactions of these proteins agrees well with the finding (Losso et al. 1986) that very efficient intermolecular crosslinking reactions involving lysine residues can fix covalently the various types of quaternary protein-protein interactions of NS proteins.

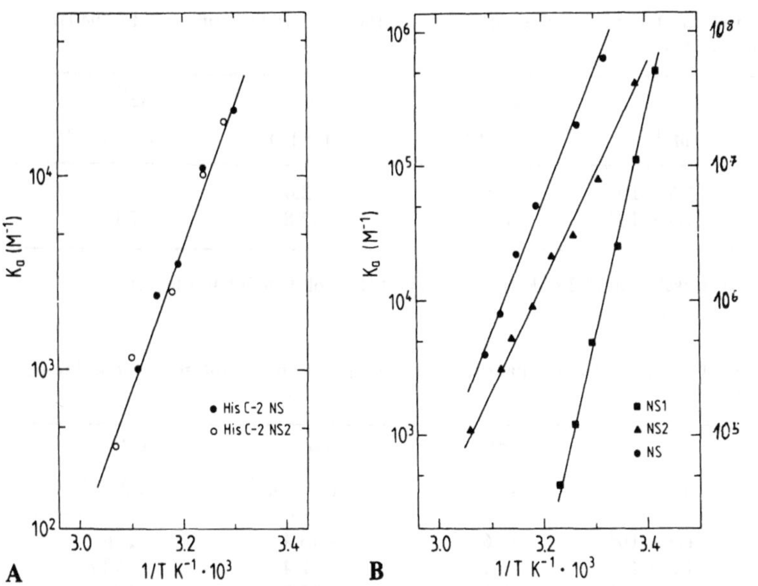

Fig. 9A,B. Van't Hoff plots for the association equilibrium monitored by the histidine and phenylalanine proton resonances. The ln K_a vs 1/T plots for the (A) His C-2 proton resonance in NS2 alone (○) or in the presence of stoichiometric amounts of NS1 (●) and (B) for the perturbed Phe resonances in NS1 (■) (ordinate valves on the right side), NS2 (▲) and NS (●) (ordinate valves on the left side). (From Paci and Gualerzi 1986)

2.5 Evaluation of the Thermodynamic Quantities of the Quaternary Interactions of the NS Proteins

Since, as shown above, the up-field perturbed Phe resonances and C-2 proton resonance of His 54 of NS2 are sensitive indicators of the quaternary structure of the NS proteins, we investigated the quaternary interactions of these proteins by following the spectral changes induced in these resonances by variation of temperature. As seen from the spectra shown in Fig. 8, increasing the temperature (from the bottom toward the top spectra) induces a progressive disappearance of the up-field shifted resonances with a corresponding intensity increase of the peaks characteristic of freely rotating Phe. In addition, above 30°C, the His C-2 proton resonance splits, giving rise to a "new" high-field resonance whose intensity increases at the expense of the original low-field one; above 55°C, only the high-field resonance remains visible. Upon cooling, these effects are fully reversed (Paci et al. 1984 and data not shown) indicating that the original chemical environment is re-attained at lower temperatures. The splitting of the His C-2 and the reshuffling of the Phe resonances indicate the existence of dynamic equilibria between forms of the proteins having the His and the Phe residues in different chemical environments. These equilibria are shifted towards the formation of the molecular species characterized by the higher field His C-2 and by the absence of perturbed Phe rings when the temperature is increased. These equilibria were followed quantitatively by integration of the resonance peaks either directly as in the case of the His C-2 proton, or indirectly by computer simulation as in the case of the Phe resonances. From the molar fraction of the two forms at equilibrium, we calculated

Table 1. Evaluation of the thermodynamic quantities for the equilibrium monitored by the His 54 C-2 proton resonance of NS2[a]

Protein	K_a (M^{-1})	ΔG^o (Kcal M^{-1})	ΔH^o (Kcal M^{-1})	ΔS^o (cal K^{-1} M^{-1})
NS2	1.6×10^5	−5.6	−32.4	−90
NS (NS1 + NS2)	4.4×10^5	−6.6	−32.8	−88

[a] The listed values were evaluated at 298 K from the Van't Hoff plot shown in Fig. 9A.

Table 2. Evaluation of thermodynamic quantities for the equilibrium monitored by the Phe-perturbed resonances[a]

Protein	K_a (M^{-1})	ΔG^o (Kcal M^{-1})	ΔH^o (Kcal M^{-1})	ΔS^o (cal K^{-1} M^{-1})
NS1	1.3×10^6	−7.6	−93.4	−288
NS2	4.8×10^5	−6.7	−22.4	− 52.8
NS (NS1 + NS2)	3.9×10^6	−8.8	−42.8	−114

[a] The listed values were evaluated at 298 K from the Van't Hoff plot shown in Fig. 9B.

the apparent K_a's of these processes as a function of temperature. The plot of ln K_a vs 1/T for the His C-2 proton in NS2 alone and in the presence of a stoichiometrically equivalent amount of NS1 is presented in Fig. 9A. The thermodynamic quantities calculated from the Van't Hoff plot are reported in Table 1. The apparent K_a's of the aggregation equilibria of NS1, NS2 and NS, evaluated from the molar fraction of the freely rotating vs perturbed Phe protons at various temperatures, are presented in Fig. 9B; the related thermodynamic quantities are listed in Table 2.

From the data presented in Tables 1 and 2, it can be concluded that all association equilibria monitored by the His and Phe proton resonances are characterized by negative ΔG, ΔH and ΔS values, indicating that all association processes occur spontaneously and involve extensive interactions, which immobilize several previously mobile side chains and which involve a large number of water molecules at the hydrophobic interface between subunits. The data also indicate that the heterologous interactions are always characterized by more negative free energy (i.e. are the most favoured interactions) and suggest that the association equilibria monitored by the Phe and His residues are probably different. Even though the precise nature of these quaternary interactions cannot be determined directly, from the comparison of their association constants with those previously estimated for the formation of dimers ($K \geqslant 10^9$ M^{-1}), tetramers ($K \cong 10^6$ M^{-1}) and larger aggregates ($K \cong 10^4$ M^{-1}) of NS (Losso et al. 1986), we can rule out that the effects studied here reflect the monomer ⇌ dimer equilibrium. Thus, the perturbation of the Phe resonances probably reflects the dimer ⇌ tetramer equilibrium, while the splitting of the His resonance probably reflects the dissociation of protein aggregates larger than tetramers.

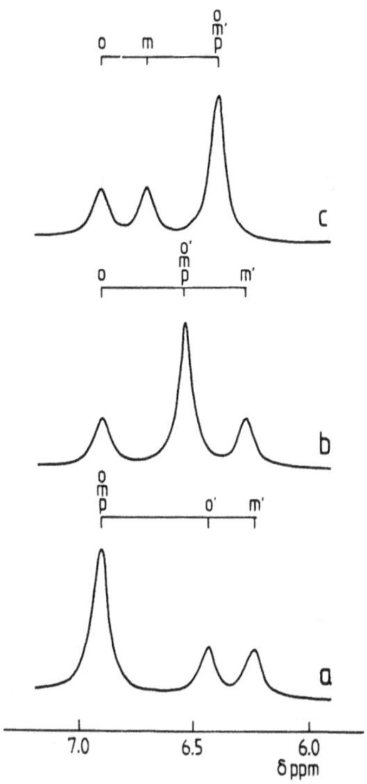

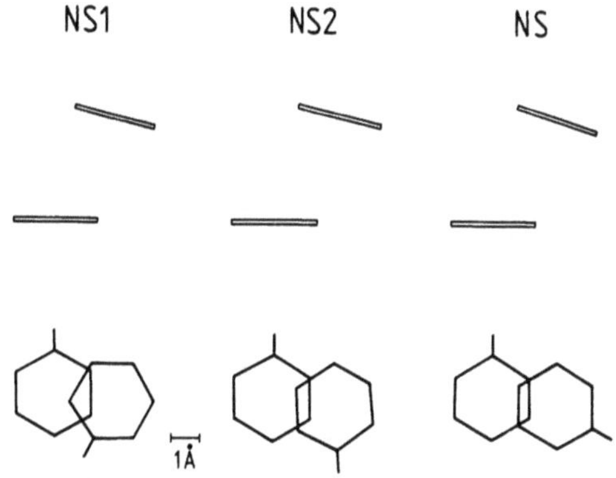

Fig. 11. Mutual orientation of the Phe rings in the homologous and heterologous interactions of NS1, NS2 and NS. The relative position and orientation of the Phe rings of NS1 and NS2 in the homologous and heterlogous interactions have been obtained from the chemical-shift differences between the freely rotating and the perturbed Phe resonances reported in Table 3. The *upper scheme* refers to an edge-on view, the *lower scheme* to a view perpendicular to the ring plane (From Paci and Gualerzi 1986)

Fig. 10a–c. Spectral shape of the perturbed phenylalanine resonances in the homologous and the heterologous interactions. Simulated spectral shapes of the resonances due to the perturbed Phe in the homologous interactions in NS1 (**a**), in NS2 (**b**) and to the heterologous interactions in NS (c). The simulations were obtained as described previously omitting the resonances of the freely rotating phenylalanines and, when appropriate, that of the C-4 proton of His. In the case of NS, simulated NS1 and NS2 spectra of normalized intensity were subtracted from the simulated NS spectrum to yield a simulated spectrum which reflects the resonances of the Phe of NS1 and NS2 engaged in heterologous interactions. The relative assignments of the various peaks are also reported: o,o′ = ortho; m,m′ = meta; p = para protons (From Paci and Gualerzi 1986)

2.6 Orientation and Mobility of the Phenylalanine Residues in the Quaternary Interactions of NS Proteins

From the integral values of the Phe resonances, we also determined the number of the perturbed and freely rotating phenylalanines. In the temperature interval examined, these numbers range from approximately 1.1 to 3.0 and from 1.9 to 0.0 for the freely rotating and perturbed phenylalanines, respectively. This indicates that each protein-protein aggregation giving rise to a perturbation of the Phe resonances involves two (out of three) phenylalanines of each polypeptide chain.

From the comparison of the aromatic spectra of NS1, NS2 and NS and from their simulation spectra, we deduced the relative orientation of the phenylalanine rings in the quaternary interactions and also obtained information concerning their mobility. The spectral shapes of the perturbed Phe resonances in the homologous and heterologous contacts, as well as the relative assignments of the various types of ring protons, are

Table 3. Chemical-shift difference between proton resonances of freely rotating phenylalanines and the corresponding perturbed Phe resonances in homologous and heterologous interactions[a]

Assigned ring protons[b]	Type of interaction NS1-NS1 ($\Delta\delta$ ppm)	NS2-NS2 ($\Delta\delta$ ppm)	NS1-NS2 ($\Delta\delta$ ppm)
o	0.3	0.3	0.4
o′	0.8	0.7	0.9
m	0.2	0.6	0.6
m′	0.9	0.9	0.9
p	∿0.4	∿0.6	∿0.9

[a] Data were obtained as described in the text comparing the experimental and the simulated spectra reported in Fig. 10.
[b] o,o′ = Ortho; m,m′ = meta; p = para protons.

Table 4. Activation energy for the rotation of the perturbed phenylalanine rings around the Cβ–Cγ bond[a]

Type of interaction	$\Delta G^{\ddagger}$ (Kcal M^{-1})	$\Delta H^{\ddagger}$ (Kcal M^{-1})	$\Delta S^{\ddagger}$ (cal K^{-1} M^{-1})
NS1-NS1	16	4	−38
NS2-NS2	15	2	−44
NS1-NS2	15	10	−18

[a] Values obtained from the chemical-shift differences between the resonances of Fig. 10 as a function of the temperature. The interconversion rate was obtained as described (Gutowsky and Holm 1956) and the activation energy was estimated using the Eyring equation.

presented in Fig. 10. The relative orientation of the stacked Phe rings (Fig. 11) was estimated (Johnson and Bovey 1959) from the differences in chemical shift between freely rotating and perturbed Phe protons in the various types of interactions (Table 3) on the assumption that no other center of magnetic anisotropy exists near the phenylalanine residues. Within the temperature range examined, we have recorded small variations of the relative positions of the perturbed Phe resonances which can be attributed to changes in the interconversion rate between the relative positions of the aromatic ring. Thus, from these data we evaluated the rate of interconversion (i.e. of rotation) (Gutowsky and Holm 1956) and, from the Eyring equation, we calculated the activation energy barrier required to overcome all the steric hindrances to the rotation of the ring around the Cβ–Cγ bond. From the results reported in Table 4, it can be seen that the mobility of the Phe rings in the aggregate is strongly impaired (interconversion rate $\cong 25\ s^{-1}$) with respect to a freely rotating ring ($10^3 - 10^5\ s^{-1}$) (Jardetzky and Roberts 1981) and that the activation energy barriers for ring rotation are very similar in NS1, NS2 and NS and are comparable to those reported for aromatic rings with reduced internal mobility due to steric hindrances. Finally, the third phenylalanine of each molecule was found to remain unstacked and substantially free to rotate.

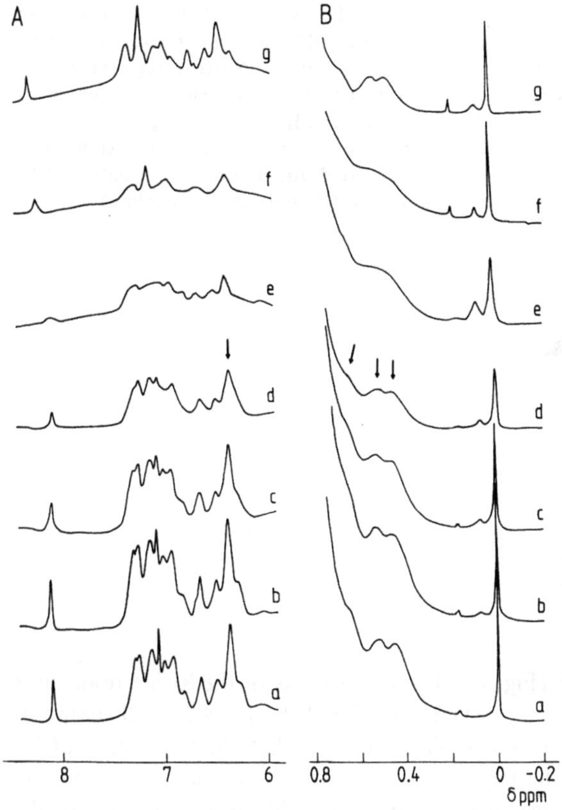

Fig. 12A,B. Effect of DNA binding on the [^{1}H]-NMR spectrum of NS. Spectra were obtained at 19° ± 1°C on 5 mg ml^{-1} protein samples (≅ 5.5 × 10^{-4} M) in 10 mM K-phosphate buffer (pH 7.5) (meter reading) containing the amounts of deuterated, fragmented *E. coli* DNA indicated below and, unless otherwise specified, 100 mM KCl in 99% D_2O. Spectra: (*a*) NS alone; (*b*) NS:DNA = 40:1 (w/w); (*c*) NS:DNA = 20:1 (w/w); (*d*) NS:DNA = 7.1 (w/w); (*e*) NS:DNA = 3.5:1 (w/w); (*f*) same as in *e* at 0.4 M KCl; (*g*) same as *e* at 1.0 M KCl. **A** Aromatic region; **B** 0.8 to 0.2 region. *Arrows:* (in **A**) a perturbed Phe resonance characteristic of NS1-NS2 heterotypic aggregates; (in **B**) ring current-shifted methyl resonances due to NS tertiary and quaternary structure (Adapted from Lammi et al. 1984)

2.7 The Interaction of NS with DNA

The interaction of NS with DNA was also investigated by NMR spectroscopy. Upon addition of increasing amounts of deuterated fragmented *E. coli* DNA to a fixed amount of NS, we observed a progressive broadening of all resonances, without any specific modification of the spectrum, except for a slight up-field shift of the Arg δCH_2 band at 3.2 ppm (see below). The broadening of the spectrum ultimately led to the reduction beyond detection of all protein resonances. This effect can be reversed by increasing the ionic strength and is not seen with the resonances due to preparation impurities which remain quite sharp. Thus, the broadening of the protein resonances is not (or only minimally) due to the viscosity increase brought about by the addition of DNA, and must be explained by the existence of dynamic exchange between free and DNA-bound proteins; the increase in the correlation time of the bound species determines an increase in the relaxation rate which, probably together with a medium-low rate of exchange, broadens the resonances of NS until they virtually disappear. The increase in the K_d of the DNA-NS complex, caused by the addition of KCl, results in an increase in the concentration of the free species in solution with the consequent reappearance of the resonance bands characteristic of NS.

As seen from the spectra presented in Fig. 12, with the addition of increasing amounts of DNA, the Phe resonances perturbed to high field because of ring stacking

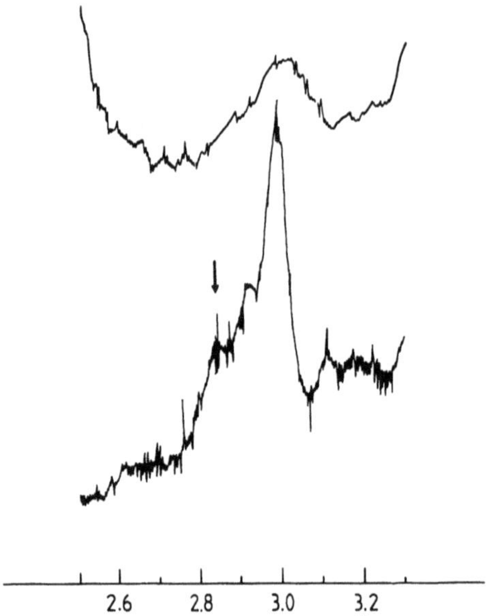

Fig. 13. Involvement of Arg residues in NS binding to DNA. Spectra: *a* a 1:1 (w/w) mixture of NS and deuterated, fragmented DNA at 100 mM KCl; *b* the same sample containing 2 M KCl. Only the δCH_2 Arg and ϵCH_2 Lys bands are shown. The incomplete reapperance of the δCH_2 Arg resonances is indicated by the *arrow* (From Lammi et al. 1984)

in the quaternary structure of NS (Fig. 12A), and the peaks of the RCSM resonances due to the tertiary and/or quaternary structure (Fig. 12B) broaden and disappear simultaneously with the other resonances. This indicates that the interaction of NS with DNA does not grossly alter the tertiary and quaternary structure of the protein (i.e. NS binds to the DNA in the aggregated form). A possible exception could be the aggregation equilibrium monitored by the His C-2 proton resonance that seems to be shifted slightly up-field upon addition of DNA and down-field when the interaction with DNA is diminished by the increase in ionic strength. It is noteworthy that when the ionic strength is increased so as to dissociate the NS-DNA complex, not all resonances reappear simultaneously, but the Arg δCH_2 band at 3.2 ppm remains relatively broad (especially at a high DNA:protein ratio). This effect is illustrated in Fig. 13, where the resonances of the Arg δCH_2 and Lys ϵCH_2 of NS are seen broadened in the presence of DNA (upper spectrum) and after addition of 2 M KCl (lower spectrum). As seen in the figure, the Arg δCH_2 band reappears incompletely, while the original line-width of the adjacent Lys ϵCH_2 band is nearly restored. This effect, which is probably due to a residual immobilization of the Arg residues, in spite of the high ionic strength, strongly suggests an involvement of Arg residues in the interaction between NS and DNA. This premise was confirmed by chemical modification experiments (Lammi et al. 1984).

2.8 Spectral Characteristics of H-NS

In addition to the spectroscopic investigation on proteins NS1, NS2 and NS described above, we have recently begun a similar study on H-NS, the larger (15 kD) histonelike protein of *E. coli.* This work, which is still at a very early stage, is made difficult by

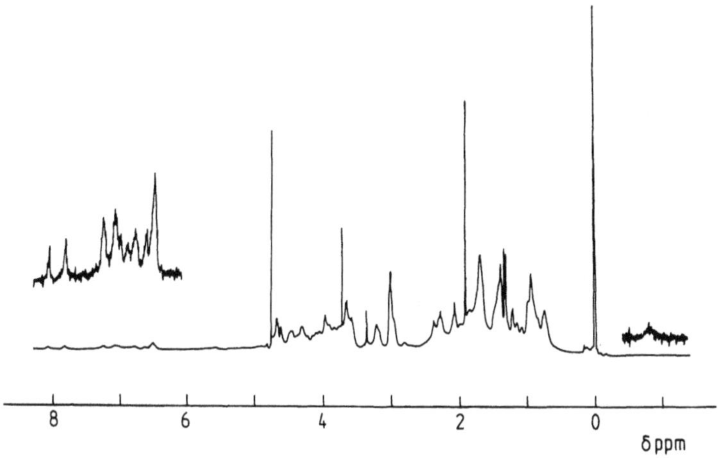

Fig. 14. 400 MHz [^{1}H]-NMR spectrum of H-NS. The spectral conditions are the same as those described in the legend of Fig. 1. The concentration of H-NS was approx. 2 mg ml^{-1}

the larger size of H-NS, by the greater complexity of its amino acid composition and, even more so, because of the lower solubility of this protein.

The 400 MHz [^{1}H]-NMR spectrum of native H-NS is shown in Fig. 14. In Fig. 15, an enlargement of the aromatic region of H-NS (a) as well as that of H-NS heated at 75°C (b) is presented. In the same figure, we also give the tentative assignments of the aromatic proton resonances. From a preliminary examination of the H-NS spectra, we can draw the following conclusions. In comparison with NS, the H-NS spectrum does not show extensive hard-to-exchange NH protons, but contains numerous up-field perturbed methyl resonances (at approximately 0.75 ppm), in addition to a discrete number of RCSM resonances. As in the case of NS and NS2, both Arg δCH_2 and Lys ϵCH_2 bands of H-NS are not symmetrical, but possess many shoulders, indicating that these residues are distributed in different chemical environments resulting from the tertiary and/or quaternary structure of the protein. As in the case of NS, several aromatic resonances are shifted up-field. It seems unlikely, however, that these are Phe ring resonances, since each polypeptide chain, which probably exists as a dimer (or a tetramer) contains only one of these residues. From the spectral changes obtained upon heating and from the line-widths and chemical shifts of the aromatic protons of H-NS, it can be concluded that all the aromatic residues of this protein are buried within the tertiary or quaternary structure and endowed with little mobility. Finally, the structure of H-NS seems to be even more heat-stable than that of NS, since higher temperatures are needed to denature the protein and complete denaturation did not occur even at 75°C, at which temperature extensive precipitation of the protein became apparent.

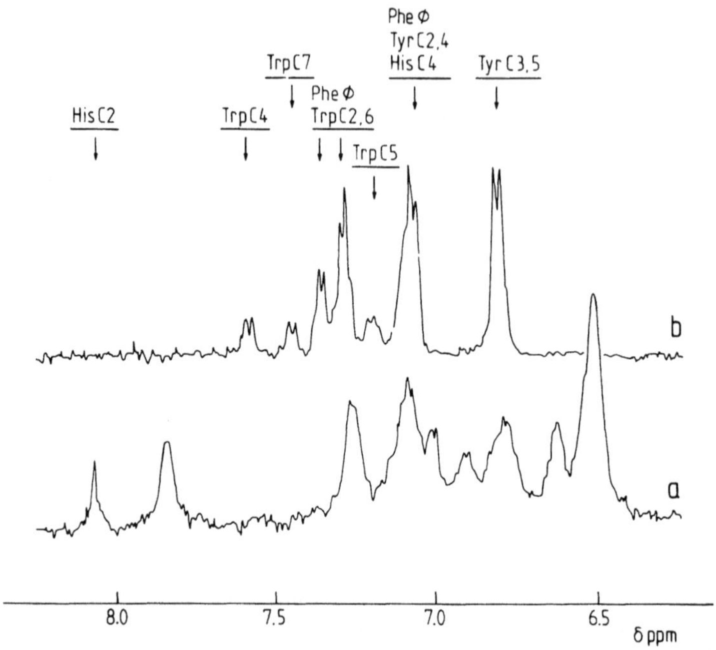

Fig. 15. Effect of heating on the aromatic region of the spectrum of H-NS. *a* Spectrum of H-NS recorded at 20°C. *b* Spectrum of the same sample after heating at 75°C. Tentative resonance assignments are given

3 Conclusions

We have shown here that the [^{1}H]-NMR spectra of NS1, NS2 and NS display several Phe ring proton resonances which are shifted up-field by stacking interactions and whose positions are characteristic for each protein. The chemical shift of the C-2 proton of His 54 of NS2 is also different in NS2 and NS. These differences in the Phe and His resonances are useful indicators of the existence of homologous and heterologous quaternary interactions. Increasing the temperature induces the following reversible effects: (a) a progressive decrease of the intensity of the up-field-shifted Phe resonances with a corresponding intensity increase of the peaks characteristic of freely rotating Phe, (b) the splitting of the His C-2 proton resonance, giving rise to a second high-field resonance whose intensity increases with temperature at the expense of the low-field one. The behavior of both Phe and His resonances reflects that expected for temperature-influenced chemical equilibria. Thus, assuming simple association equilibria for the transitions monitored by the spectral changes, we have calculated the pertinent thermodynamic quantities. It was found that heterologous interactions are characterized by the most negative ΔG (i.e. are the most favored interactions).

From the comparison of the association constants determined by NMR with those estimated from cross-linking experiments, we conclude that the perturbation of the

Phe resonances probably reflects the dimer ⇌ tetramer equilibrium, while the splitting of the His resonance probably monitors the dissociation of protein aggregates larger than tetramers. If, as it is likely, the Phe residues of NS1 and NS2 are located on the protein surfaces involved in the dimeric interaction (Tanaka et al. 1984) and if we interpreted correctly the nature of the quaternary interaction monitored by the perturbed Phe resonances, we must conclude that a stacking interaction occurring at the dimeric interface monitors the formation of tetramers. Thus, it would stem from this premise that a stacking interaction between Phe rings of different protein subunits does not take place (or at least not to the extent observed by NMR) at the dimeric interface until it is triggered by the interaction between dimers yielding tetramers. Our results also showed that two of the three phenylalanines present in each polypeptide chain are involved in intermolecular stacking interactions with slightly different orientations depending on the type of quaternary interactions (i.e NS1-NS1, NS2-NS2 or NS1-NS2). The third phenylalanine of each polypeptide remains unstacked and substantially free to rotate. The mobility of these rings in the aggregate and the activation energy barriers for ring rotation were also estimated.

Upon addition of increasing amounts of DNA, the resolution of the [^{1}H]-NMR spectra of NS is progressively lost due to the broadening of all resonance lines. Since all resonances broaden simultaneously without prior selective loss of the spectral features due to the tertiary and quaternary structure of the protein, we concluded that NS binds to DNA in the aggregated form and without gross alteration of its tertiary structure.

Finally, from the comparison of the properties of the prokaryotic histonelike proteins with those of genuine histones, we conclude that these two groups of proteins display both important similarities and differences. Like eukaryotic histones, the NS proteins form specific molecular aggregates among themselves that involve primarily hydrophobic interactions resulting in hard-to-exchange NH groups and in upfield-shifted Phe resonances similar to those observed in the complex between histones H3 and H4 (Böhm et al. 1977; Crane-Robinson 1979). Furthermore, the thermodynamic quantities estimated here for the equilibria monitored by the perturbed Phe and the His C-2 resonances are strikingly similar to those calorimetrically determined for the equilibria leading to the assembly of nucleosomal octamers: (H2A-H2B) + $(H3\text{-}H4)_2$ ⇌ hexamer and hexamer + (H2A-H2B) ⇌ octamer (Benedict et al. 1984).

Unlike eukaryotic histones, NS1 and NS2, isolated or in combination, do not show any evidence for the presence of distinct structural domains (i.e. hydrophobic-globular vs hydrophilic-random coil regions) and are virtually insensitive to large variations in ionic strength.

Acknowledgements. We would like to express our gratitude to Ms. E. Philippi for her enthusiastic help in the preparation of the drawings and Ms. J Belart for the careful and patient typing of the manuscript. This work was supported in small part by funds of the Italian Ministry of Public Education.

References

Benedict RC, Moudrianakis EN, Ackers GK (1984) Interactions of the nucleosomal core histones: a calorimetric study of octamer assembly. Biochemistry 23:1214–1218

Böhm L, Hayashi H, Cary PD, Moss T, Crane-Robinson C, Bradbury EM (1977) Sites of histone-histone interaction in the H3-H4 complex. Eur J Biochem 77:487–493

Crane-Robinson C (1979) High resolution NMR studies of histones. In: Berliner LJ, Reuben J (eds) Biological magnetic resonances, vol 1. Plenum, New York, pp 33–90

Gualerzi C, Pon CL (1979) Radioactive chemical labeling of ribosomal proteins and translational factors in vitro. Methods Enzymol 59:782–795

Gutowsky HS, Holm RH (1956) Rate processes and nuclear magnetic resonance spectra. Hindered internal rotation of amides. J Chem Phys 25:1228–1234

Hawkins AR, Wootton JC (1981) A single DNA-binding protein from *Pseudomonas aeruginosa* homologous to proteins NS1 and NS2 (HU proteins) of *Escherichia coli* and other bacteria. FEBS Lett 130:275–278

Hirshbein L, Guillen N (1982) Characterization, assay and use of isolated bacterial nucleoids. Methods Biochem Anal 28:297–328

Jardetzky O, Roberts GCK (1981) NMR in molecular biology. Academic Press, New York

Johnson CE, Bovey FA (1959) Calculation of nuclear magnetic resonance spectra of aromatic hydrocarbons. J Chem Phys 29:1012–1017

Kyte J, Doolittle RF (1982) A simple method for displaying the hydropathic character of a protein. J Mol Biol 157:105–132

Lammi M, Paci M, Gualerzi CO (1984) Proteins from the prokaryotic nucleoid. The interaction between protein NS and DNA involves the oligomeric form of the protein and at least one Arg residue. FEBS Lett 170:99–104

Losso MA, Miano A, Gianfranceschi GL, Gualerzi CO (1982) Proteins from the prokaryotic nucleoid II. Inhibition of DNA transcription by NS1 and NS2 (HU proteins). Biochem Int 5:423–427

Losso MA, Pawlik RT, Canonaco MA, Gualerzi CO (1986) Proteins from the prokaryotic nucleoid. A protein-protein crosslinking study on the quaternary structure of *Escherichia coli* DNA-binding proteins NS (HU). Eur J Biochem 155:27–32

Mende L, Timm B, Subramanian AR (1978) Primary structures of two homologous ribosome-associated DNA-binding proteins of *Escherichia coli.* FEBS Lett 96:395–398

Miano A, Losso MA, Gianfranceschi GL, Gualerzi CO (1982) Proteins from the prokaryotic nucleoid I. Effect of NS1 and NS2 (HU) proteins on the thermal stability of DNA. Biochem Int 5: 415–422

Paci M, Gualerzi C (1986) Proteins from the prokaryotic nucleoid. ^{1}H NMR study on the quaternary structure of *Escherichia coli* DNA-binding protein NS (HU). Biochemistry 25:2765–2769

Paci M, Pon CL, Losso MA, Gualerzi C (1984) Proteins from the prokaryotic nucleoid. High-resolution ^{1}H-NMR spectroscopic study of *Escherichia coli* DNA-binding proteins NS1 and NS2. Eur J Biochem 138:193–200

Tanaka I, Appelt K, Dijk J, White SW, Wilson KS (1984) 3-Å resolution structure of a protein with histone-like properties in prokaryotes. Nature 310:376–381

Varshavsky AJ, Bakayev VV, Nedospasov SA, Georgiev GP (1977) On the structure of eukaryotic, prokaryotic and viral chromatin. Cold Spring Harbor Symp Quant Biol 42:457–473

Characterization of *Bacillus subtilis* Nucleoid DNA-Binding Proteins

F. LE HÉGARAT, V. SALTI, and L. HIRSCHBEIN[1]

The tertiary structure of the bacterial genome has now been studied to some extent, mostly in the Gram-negative bacterium *Escherichia coli,* and various small basic proteins called "histonelike proteins" have been described by several authors. In the Gram-positive sporulating bacterium *Bacillus subtilis,* proteins together with RNA are also essential for the structural integrity of the isolated nucleoid. This work was focused on the identification and characterization of a group of nucleoid-associated proteins displaying some properties shown by the histones, i.e. heat stability, acid solubility and abundance in the cell.

Nucleoids could be isolated at low ionic strength (0.1 M NaCl) and low lysozyme concentration. The associated proteins were separated by gel electrophoresis, transferred to nitrocellulose filters and characterized by their ability to bind radioactive DNA. Among several DNA-binding proteins, four small, heat-stable, acid-soluble proteins (HASP) of respectively 8.5, 12, 23, and 26 kD were identified as being abundant in the *B. subtilis* nucleoids. The HASP/DNA ratio in the nucleoids is about 12 molecules per kilobase of DNA, instead of 40 for the histones in chromatin. These four HASP are able to bind homologous chromosomal or plasmid DNA and also heterologous DNA from the λ bacteriophage or from calf thymus (native or denatured DNA). Their affinity for DNA, as measured by the sensitivity to the ionic strength of the protein-DNA complex (0.1–0.4 M NaCl), is moderate. This set of proteins was not detected in purified membranes. Partially purified fractions containing the HASP were obtained from an S 100 extract prepared from *B. subtilis* vegetative cells, by several ammonium sulfate precipitation and chromatographic steps.

The possible involvement of the HASP in the compaction of the *B. subtilis* genome and their relationship with the *E. coli* histonelike proteins will be discussed.

Introduction

In bacterial chromatin the DNA is folded in a way that ensures replication and expression of the genome. Information about the folding of the prokaryotic DNA has come mostly from studies conducted on the effect of physical, chemical and enzymatic

1 Institut de Microbiologie, Bat. 409, Université Paris-Sud, 91405 Orsay Cedex, France

Bacterial Chromatin
(ed. by C.O. Gualerzi and C.L. Pon)

treatments of purified nucleoids (for a review see Hirschbein and Guillen 1982). The models proposed for the organization of the chromosome point out the folding of the DNA in loop-shaped domains of supercoiling (Worcel and Burgi 1972; Pettijohn and Hecht 1973; Kleppe et al. 1979; Hirschbein and Guillen 1982). These domains are dispersed or confined, depending on the local state of gene expression (Ryter and Chang 1975). Supercoils induced by the DNA gyrase are stabilized either by a nucleosomelike structure or by RNA-DNA interactions. In these dynamic models, the proteins considered to interact with the DNA, either directly or through RNA-protein complexes are RNA polymerase (Kleppe et al. 1979), cell envelope proteins (Leibowitz and Schaechter 1975) and histonelike proteins. Such basic DNA-binding proteins (DBP) have been characterized both in Gram-negative (Rouvière-Yaniv and Gros 1975; Varshavski et al. 1977; Hübscher et al. 1980; Lathe et al. 1980) and in Gram-positive bacteria (Nakayama et al. 1981; Imbert et al. 1982; Dijk et al. 1983; Salti et al. 1985). Among these DBP, the histonelike protein HU from *E. coli* gives rise to nucleosome-like structures observed under electron microscopy after reconstitution experiments with SV40 DNA molecules (Rouvière-Yaniv et al. 1979). However, little is known about their role in the organization of the genome. On the other hand, when *Bacillus subtilis* nucleoids are treated with sarcosyl (Guillen et al. 1978), DNA unfolding takes place, indicating that membrane proteins stabilize the folded structure of the genome. At present, such membrane DBP have not been identified.

In an attempt to obtain more information on the relationship between nucleoid DBP and the genome conformation in vivo, we have further characterized a group of heat-stable, acid-soluble nucleoid histonelike proteins (HASP) and have developed a purification method for these HASP. The identification of DBP has also been undertaken on purified membranes from *B. subtilis*.

Materials and Methods

1. Bacterial Strain and Growth

Bacillus subtilis 168 *trpC2* was grown at 37°C in nutrient broth (Schaeffer et al. 1965) to an optical density (OD_{570}) of 0.8. For radioactive labelling of DNA, (methyl-^3H) thymidine (1 μCi ml^{-1}, specific activity 48 Ci $mmol^{-1}$; 1.78 TBq $mmol^{-1}$) was added at the time of inoculation.

2. Preparation of "Low Ionic Strength Nucleoid" Proteins

The "low ionic strength nucleoids" were purified according to the procedure of Guillen et al. (1978), modified as described previously (Salti et al. 1985). Nucleoid proteins were obtained by treating the nucleoid preparation with DNase I (1 μg ml^{-1}) in the presence of 10 mM $MgCl_2$ for 3 h at 4°C. HASP were isolated from nucleoid proteins by a two-step procedure as described by Salti et al. (1985). For each preparation, 10 ml of nucleoid solution were first heated to 100°C for 10 min. The thermostable fraction, isolated by centrifugation, was then treated with HCl at a final concentration of 0.25 M for 15 min. The protein concentration was measured by the Lowry method.

3. Preparation of Membrane Proteins

Membrane vesicles were obtained according to the method of Kaback (1971), modified by Bohin et al. (1976).

4. Electrophoretic Analysis of Proteins

Protein samples were precipitated by 5% TCA, neutralized with ammonia and dissolved in 10 mM sodium phosphate buffer, pH 7, containing 8 M urea. The SDS polyacrylamide slab gel electrophoresis was carried out using the system of Laemmli (1970) with 0.1% SDS, 4 M urea and 15% acrylamide plus 0.4% bisacrylamide. The transfer of proteins was performed in a Bio Rad Transblot apparatus (Towbin et al. 1979). The remaining proteins on the gel were stained with Coomassie brilliant blue.

5. DNA-Binding Assays

Bacillus subtilis chromosomal DNA was prepared according to Saito and Muira (1963). The DNA was further purified by a CsCl/ethidium bromide centrifugation step. *B. subtilis* plasmid pGY102 DNA was kindly supplied by Dr. J.C. Le Hégarat (Huerre and Le Hégarat 1979). DNA samples were labelled with [^{32}P] to high specific activity (0.5–1 × 10^8 cpm μg^{-1}) using the Amersham Nick Translation kit. The DNA-binding experiments were done by the method of Bowen et al. (1980).

6. Purification of DNA-Binding Proteins

All the following steps were conducted at 4°C. 40 g (wet weight) of exponentially growing cells were suspended in 100 ml of solution A (10 mM Tris-HCl pH 7.9, 1 mM EDTA, 200 mM KCl, 0.1 mM DTT, 5% glycerol, 3 mM PMSF). The crude extract made with the French Press was incubated 20 min with hemoglobin-Sepharose (2 g for 10 g of cells) and centrifuged to obtain the S100 extract. The S100 extract was then adjusted to 30% saturation of ammonium sulfate by addition of 176 g solid crystal to 1 liter solution. The precipitate was collected by centrifugation at 12,000 rpm for 30 min using a JS13 rotor in a J21 Beckman centrifuge. The fraction resuspended in solution B (10 mM Tris-HCl pH 7.9, 1 mM EDTA, 50 mM KCl, 0.1 mM DTT, 10% glycerol, 3 mM PMSF) was kept at –20°C for further purification (Fraction 1). The supernatant of the centrifugation was again adjusted to 50% saturation of ammonium sulfate by adding 127 g solid crystals of ammonium sulfate. The protein precipitate was collected as above and resuspended in the same buffer (Fraction 2).

6.1 Phosphocellulose Chromatography

Fraction 1 and Fraction 2 were dialyzed against solution B, concentrated and loaded onto a phosphocellulose column equilibrated with the same buffer. The proteins bound to the phosphocellulose were eluted with solution B containing 1 M KCl.

Table 1. Comparison of nucleoids purified at three different ionic strengths

NaCl concentration of the lysate	Sedimentation rate (S)	Lysate DNA[a] (%)	Bacterial phospholipid[b] (%)	Total cell protein (%)	Presence HU_{BS}	Protein (w/w) DNA
1 M[c]	1600–2000	80–95	2– 5	3– 5	No	4–5
0.2 M[c]	3000–4500	80–95	10–12	7–10	No	6–8
0.115 M	4900–5400	80–95	11–13	12–15	Yes	8.3[d]

[a] Measured in the pooled fraction.
[b] Determined by the method of Bligh and Dyer (1959).
[c] Described in Guillen et al. (1978).
[d] Mean value.

6.2 DSCT DNA-Cellulose Chromatography

Double-stranded calf thymus (DSCT) DNA-cellulose was prepared by the technique of Litman (1968) as modified by Alberts and Herrick (1971). Protein fractions eluted from phosphocellulose were dialyzed against solution C (10 mM Tris-HCl pH 7, 1 mM EDTA, 50 mM NaCl, 0.1 mM DTT, 10% glycerol, 3 mM PMSF) and applied to the DSCT DNA-cellulose column. After a thorough washing, the bound proteins were eluted in three steps with solution C containing respectively, 0.2, 0.4 and 1 M NaCl.

Results

Properties of Nucleoids Isolated at Low Ionic Strength

Nucleoids have been isolated from *B. subtilis* by lysing the cells in three different conditions of ionic strength as described in Salti et al. (1985). The mildest extraction procedure was designed to preserved nucleoid proteins possibly involved in the maintenance of the tertiary structure of the genome. The HU_{BS} protein, the *B. subtilis* equivalent of the *E. coli* HU protein (Fleury-Guerout 1978), is found in the nucleoid supernatant when the NaCl concentration of the cell lysate is over 0.15 M. Therefore, the cell lysis was performed at 0.115 M NaCl. Furthermore, competition for DNA binding was reduced by lowering the amount of lysozyme used to lyse the bacteria.

The properties of these "low ionic strength nucleoids" have been compared with those of the two other types of nucleoids (Table 1). The protein and lipid contents are clearly increased at low ionic strength. The presence of a larger nucleoid-associated membrane fraction (Charret et al. 1980) can account for these quantitative differences. However, a striking qualitative change has also occurred: the HU_{BS} protein is only detected in the 0.115 M nucleoids. Therefore, the low ionic strength nucleoids were used as starting material for further nucleoid protein isolation and characterization.

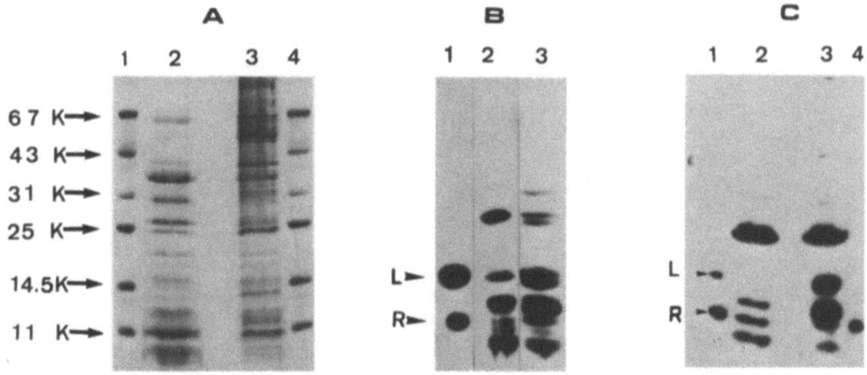

Fig. 1A–C. Binding of nucleoid proteins to homologous DNAs. **A** SDS-PAGE (15% polyacrylamide). *Lane 1* Marker proteins: BSA (67 kD), ovalbumin (43 kD), DNase (31 kD), chymotrypsinogen (25 kD), RNase (14.5 kD), lysozyme (11 kD); *lane 2* HASP from *B. subtilis* nucleoids; *lane 3* total nucleoid proteins; *lane 4* marker proteins. **B** and **C** Autoradiograms of the nitrocellulose filters incubated with [^{32}P]-labelled DNA. **B** *B. subtilis* chromosomal DNA. *Lane 1* Marker proteins: RNase (R: 14.5 kD), lysozyme (L: 11 kD); *lane 2* nucleoid HASP exposed for 4 h; *lane 3* total nucleoid proteins exposed for 40 min. **C** Plasmid pGY102 DNA. *Lane 1* RNase and lysozyme; *lane 2* nucleoid HASP; *lane 3* total nucleoid proteins; *lane 4* lysozyme. The autoradiograms were exposed for 4 h

Isolation of Nucleoid Proteins

The nucleoid protein fraction was obtained by treating low ionic strength nucleoid by pancreatic DNase as described in Methods. At least 40 polypeptides, ranging from 7 to 120 kD, could be detected by SDS-PAGE of this fraction (Fig. 1A, lane 3). After electrophoresis the fraction was transferred to a nitrocellulose filter by electroblotting and assayed for [^{32}P]-DNA binding. A small proportion of the nucleoid proteins were able to bind either chromosomal or plasmid *B. subtilis* DNA (Fig. 1B and C, lanes 3). A more detailed analysis of the DNA-binding components was achieved by DNA-cellulose affinity chromatography of the nucleoid protein fraction. About 4% in weight of the nucleoid proteins are bound to the calf thymus DNA-cellulose. By assaying this fraction for binding of [^{32}P]-*B. subtilis* chromosomal DNA, approximately 20 polypeptides from 8 kD to 48 kD were detected (Fig. 2B). Being present at a low level in the nucleoids, many DNA-binding proteins appear to be detected only after an enrichment by the affinity chromatographic step. Among these nucleoid DBP, we attempted to characterize two classes of structural components, histonelike proteins and membrane bifunctional proteins.

Characterization of a Heat-Stable, Acid-Soluble Protein Fraction (HASP)

Bacterial histonelike proteins are presumed to have specific properties, such as DNA binding, abundance, heat resistance, a positive charge and a small size. These criteria were applied to the bulk of nucleoid proteins. Heat-stable, acid-soluble proteins (HASP) were prepared (see Methods) and assayed for [^{32}P]-DNA binding.

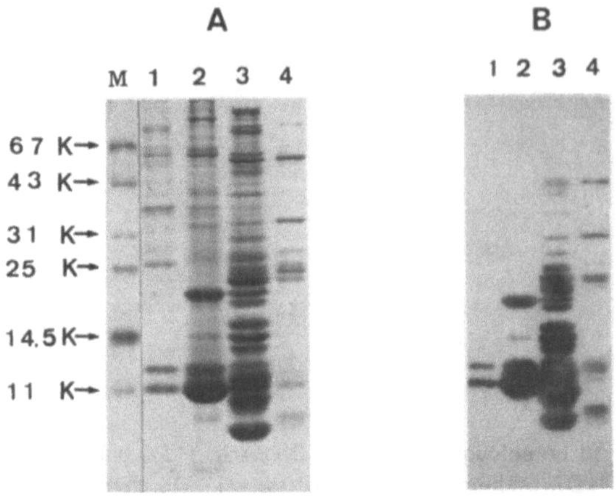

Fig. 2A,B. DSCT DNA-cellulose affinity chromatography of *B. subtilis* total nucleoid proteins. A sample of *B. subtilis* total nucleoid proteins was submitted to DNA-cellulose affinity chromatography. The fractions removed from the column by increasing the NaCl concentration of the elution buffer in four steps were subjected to SDS-PAGE (15% polyacrylamide) and the proteins stained with Coomassie Blue (A). Electroblotting was then used to transfer the protein to a nitrocellulose filter for incubation with [^{32}P]-labelled *B. subtilis* chromosomal DNA. The autoradiogram (B) was exposed for 2 h. Lane numbering same for A and B: *M* marker proteins (see legend to Fig. 1); *lane 1* 0.1 M NaCl eluate; *lane 2* 0.2 M NaCl eluate; *lane 3* 0.4 M NaCl eluate; *lane 4* 1 M NaCl eluate

The electrophoretic pattern of the HASP fraction (Fig. 1A, lane 2) revealed a small number of polypeptides among which four major species of respectively, 8.5, 12, 23, and 26 kD were able to bind either chromosomal or plasmid *B. subtilis* DNA (Fig. 1B and C, lanes 2). Several minor bands could also be detected. Their DNA-binding activity varied between experiments (for example, a 14.5 kD species seen in Fig. 1B, lane 2). One band co-migrates with the lysozyme molecular marker. It corresponds to the exogenous lysozyme used to lyse the cells. In *B. subtilis,* four major HASP meet the criteria defined above and bind indifferently native or denatured heterologous DNA (data not shown). Therefore, they are good candidates for organizing the *B. subtilis* chromosome into bacterial chromatin.

Identification of Membrane DNA-Binding Proteins

These proteins were identified by isolating pure membrane proteins, examining them for DNA-binding activity and then by comparing their binding pattern to that of the nucleoid proteins. The membrane proteins were purified by preparing membrane vesicles following the technique of Kaback (1971) modified by Bohin et al. (1976). This procedure avoids the trapping of unspecific material inside the vesicles by using large volumes of buffer. The DNA-binding experiments were achieved with [^{32}P]-labelled *B. subtilis* chromosomal DNA as described above. The DNA-binding patterns of the membrane proteins (Fig. 3B, lanes 3 and 4) and of the nucleoid proteins (Fig. 3B, lane 2) were clearly different. The membrane proteins which bound most intensively the DNA had respective molecular weights around 42, 14, 11.5, and 11 kD. Four

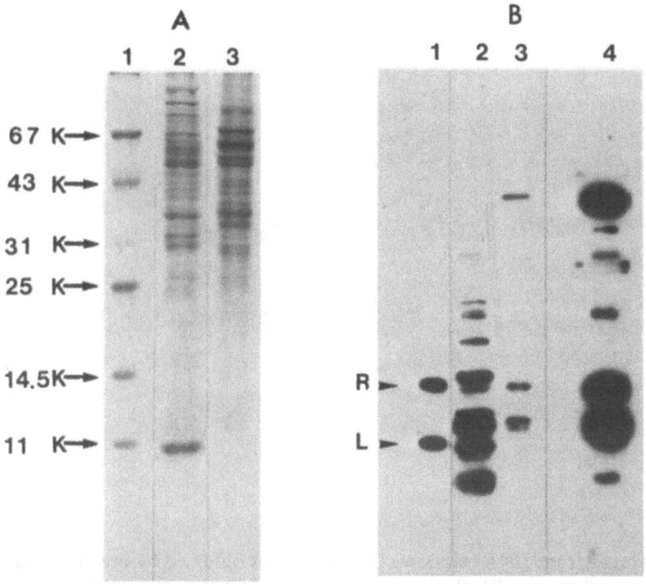

Fig. 3A,B. Binding of *B. subtilis* cytoplasmic membrane proteins to homologous chromosomal DNA. **A** SDS-PAGE (15% polyacrylamide). *Lane 1* Marker proteins (see legend to Fig. 1); *lane 2* total nucleoid proteins; *lane 3* membrane proteins. **B** Autoradiograms of the nitrocellulose filter incubated with [^{32}P]-labelled *B. subtilis* chromosomal DNA. *Lane 1* RNase and lysozyme; *lane 2* total nucleoid proteins; *lane 3* membrane proteins. The autoradiogram was exposed for 6 h; *lane 4* autoradiogram of the membrane proteins exposed for 60 h

other species binding less DNA and of molecular weights 35, 30, 21, and 7.5 kD could also be seen. Regarding their molecular weights, these eight proteins can be related to corresponding nucleoid DBP.

Purification of DBP

In our first attempts to purify nucleoid-associated DBP, low ionic strength nucleoids were used as starting material. In the present work, we turned to *B. subtilis* whole cells, so as to increase the yield of pure proteins. The purification scheme was devised in order to isolate DBP by phosphocellulose and DNA-cellulose affinity chromatography and then to identify the positively charged, heat-stable species by heat and acid treatments. Hemoglobin-Sepharose was used to remove proteases from the crude extract and protease inhibitors were added at each step of the procedure (see Methods). The DNA-binding activity of the purified fractions was examined by DNA-binding assays as described above.

A notable enrichment in several abundant HASP was obtained after phosphocellulose and DNA-cellulose chromatography as shown in Figs. 4 and 5. During the purification of Fraction 1 (see Fig. 4, legend) a 12 kD HASP was found as an almost unique species in the 0.2 M NaCl DNA-cellulose eluate (Fig. 4A and B, lanes 2, 3, 4). However, the 8.5 kD and 26 kD HASP were eluted from DNA cellulose at 0.4 M NaCl together with several other HASP not found in the nucleoids (Fig. 4A and B, lanes 5,

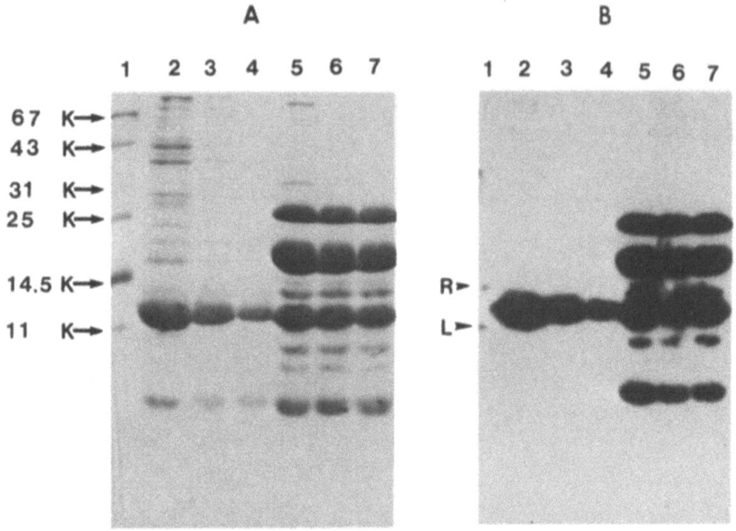

Fig. 4A,B. Fraction 1 HASP eluted from DNA-cellulose. Fraction 1 was obtained by precipitating at 30% saturation of ammonium sulfate the S100 extract of *B. subtilis* 168 M exponentially growing cells. The purification was conducted as follows (for details see Methods). Fraction 1 was first dialyzed and submitted to phosphocellulose chromatography. Proteins eluted at 1 M KCl from phosphocellulose were dialyzed and then applied to a DSCT DNA-cellulose column. Two fractions were successively eluted from the DNA-cellulose with solution 3 containing 0.2 M and 0.4 M NaCl. Protein samples of 200 μg from each fraction were heated at 100°C for 10 min and acidified with HCl at a final concentration of 0.25 M for 15 min in order to obtain the HASP fraction (see Methods). SDS-PAGE was performed on the fractions. Proteins were then transferred to a nitrocellulose filter and tested for DNA binding with [^{32}P]-labelled *B. subtilis* chromosomal DNA. **A** SDS-PAGE (15% polyacrylamide). **B** Autoradiogram exposed for 30 min. Lanes (numbering same for **A** and **B**): *lane 1* marker proteins (see legend to Fig. 1); *lane 2* untreated 0.2 M NaCl eluate; *lane 3* heat-treated 0.2 M NaCl eluate; *lane 4* heat/acid-treated 0.2 M NaCl eluate; *lane 5* untreated 0.4 M NaCl eluate; *lane 6* heat-treated 0.4 M NaCl eluate; *lane 7* heat/acid-treated 0.4 M NaCl eluate

6, 7). In contrast, the similar purification of Fraction 2 lead to a 0.2 M NaCl DNA-cellulose eluate devoid of HASP though rich in DNA-cellulose-bound proteins (Fig. 5A and B, lanes 2, 3, 4). The 0.4 M NaCl DNA-cellulose eluate bears many similarities with its Fraction 1 equivalent, specially the presence of a 26 kD HASP (Fig. 5, lanes 5, 6, 7). The purification of the 8.5, 12, and 26 kD proteins are in progress.

Discussion

In order to obtain more information on the structure of the bacterial genome, DBP associated with the nucleoid of *B. subtilis* were further characterized and a subset of heat-stable, acid-soluble proteins (HASP) was identified. Four of them have low molecular weights: 8.5, 12, 23, and 26 kD. They bind homologous and heterologous,

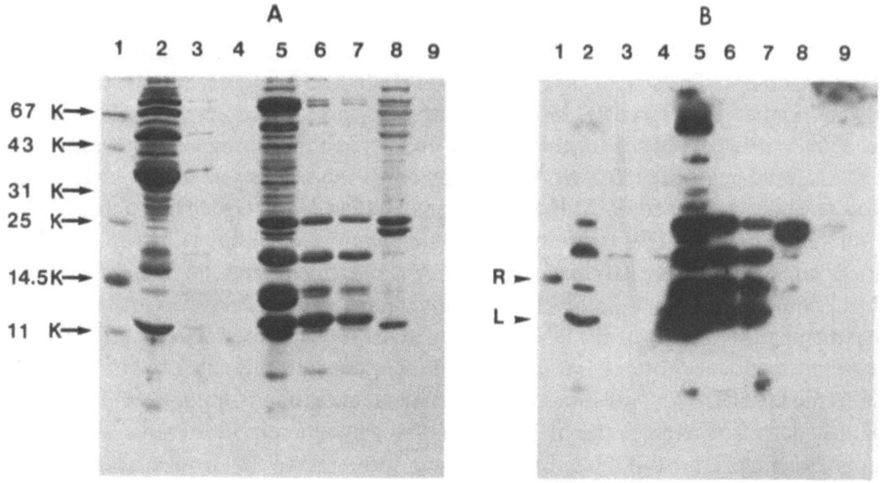

Fig. 5A,B. Fraction 2 HASP eluted from DNA-cellulose. The supernatant of the Fraction 1 was adjusted to 50% saturation of ammonium sulfate. The precipitate (Fraction 2) was submitted to the same purification steps as Fraction 1. Proteins attached to the DSCT DNA-cellulose were eluted in three steps with solution 3 containing respectively, 0.2, 0.4 and 1 M NaCl. HASP were obtained from 200 μg proteins of each fraction (see legend to Fig. 4). **A** SDS-PAGE (15% polyacrylamide). **B** Autoradiogram exposed for 30 min. Lanes (numbering same for **A** and **B**): *lane 1* marker proteins; *lane 2* untreated 0.2 M NaCl eluate; *lane 3* heat-treated 0.2 M NaCl eluate; *lane 4* heat/acid-treated 0.2 M NaCl eluate; *lane 5* untreated 0.4 M NaCl eluate; *lane 6* heat-treated 0.4 M NaCl eluate; *lane 7* heat/acid-treated 0.4 M NaCl eluate; *lane 8* untreated 1 M NaCl eluate; *lane 9* heat-treated 1 M NaCl eluate

native or denatured DNA equally well, and are relatively abundant (33% of the nucleoid HASP). Altogether, they meet the criteria adduced for histonelike proteins. The most abundant one, (HU_{BS}), is immunologically related to the HU protein of *E. coli* (data not shown). It would be interesting to ascertain the homology of HU_{BS} with small basic proteins found in *B. globigii* (Imbert et al. 1982) and *B. stearothermophilus* (Djijk et al. 1983; Dijk and Reinhardt, this volume).

The other three major HASP had not been described previously. At the moment, it is not known whether or not they are related to proteins described in *E. coli* as the HLP protein (Lathe et al. 1980), the H protein (Hübscher et al. 1980), the H1 protein (Spassky et al. 1984; Rimsky and Spassky, this volume) or H-NS (Gualerzi et al., this volume).

With respect to bacterial chromatin, the main question is how DNA is arranged with proteins and other components inside the living cell. Nucleosomelike beaded structures have been detected in electron micrographs of gently lysed cells (Griffith 1976; Materman and Van Gool 1978; Sjåstad et al. 1982). Their diameters varied from 20 to 150 nm. In recent findings obtained by quick freezing followed by thin sectioning, Hobot et al. (1985) observed in *E. coli* and *B. subtilis* nucleoids a variety of fine structures whose sizes are compatible with DNA superstructures, such as supercoils, aggregates and nucleosomes. In *B. subtilis* purified nucleoids, the ratio of the four major HASP to DNA is 0.23 μg μg^{-1} DNA, while it is 1/1 for the histones in eukaryotic chromatin. However, the amount of HASP in the nucleoid could be underevaluated in our calculations by a depletion due to a carrier effect of the denatured proteins during the heat treatment. The size of the nucleosomes could also be different in prokaryotes from that found in eukaryotes. Moreover, several types of proteins, basic and non-basic, could participate in the organization of prokaryotic nucleosome structures.

The bacterial genome is known to be attached to the cell envelope (for review see Leibowitz and Schaechter 1975; Hirschbein and Guillen 1982; Ogden and Schaechter, this volume). However, the nature of the binding site is still unknown. In *B. subtilis,* purified nucleoids are destabilized by detergent treatment and no membrane-free nucleoid could be obtained, indicating that membrane proteins are important factors in maintaining the folding of the DNA (Guillen et al. 1978). In our attempt to identify putative bifunctional proteins, we found eight membrane proteins able to bind *B. subtilis* chromosomal DNA. This type of in vitro characterization had not been described previously despite several studies of DNA-membrane attachment in bacteria. In *E. coli,* the photochemical cross-linking of membrane proteins to BUdr-substituted DNA enabled Protalier and Worcel (1976) to detect an enrichment of the inner membrane proteins in 56 kD and 80 kD species after UV-irradiation. Direct evidence of attachment of the *E. coli* replication initiation point to the outer membrane has been found (Kusano et al. 1984) involving two proteins of 55 kD and 75 kD (Hendrickson et al. 1982). In *B. subtilis,* specific membrane proteins (P15, P22, P28, and P35) appeared to be associated with a DNA fragment belonging to the origin of the chromosome (Sueoka and Hammers 1974). In a DNA-membrane complex isolated by a combination of centrifugation and electrophoresis, Toyoda et al. (1978) detected several proteins (29, 27, and 24 kD) that remained tightly bound to the DNA after sonication of the complex. A membrane-bound DNA sequence containing the *purA* marker was recently cloned by Sargent and Bennett (1985). Comparing the affinity of membrane DBP for such fragments and for the bulk of the DNA should provide a useful method for determining the molecular composition of membrane-binding sites on the *B. subtilis* chromosome. Consequently, after purification, their functional significance should be facilitated.

Acknowledgement. This study was supported by a grant from the CNRS (LA 136). We thank Y. Fontaine and J. Le Dérout for skillful assistance.

References

Alberts B, Herrick G (1971) In: Colowick SP, Kaplan NO (eds) Methods in enzymology, vol 21D. Academic Press, New York, pp 198–217

Bligh EG, Dyer MJ (1959) A rapid method of total lipid extraction and purification. Can J Biochem Physiol 37:911–917

Bohin JP, Bohin A, Schaeffer P (1976) Increased nitrate reductase A activity as a sign of membrane alteration in early blocked asporogenous mutants of *Bacillus subtilis.* Biochimie (Paris) 58:99–108

Bowen B, Steinberg J, Laemmli UK, Weintraub H (1980) The detection of DNA-binding proteins by protein blotting. Nucleic Acids Res 8:1–18

Charret R, Guillen N, Fleury AM, Le Hégarat H, Hirschbein L (1980) Organization of the isolated *Bacillus subtilis* nucleoid. Biol Cell 38:105–114

Dijk J, White SW, Wilson KS, Appelt K (1983) On the DNA binding protein II from *Bacillus stearothermophilus.* J Biol Chem 258:4003–4006

Fleury-Guérout AM (1978) Caractérisation des protéines associées au nucléoide de *Bacillus subtilis.* Thèse de doctorate de 3ème cycle, Université Paris-Sud

Griffith JD (1976) Visualization of procaryotic DNA in a regularly condensed chromatin-like fiber. Proc Natl Acad Sci USA 73:563–567

Guillen N, Le Hégarat F, Fleury AM, Hirschbein L (1978) Folded chromosomes of vegetative *Bacillus subtilis:* composition and properties. Nucleic Acids Res 5:475–489

Hendrickson WG, Kusano T, Yamaki H, Balakhrishnan R, King M, Murchie J, Schaechter M (1982) Binding of the origin of replication of *Escherichia coli* to the outer membrane. Cell 30:915–923

Hirschbein L, Guillen N (1982) Characterization, assay and use of isolated bacterial nucleoids. Methods Biochem Anal 28:297–328

Hobot JA, Villiger W, Escaig J, Maeder M, Ryter A, Kellenberger E (1985) Shape and fine structure of nucleoids observed on sections of ultrarapidly frozen and cryosubstituted bacteria. J Bacteriol 162:960–971

Hübscher U, Lutz H, Kornberg A (1980) Novel histone H2A-like protein of *Escherichia coli.* Proc Natl Acad Sci USA 77:5097–5101

Huerre C, Le Hégarat JC (1979) Etude des clones de *Bacillus subtilis* resistants à la tétracycline obtenus par transformation avec des plasmides recombinés *in vitro.* Microbiologie 1979, (Edns du CNRS) Paris, p 179

Imbert R, Bachinger H, Bickle TA (1982) Purification and characterization of a small DNA-binding protein, HB, from *Bacillus globigii.* Eur J Biochem 122:627–632

Kaback HR (1971) In: Jacob B (ed) Methods in enzymology, vol 22. Academic Press, New York, pp 99–120

Kleppe K, Ovrebö S, Lossius I (1979) The bacterial nucleoid. J Gen Microbiol 112:1–13

Kusano T, Steinmeth D, Hendrickson WG, Murchie J, King M, Benson A, Schaechter M (1984) Direct evidence for specific binding of the replicative origin of the *Escherichia coli* chromosome to the membrane. J Bacgeriol 158:313–316

Laemmli UK (1970) Cleavage of structural proteins during the assembly of the head of bacteriophage T4. Nature 227:680–685

Lathe R, Buc H, Lecocq JP, Bautz EKF (1980) Prokaryotic histone-like protein interacting with RNA polymerase. Proc Natl Acad Sci USA 77:3548–3552

Leibowitz PJ, Schaechter M (1975) The attachment of the bacterial chromosome to the cell membrane. Int Rev Cytol 41:1–28

Litman R (1968) A deoxyribonucleic acid polymerase from *Micrococcus luteus* (*Micrococcus lysodeikticus*) isolated on deoxyribonucleic acid-cellulose. J Biol Chem 243:6222–6233

Materman EC, Van Gool AP (1978) Compact *Escherichia coli* nucleoids in a highly supercoilsed conformation. J Bacterial 135:703–706

Nakayama T, Kurogi Y, Irijura M (1981) Purification of polypeptide P6 with a molecular weight of 6,000 from the vegetative chromosomes of *Bacillus subtilis.* J Bacteriol 145:953–957

Pettijohn DE, Hecht RM (1973) RNA molecules bound to the folded bacterial genome stabilize DNA folds and segregate domains of supercoiling. Cold Spring Harbor Symp Quant Biol 38: 31–34

Portalier R, Worcel A (1976) Association of the folded chromosome with the cell envelope of *E. coli:* Characterization of the proteins at the DNA-membrane attachment site. Cell 8:245–255

Rouvière-Yaniv J, Gros F (1975) Characterization of a novel, low-molecular-weight DNA-binding protein from *E. coli.* Proc Natl Acad Sci USA 72:3428–3432

Rouvière-Yaniv J, Yaniv M, Germond JE (1979) *Escherichia coli* DNA-binding protein HU forms nucleosome-like structure with circular double stranded DNA. Cell 17:265–274

Ryter A, Chang A (1975) Localization of transcribing genes in the bacterial cell by means of high resolution autoradiography. J Mol Biol 98:797–810

Saito H, Muira K (1963) Preparation of transforming deoxyribonucleic acid by phenol treatment. Biochim Biophys Acta 72:619–629

Salti V, Le Hégarat F, Hirschbein L (1985) Isolation and characterization of small heat-stable acid-soluble DNA-binding proteins from *Bacillus subtilis* nucleoids. J Gen Microbiol 131:581–590

Sargent MG, Bennett MF (1985) Amplification of a major membrane-bound DNA sequence of *Bacillus subtilis*. J Bacteriol 161:589–595

Schaeffer P, Ionesco H, Ryter A, Balassa G (1965) La sproulation de *Bacillus subtilis*, étude génétique et physiologique. Col Int CNRS 124:553–563

Sjåstad K, Fadnes P, Kruger PG, Lossius I, Kleppe K (1982) Isolation, properties and nucleolytic degradation of chromatin from *Escherichia coli*. J Gen Microbiol 128:3037–3050

Spassky A, Rimsky S, Garreau H, Buc H (1984) H1a an *E. coli* DNA-binding protein which accumulates in stationary phase, strongly compacts DNA *in vitro*. Nucleic Acids Res 12:5321–5340

Sueoka N, Hammers JM (1974) Isolation of DNA-membrane complex in *Bacillus subtilis*. Proc Natl Acad Sci USA 71:4787–4791

Toyoda H, Yamaguchi K, Yoshikawa H (1978) Chromosome-membrane association in *Bacillus subtilis* IV. Further purification of DNA-membrane complex by using a combination of centrifugation and electrophoresis. J Bacteriol 136:799–802

Towbin H, Staehelin T, Gordon J (1979) Electrophoretic transfer of proteins from polyacrylamide gels to nitrocellulose sheets: procedure and some applications. Proc Natl Acad Sci USA 76: 4350–4354

Varshavsky AJ, Nedospasov SA, Bakayev VV, Bakayeva TG, Georgiev GP (1977) Histone-like proteins in the purified *Escherichia coli* deoxyribonucleoprotein. Nucleic Acids Res 4:2725–2745

Worcel A, Burgi E (1972) On the structure of the folded chromosome of *Escherichia coli*. J Mol Biol 71:127–147

Escherichia coli Protein H1 a Strongly Compacts DNA in Vitro

S. RIMSKY and A. SPASSKY[1]

E. coli H1 protein was first isolated in an early step during the preparation of RNA polymerase (Jacquet et al. 1971; Cukier-Kahn et al. 1972). Since then, evidence that prokaryotic chromosomes are folded in a nucleosomelike structure has led to a search for histonelike proteins and to the development of methods to lyse bacterial cells mildly. Using these techniques, several authors have isolated deoxyribonucleoprotein complexes (DNP) and purified a number of these DNA-binding proteins. Among them, a 15–17 kD protein has been characterized, which is called either protein 1 (Varshavsky et al. 1977) BH1 or B1 (Bakaev 1981) or 16 K (Laine et al. 1984). On the basis of their physicochemical properties, these proteins have been shown to be identical to the protein H1 (Bakaev 1981; Laine et al. 1984). Protein HNS (Lammi et al. 1983; Gualerzi et al., this volume) also appears to be identical to H1 (Gualerzi C., personal communication).

By gentle lysis of *E. coli* cells, using low concentrations of lysozyme and sodium dodecyl sulfate, we have isolated a stable DNP complex of the plasmid pMB9 and a few discrete proteins: essentially core RNA polymerase and 30, 15 and 9 kD species (Busby et al. 1979). The 15 kD species has been identified as H1 protein and the 9 kD as Hu (Rouvière-Yaniv and Gros 1975). Furthermore, the 15 kD plasmid protein, as well as the H1 protein purified by the original method (Jacquet et al. 1971), consists of at least three species H1a, H1b, H1c, differing in their isoelectric points (Fig. 1). In order to determine whether these three species are artefacts of purification, we analyzed radioactively labelled crude cellular extracts (prepared using the same lysis procedure) by two-dimensional electrophoresis. The same three species can be identified in these crude cellular extracts. We have determined the relative amounts of the three species, as a function of cellular growth, by cutting spots from the gel, and counting their [^{35}S]-content. Figure 2 shows the number of copies per cell of H1a, H1b, H1c as a function of cellular growth. The concentration of H1b and H1c does not vary significantly, while there is a marked accumulation of H1a in stationary phase bacteria.

Previous purification of H1 gave rise to the three 15 kD proteins. We have since designed a new purification in which we obtain H1a free of H1b and H1c. The first steps of the H1 preparation devised by Jacquet et al. were retained, up to the step in which the solution was heated to 100°C for 15 min. This step was replaced by affinity chromatography on a phosphocellulose column. By this procedure, protein H1a is

1 Institut Pasteur, 28, rue du Dr. Roux, 75724 Paris 15, France

Bacterial Chromatin
(ed. by C.O. Gualerzi and C.L. Pon)

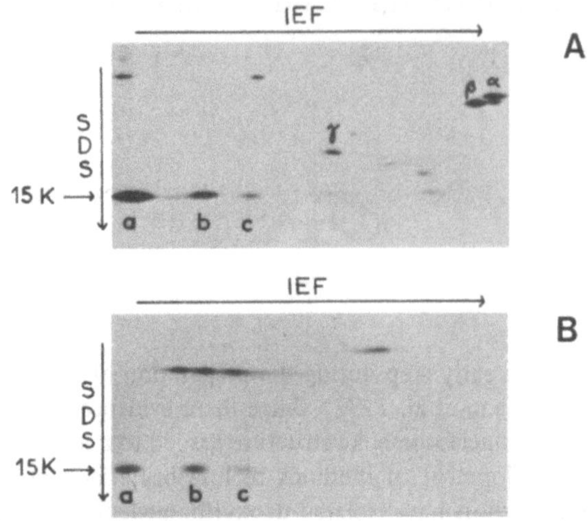

Fig. 1A,B. Two-dimensional gel electrophoresis. Comparison between unlabelled purified H1 protein (A) and the [^{35}S]-labelled proteins present in a plasmid-protein complex (**B**). 10 ml of *E. coli* strain HB101 containing pMB9 plasmid were grown to an absorbance of 0.6 at 600 nm in low sulfate medium containing 0.1 mCi [^{35}S]-sulfate. Cells were lysed and plasmid-protein complexes were purified by the procedure described previously (Busby et al. 1979). 20 μl of plasmid-protein complex (1.3 μg protein/μg DNA) were mixed with pure unlabelled H1 protein and a set of non-radioactive markers (α, β, γ, part of a set of muscle protein extract obtained from R. Whalen, Institut Pasteur). This mixture was analyzed by isoelectric focusing in the first dimension and SDS electrophoresis at the second dimension. Unlabelled proteins are revealed by Coomassie blue staining (A) and [^{35}S]-labelled ones by fluorography (**B**). It has been verified that all the 15 K protein initially present in the mixture is accounted for by the spots present on this gel

eluted between 0.35 and 0.45 molar salt, and it is then purified further on a DEAE-cellulose column. Protein H1a has been obtained to at least 98% purity (Fig. 3).

Using this purified protein, a number of studies in vitro have been performed.

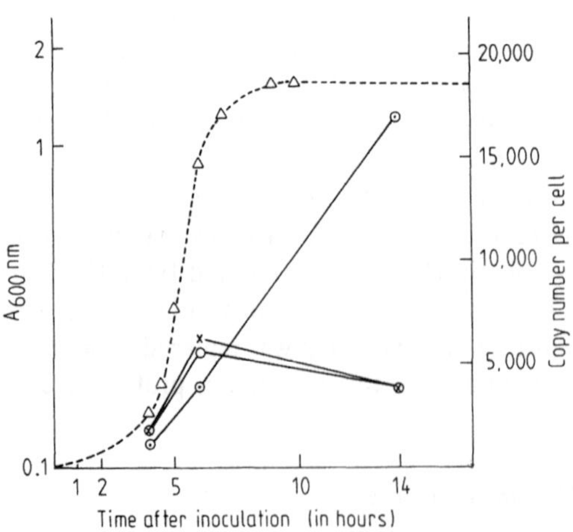

Fig. 2. Content of H1a, H1b, and H1c in crude cellular extracts at various growth phases. *E. coli* MRE600 cells were grown in low sulfate medium, and the growth was monitored by measuring the turbidity (- - △ - -). At the times indicated, 10 ml aliquots of culture were harvested, cellular extracts were prepared, and the content per cell of H1a (⊙) H1b (○), and H1c (x) was determined

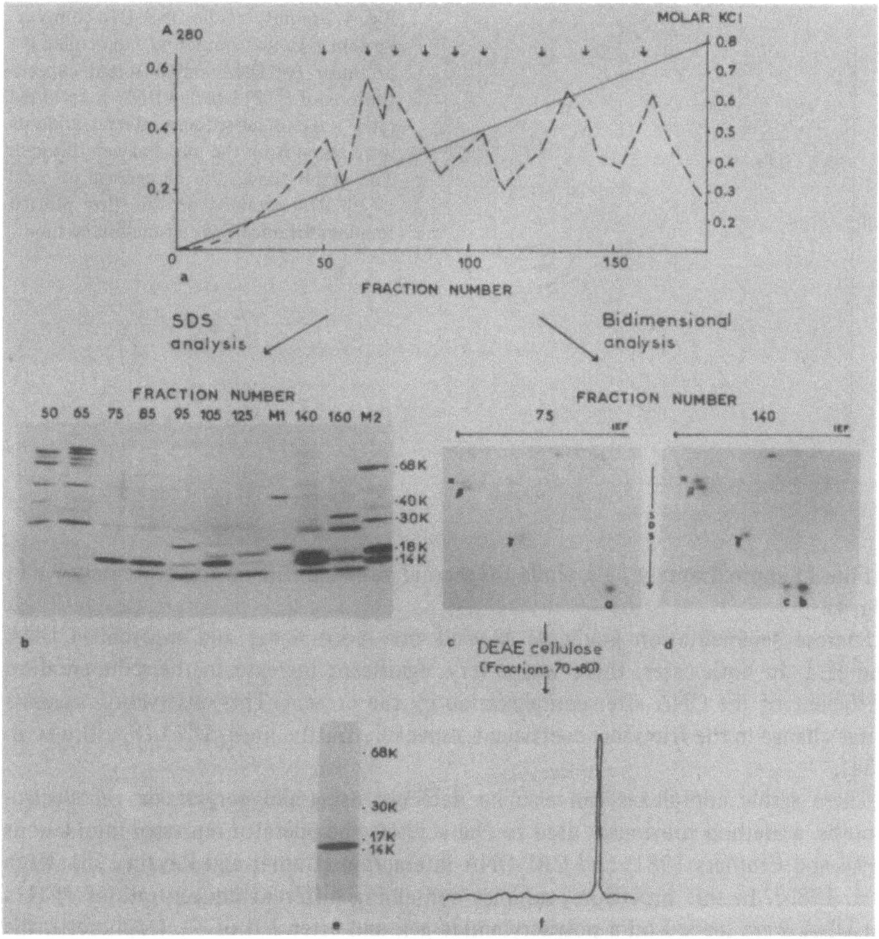

Fig. 3a–f. Purification of H1a. First steps of H1a purification are identical to H1 purification described earlier (Jacquet et al. 1971; Cukier-Kahn et al. 1972) until the heat step is replaced by the phosphocellulose step in phosphate buffer (potassium phosphate 40 mM pH 8, KCl 0.1 M, EDTA 10^{-3} M, glycerol 5%, DTT 10^{-4} M). **a** Phosphocellulose column elution profile by a 0.1–0.8 M linear KCl gradient. **b** Analysis of the different fractions on a SDS 20% acrylamide gel. Two-dimensional analysis gel of fractions 75 (**c**) and 140 (**d**). Fractions 70 to 80, containing H1a (80%) are pooled and loaded on DEAE cellulose in Tris 20 mM pH 8, NaCl 0.1 M, EDTA 10^{-3} M, DTT 10^{-4}. After washing, retained proteins are fractionated by a 0.1–0.5 M NaCl linear gradient. At 0.35 M NaCl, H1a is eluted pure at 98% as shown by the SDS gel (**e**) and the scan of this gel (**f**)

It was shown that H1a has the same affinity for linear as for supercoiled DNA, as measured by nitrocellulose membrane filter assay (Spassky et al. 1984). The filtration method has also been used to check the stability of the DNA-H1a complex. H1a was mixed with radioactive DNA in a 20/1 weight ratio and an excess of unlabelled DNA was then added (Busby et al. 1981). The dissociation of the protein from the DNA was then measured. It appears that the dissociation of the complex is biphasic, suggesting two types of complexes of different stability. The first phase has a

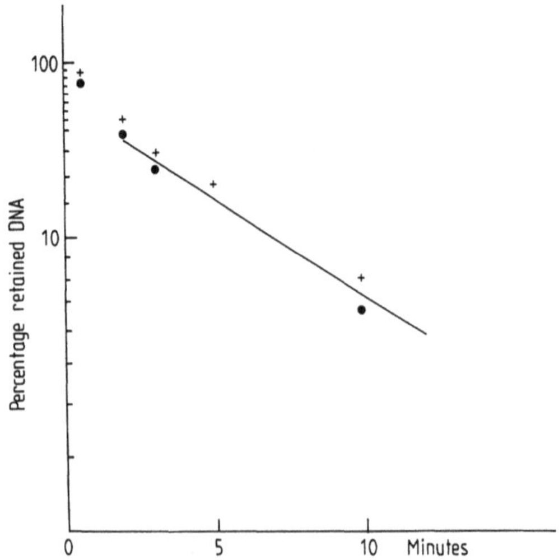

Fig. 4. Stability of the DNA-H1a complex. Protein H1a was bound to supercoiled (+) or linear (•) DNA and different concentrations of [^{32}P]-labelled DNA were added (at 1 = 0); at subsequent interval, aliquots were taken from the mix and were filtered. The graph shows the percentage of total [^{32}P]-DNA retained on the filter plotted semilogarithmically as a function of time

half-life of approximately 15 s, while the second has a half-life of approximately 7 min (Fig. 4).

Sucrose sedimentation gradients showed that both linear and supercoiled DNA bind H1a. In both cases, there was a very significant increase in the sedimentation coefficient of the DNA after complexation by the protein. This observation suggests a large change in the frictional coefficient, especially for the linear DNA (Spassky et al. 1984).

These stable complexes can also be detected using polyacrylamide gel electrophoresis, a method previously used to characterize the operator-repressor interactions (Fried and Crothers 1981) and CRP-DNA interaction (Garner and Revzin 1981; Kolb et al. 1983). In this procedure, samples containing different concentrations of H1a and DNA were loaded on a polyacrylamide gel, and after 2 h of electrophoresis, the position of the DNA was revealed by ethidium bromide staining, or by autoradiography (Figs. 5 and 6). Surprisingly, we observed that the disappearance of the free-DNA band is accompanied by the progressive appearance of a single band of DNA-protein complex (complex A) – no intermediate band can be seen. At higher concentrations of H1a, a second band (complex B) progressively replaced the first one. In order to determine whether these discrete bands correspond to stoichiometric complexes, we performed the experiment by mixing [^{125}I]-H1a with labelled [^{32}P]-DNA fragment, and then measuring the radioactive iodine and phosphorus content in each band. We deduced that the stoichiometry is constant. Six different experiments have shown that one monomer of H1a is bound per 35 ± 10 base pair in complex A. The accurate determination of DNA and protein concentrations in the complex B has been difficult because of the poor resolution of the band. But, the stoichiometric ratio is approximatively the same, and on this basis, we assume that complex B is an aggregate of complex A.

The production of only one complex was unexpected. To explain such a result, it can be assumed that the binding is a cooperative process, and that the most stable structure is made of a precise number of H1a monomers per base pair.

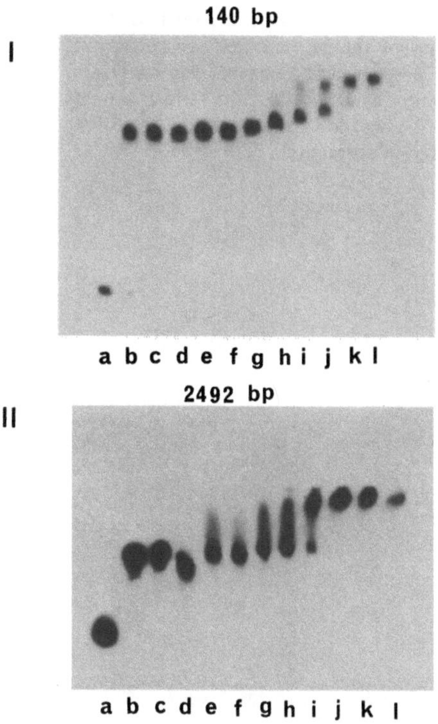

Fig. 5. Visualization of the two complexes A and B on an electrophoretic gel. 15 μl samples containing 18 μg ml^{-1} [^{32}P]-labelled DNA [dimension of fragment used is 140 base pairs) (I) or 2492 pb (II)] and increasing concentrations of protein H1a: (*Lane a* no H1a; *b* 10 μg ml^{-1}; *c* 15 μg ml^{-1}; *d* 20 μg ml^{-1}; *e* 25 μg ml^{-1}; *f* 30 μg ml^{-1}; *g* 35 μg ml^{-1}; *h* 40 μg ml^{-1}; *i* 50 μg ml^{-1}; *j* 80 μg ml^{-1}; *k* 100 μg ml^{-1}; *l* 150 μg ml^{-1}) were applied to a polyacrylamide gel. This figure shows autoradiograms of the gels

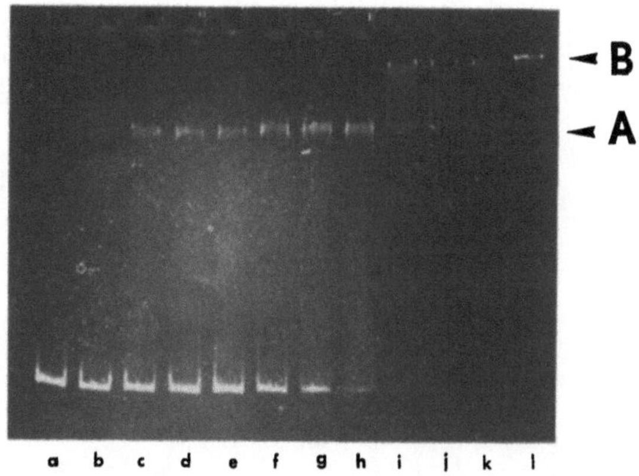

Fig. 6. Visualization of the two complexes A and B on an electrophoretic gel. 15 μl samples containing 18 μg ml^{-1} [^{32}P]-labelled DNA (dimension of fragment used is 203 base pairs) and increasing concentrations of protein H1a: (*Lane a* no H1a; *b* 17 μg ml^{-1}; *c* 22 μg ml^{-1}; *d* 25 μg ml^{-1}; *e* 28 μg ml^{-1}; *f* 32 μg ml^{-1}; *g* 36 μg ml^{-1}; *h* 43 μg ml^{-1}; *i* 46 μg ml^{-1}; *j* 51 μg ml^{-1}; *k* 56 μg ml^{-1}; *l* 63 μg ml^{-1}) were applied to a polyacrylamide gel. In this figure DNA has been revealed after staining by ethidium bromide

The other striking feature revealed by this experiment is that the migration of the complex is slowed down more than would be expected on the basis of the increase in molecular weight or the neutralization of charge, since the isoelectric point of the protein is 7.5. This behaviour is analogous to that observed with CRP, the cyclic adenosine monophosphate receptor protein, complexed at its specific site. In the latter case, the delayed electrophoretic migration has been attributed to the formation of a bend in the DNA induced by the binding of protein (Kolb et al. 1983; Wu and Crothers 1984). As shown on Figs. 5 and 6, the phenomenon is independent of the sequence or the length of the fragment.

The large change in the frictional coefficient revealed both by the sucrose sedimentation experiment and by the delayed electrophoretic migration, led us to conclude that the protein H1a drastically compacts the DNA (Spassky et al. 1984).

To analyze further the H1a-induced structural changes in the DNA, we partially digested complex A with DNase I. We found (Fig. 7) that the protected and exposed sites are completely different in free DNA and in complexed DNA, thus demonstrating that H1a alters the structure of the DNA double helix (Lomonossoff et al. 1981; Drew

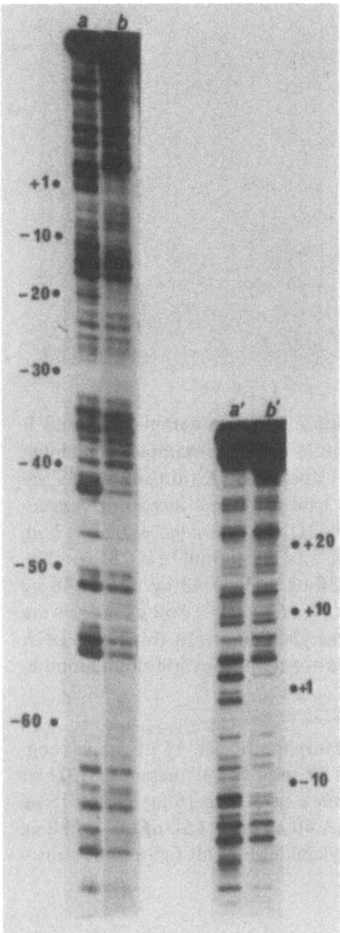

Fig. 7. Footprinting analysis of a 140 base pairs DNA fragment. Autoradiograms of the gel in which the pattern of digestion of the 5′end-labelled fragment was analyzed. *Lane a and a′*: digestion by DNase I of the DNA fragment alone. *Lane b and b′*: digestion by DNase I of the DNA fragment complexed with protein H1a

1984). We obtained the same results using a variety of DNA fragments. No consensus sequence can be found, but the reproducibility of the results shows that the binding of H1a is not aleatory. It is likely that the protein recognizes a particular secondary structure.

Finally, the binding of H1a inhibits the initiation of transcription. We have analyzed this phenomenon using the lactose UV5 promoter in an abortive initiation assay devised by McClure (1980). We have found that the promoter and H1a complex can reversibly associate with RNA polymerase, leading to a closed complex. However, the subsequent step, the isomerization from the closed to the open complex, is slowed down. This implies that RNA polymerase and H1a can form a ternary complex at a promoter. It has been shown elsewhere that in the absence of H1a, this isomerization most probably reflects the specific and stable binding of RNA polymerase at the UV5 promoter. The −35 and −10 regions of this promoter are not easily placed in register during open complex formation. By hindering the rotational flexibility of the DNA, H1a could make this fit even more difficult.

DNA compaction, induced by histones and HU protein, is accompanied by an unwinding of the helix DNA (Germond et al. 1975; Rouvière-Yaniv and Yaniv 1979). We found that only a very slight unwinding of the helix is induced by the binding of H1a. We suggest that H1a could cause alternate positive and negative changes in writhing, which would then lead to a small effect on the degree of superhelicity. It is also possible that the binding of H1a to the DNA results in a change of pitch of the double helix, but that this effect is cancelled out by a change in writhing.

In conclusion, *E. coli* H1a protein compacts significantly the double-stranded DNA in vitro. It could efficiently participate in the packaging of DNA in the bacterial nucleoid.

References

Bakaev VV (1981) Structure of the chromosomal deoxyribonucleoproteins XI. Organization of deoxyribonucleoprotein complex in the bacterial cell. Mol Biol (Mosc) 15:1045–1056

Busby S, Kolb A, Buc H (1979) Isolation of plasmid-protein complexes from *E. coli*. Eur J Biochem 99:105–111

Busby S, Spassky A, Buc H (1981) On the binding of tRNA to *E. coli* RNA polymerase. Interactions between the core enzyme, DNA and tRNA. Eur J Biochem 118:443–451

Cukier-Khan R, Jacquet M, Gros F (1972) Two heat-resistant, low molecular weight proteins from *E. coli* that stimulate DNA-directed RNA synthesis. Proc Natl Acad Sci USA 69:3643–3647

Drew HR (1984) Structural specificities of five commonly used DNA nucleases. J Mol Biol 176: 535–557

Fried M, Crothers DM (1981) Equilibria and kinetics of lac repressor-operator interactions by polyacrylamide gel electrophoresis. Nucleic Acids Res 9:6505–6525

Garner MM, Revzin A (1981) A gel electrophoresis method for quantifying the binding of proteins to specific DNA regions: application to components of the *E. coli* lactose operon regulatory system. Nucleic Acids Res 13:3047–3060

Germond JE, Hirt B, Oudet P, Gross-Bellard M, Chambon P (1975) Folding of the DNA double helix in chromatin like structures from simian virus 40. Proc Natl Acad Sci USA 72:1843–1847

Jacquet M, Cukier-Kahn R, Pla J, Gros F (1971) A thermostable protein factor acting *in vitro* DNA transcription. Biochem Biophys Res Commun 45:1597–1607

Kolb A, Spassky A, Chapon P, Blazy B, Buc H (1983) On the different binding affinities of CRP at the *lac, gal* and *mal* T promoteur regions. Nucleic Acids Res 11:7833–7852

Laine B, Sautière P, Spassky A, Rimsky S (1984) A DNA-binding protein from *E. coli*. Isolation, characterization and its relationship with proteins H1 and B1. Biochem Biophys Res Commun 119:1147–1153

Lammi M, Paci M, Pon CL, Losso MA, Miano A, Pawlik RT, Gianfranceschi GL, Gualerzi CO (1983) In: Hübscher H, Spadari S (eds) Protein involved in DNA replication. Plenum, New York

Lomonossoff GP, Butler PJG, Klug A (1981) Sequence-dependent variation in the conformation of DNA. J Mol Biol 149:745–760

McClure W (1980) Rate-limiting steps in RNA chain initiation. Proc Natl Acad Sci USA 77:5634–5638

Rouvière-Yaniv J, Gros F (1975) Characterization of a novel, low-molecular-weight DNA binding protein from *E. coli*. Proc Natl Acad Sci USA 72:3428–3432

Rouvière-Yaniv J, Yaniv M (1979) *E. coli* DNA binding protein HU forms nucleosome-like structure with circular double strand DNA. Cell 17:265–274

Spassky A, Rimsky S, Garreau H, Buc H (1984) H1a, an *E. coli* DNA-binding protein which accumulates in stationary phase, strongly compacts DNA in vitro. Nucleic Acids Res 12:5321–5340

Varshavsky AJ, Nedospasov SA, Bakayev VV, Bakayeva TG, Georgiev GP (1977) Histone-like proteins in the purified *E. coli* deoxyribonucleoprotein. Nucleic Acids Res 4:2725–2726

Wu HM, Crothers DM (1984) The locus of sequence-directed and protein-induced DNA binding. Nature 308:509–513

The Archaebacterial Histone "HTa"

D.G. SEARCY[1]

1 Introduction

The extreme thermo-acidophile *Thermoplasma acidophilum* was first described by Darland et al. (1970). It was isolated from a "spontaneously burning" tailings pile of low-grade coal refuse outside of the Friar Tuck Mine in Indiana, U.S.A. Optimal culture conditions for the organism are 59°C and pH 1–2. It has also been isolated from a natural acidic hot spring in Japan (T. Oshima, personal commun.). *T. acidophilum* is clearly an Archaebacterium, but apparently is not closely related to other known Archaebacteria (Stetter and Zillig 1985).

Although it is thermophilic, *T. acidophilum* was originally described to have DNA containing only 25% G+C (guanosine plus cytosine) (Darland et al. 1970). Since a low G+C content would make the DNA more easily denatured by heat, and yet the organism is a thermophile, we anticipated that a histonelike protein might be present that would stabilize the DNA. Such a protein was indeed found, although later it was discovered that the G+C content of the DNA is higher than reported originally (Searcy and Doyle 1975).

This paper reviews our work on the *T. acidophilum* protein in the context of its relationship to other reported bacterial and eukaryotic histones. Also included are some recent observations on the unusual fluorescence properties of this protein.

1.1 *T. acidophilum* Chromatin

The nucleoprotein of the Archaebacterium *T. acidophilum* is different from that of other known prokaryotes. When the nucleoprotein was isolated from this organism by either gel filtration or centrifugation in 0.2 M NaCl, there was an abundant histone-like protein (HTa) tightly bound to the DNA (Searcy 1975). The protein remained bound to DNA in ionic concentrations up to 0.7 M NaCl (DeLange et al. 1981a), which is more than ten times above the physiological intracellular ionic concentration of about 50 mM K^+ (Searcy 1976). When the total chromosomal proteins were analyzed on SDS-polyacrylamide gels, HTa was the only protein band visible unless the gel was deliberately overloaded (Searcy and Stein 1980). The protein was bound to DNA in a ratio of about 0.42 (w/w). Overall, the situation resembles that of eukaryotic

1 Department of Zoology, University of Massachusetts, Amherst, MA 01003, USA

Bacterial Chromatin
(ed. by C.O. Gualerzi and C.L.Pon)

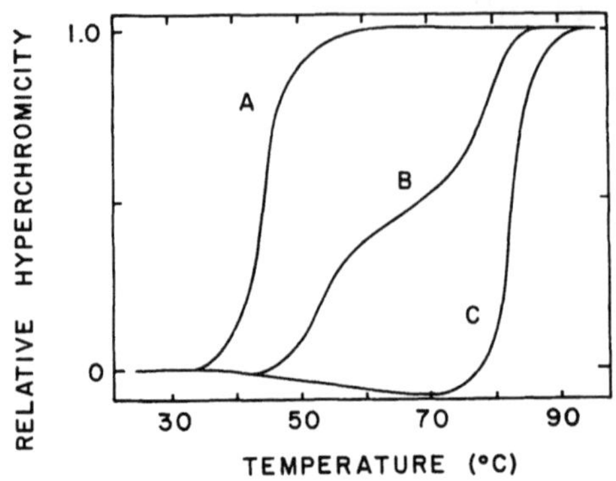

Fig. 1. Thermal denaturation of DNA or DNA complexed with HTa. Calf thymus DNA was analyzed either alone (*A*), half saturated with the protein HTa (*B*), or fully saturated (*C*). The preparation was dissolved in 0.25 mM Na_3-EDTA, pH 8. The Y-axis is the normalized increase in absorbance at 260 nm. (Adapted from Stein and Searcy 1978)

chromatin, where histones are typically abundant, tightly bound, and condense DNA into chains of globular particles.

In contrast, when *Escherichia coli* deoxynucleoprotein was purified in 0.2 M NaCl by the same procedure, there were no proteins bound to the DNA in similar abundance (Searcy 1975). Previous studies of the *E. coli* protein HU (also named NS) have typically used buffers of low ionic concentration. Such conditions can result in the binding of nonchromosomal proteins to the DNA (Johns and Forrester 1969). Even at low ionic strength the putative *E. coli* histone was not conspicuously abundant in the purified nucleoprotein unless an additional step, such as heat, was used to denature and remove the other proteins present (Rouvière-Yaniv 1978). The *E. coli* protein HU is not abundant in the chromosome in a way similar to that of *T. acidophilum* HTa.

In other studies putative eubacterial histonelike proteins have been identified by their affinity for DNA-cellulose rather than by their presence in the chromosome. This can lead to misleading results. For example, in the organism *Sulfolobus acidocaldarius* we found that in cell extracts the most tightly binding protein to DNA-cellulose was not a chromosomal protein (Green et al. 1983). Affinity for DNA-cellulose is not a good criterion for the identification of chromosomal proteins.

The explanation for the occurrence of the histonelike protein in *T. acidophilum* appears to be that it is an adaptation to heat in the absence of a cell wall. Without a cell wall, the organism must be close to equilibrium osmotically between its cytoplasm and the external environment. Typically, *T. acidophilum* is about 50 mM K^+ inside, and can be as low as 18 mM K^+ in more dilute culture conditions (Searcy 1976). Since low ionic concentrations are destabilizing to the double-helical structure of DNA, *T. acidophilum* lives at temperatures where it is at risk of DNA denaturation. The protein HTa can stabilize DNA by up to 40°C, as shown in Fig. 1 (see also Stein and Searcy 1978). Therefore, we think that the protein HTa evolved as an adaptation to protect and stabilize the DNA. Since it precipitates the DNA and replaces other cations normally present, it may serve also an osmotic function by reducing the intracellular osmotic pressure.

In contrast, other thermophilic prokaryotes typically have rigid cell walls, and therefore can have high salt concentrations inside without bursting. For example, *Methanobacterium thermoautotrophicum* contains 1.1 M K^+ (Jarrell et al. 1984). A high salt concentration by itself is sufficient to protect DNA against thermal denaturation. Presumably, walled organisms have found it easier to adapt to heat by increasing intracellular salt than to evolve a tightly-binding histonelike protein.

1.2 Relation of HTa to Eukaryotic Histones and to the *E. coli* Protein HU

The amino acid sequence of HTa has been determined (DeLange et al. 1981b). There is a significant amount of sequence homology with eukaryotic histones. For example, the sequence of the N-terminal 24 amino acid residues of the protein was 25% homologous with a sequence in calf histone H2A (Searcy and DeLange 1980). This was statistically significant at the level of $P < 0.003$. Examining only the remaining nonidentical amino acid residues in these same sequences, the genetic distance in terms of base-pair substitutions was also significantly less than the random expectation ($P < 0.004$). Since these two tests are independent, the overall probability can be calculated as the product of the two tests ($P < 10^{-5}$).

The above sequence in calf histone H2A has also been found in other eukaryotic histones, including H2B, H3, and H4 (Reeck et al. 1978). Since this sequence is not particularly basic, it seems unlikely to be in extensive contact with the DNA. Therefore, it should not be the product of evolutionary convergence due to the steric requirements of DNA binding (in contrast, for example, to repeating patterns in certain sperm protamines). Instead, the sequence is probably the vestige of an ancestral gene. If so, then *T. acidophilum* and eukaryotic cells are evolutionarily related. Such a relationship is suggested also by a variety of other observations (Searcy et al. 1981; Searcy 1982).

At the same time, the amino acid sequence of the protein HTa is 27% homologous with the *E. coli* histonelike protein HU (DeLange et al. 1981b). Thus, the thermoplasma protein HTa is the link that can relate eukaryotic histones to the bacterial histonelike proteins. Nevertheless, as described earlier, the protein HU appears not to be bound abundantly to the chromosome, and may have a function different from that of eukaryotic histones.

1.3 Nucleoprotein Subunit Structure in *T. acidophilum*

In addition to sequence homology, the protein HTa resembles eukaryotic histones in that it can protect associated DNA from digestion by staphylococcal nuclease (Searcy and Stein 1980), unlike the *E. coli* protein HU (Rouvière-Yaniv et al. 1979). With *T. acidophilum* nucleoprotein, nuclease digestion resulted in protected segments of DNA each 40 base pairs in length. By electron microscopy, globular nucleoprotein particles 5.5 nm in diameter were seen. With the addition of each nucleoprotein particle, the overall free contour length of the DNA molecule was reduced by 40 base pairs. In closed circular DNA molecules there was approximately one supercoil introduced into the DNA by each nucleoprotein particle (D. Stein and D. Searcy, unpublished). Both protein cross-linking studies and measurements of the ratio of protein to DNA indicated that each HTa tetramer was associated with 40 base pairs of DNA (Searcy and Stein 1980).

The observations suggested the structural model shown in Fig. 2. Forty base pairs of DNA are wrapped around a globular core of protein. Overall, the structure is similar to that of eukaryotic nucleosomes except that there is only about one-third as much

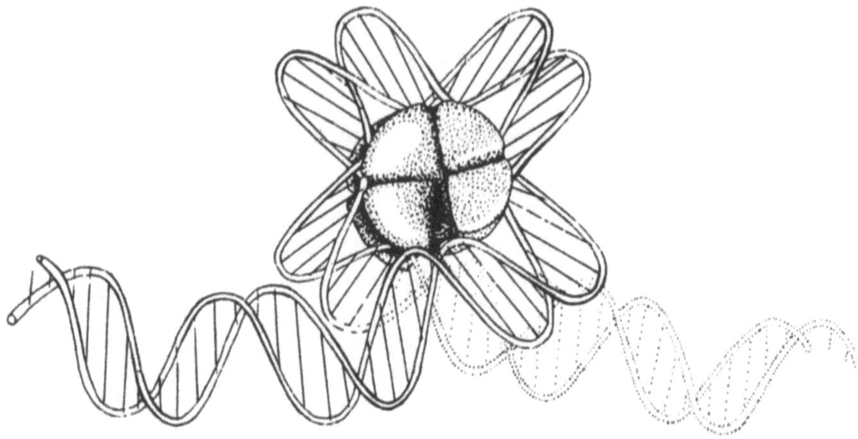

Fig. 2. Drawing of the *T. acidophilum* nucleoprotein particle. Each particle consists of 40 base pairs of DNA coiled around a tetrameric core of the protein HTa. [Reprinted from Searcy (1982) copyright Elsevier Biomedical Press]

DNA. The remarkably small size of these particles has been confirmed by low-angle X-ray scattering analysis (Notbohm 1982). Indeed, the DNA must be bent more sharply than is theoretically possible.

The mechanism of how the DNA is coiled is under investigation. Recent observations by circular dichroism and by absorption difference spectrophotometry suggest that the DNA is distorted into an "A"-like conformation, in contrast to the "B" form of DNA that is usual in solution. In addition, there are tyrosine residues that interact closely with the DNA, as described below, and these might stabilize kinks in the DNA by inserting into the double helix.

2 Recent Physical and Chemical Studies

2.1 Fluorescence Properties of HTa

Since the protein HTa lacks cysteine, histidine, and tryptophan, the single tyrosine and five phenylalanine residues in each protein molecule are particularly accessible to study by fluorescence techniques.

At excitation wavelengths that excite primarily the phenylalanine residues (252 nm), the corrected emission spectrum of the protein showed that most of the fluorescence emission comes only from the tyrosine residue. Moreover, the intensity of the tyrosine fluorescence was 210% of that expected of tyrosine alone. The explanation is that there appears to be fluorescent energy transfer from the phenylalanines to the tyrosine.

Unexpectedly, when excited at a wavelength of 252 nm the tyrosine residue had two fluorescence lifetimes (see Table 1). The shorter lifetime was 3.9 ns, and can be attributed to the normal direct route of excitation and emission of the tyrosine residue. The second lifetime was 12.1 ns, and is due to energy transfer from the

Table 1. Fluorescence lifetime data

Preparation	Excitation wavelength (nm)	Residue excited	Emission wavelength (nm)	Residue detected	Lifetime (ns)
HTa	282	tyr	330	tyr	4.0
	252	tyr (27%) phe (73%)	280	phe	12.0
	252	tyr (27%) phe (73%)	330	tyr	3.9 (35%) 12.1 (65%)
Tyrosine	282	tyr	330	tyr	3.3
	252	tyr	330	tyr	3.3
Phenylalanine	252	phe	280	phe	7.0

Conditions: HTa (160 μM), tyrosine (160 μM), or phenylalanine (800 μM) was dissolved in 0.75 M NaCl, 0.01 M sodium cacodylate, pH 7.0. Lifetimes were measured at 20°C by the single-photon technique using an Edinburgh Instruments Ltd. spectrofluorometer.

phenylalanine. This longer lifetime is the excited-state decay of the phenylalanine residues "seen" through the tyrosine residue. It is observable because the tyrosine comes into a steady-state relationship between accepting energy from the excited phenylalanines and re-emission of that energy by fluorescence.

In the native protein the fluorescence lifetimes of both the tyrosyl and phenylalanyl residues are longer than those of the free amino acids. This suggests that in the protein both types of amino acid residues are well shielded from quenching effects. The long lifetimes depend upon the native conformation of the protein, since when the protein was denatured in 6 M guanidinium chloride both lifetimes decreased to values more nearly typical of those of other proteins. In parallel with the lifetimes, the net intensity of the fluorescence in 6 M guanidinium chloride decreased also to 49% of that of the native protein.

2.2 Phosphate Binding by HTa

While doing the above studies, we observed that dilute phosphate buffers quenched the tyrosine fluorescence of the protein. This was unexpected because normally phosphate is considered to be a weak quenching agent. Further analysis showed that phosphate ions are bound by the protein HTa. The affinity constant for phosphate depended upon the concentrations and types of other anions present. For example, at pH 7.8 in 5 mM Cl^- the binding constant for phosphate was 1080 M^{-1}, but decreased to 1.75 M^{-1} when 2 M NaCl was added. As shown in Fig. 3, the Scatchard plots of the binding data were nonlinear at low concentrations of phosphate. This indicates that when phosphate is bound there is a cooperative effect, probably associated with a conformational change in the protein. Consistent with this possibility, when phosphate was bound there were also changes in the ultraviolet absorption spectrum of the

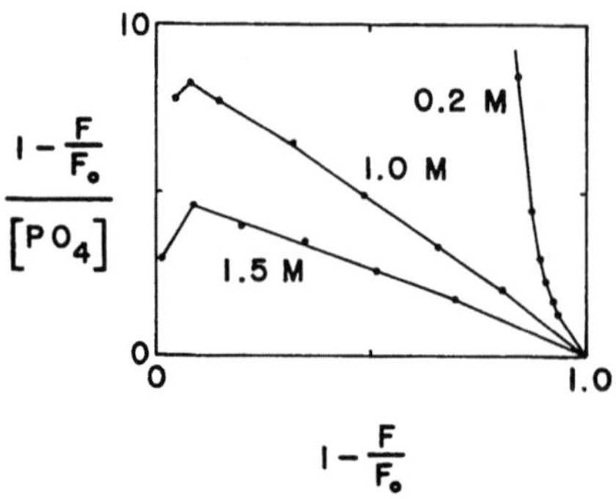

Fig. 3. Scatchard-type plot of the effect of phosphate on the fluorescence of the protein HTa. The protein was dissolved in 10 mM Tris-chloride buffer, pH 7.8, with either 0.2 M, 1.0 M, or 1.5 M NaCl, as indicated in the figure

protein. These included a red-shift of the phenylalanine absorption bands, and suggested a more compact folding of the protein core. Presumably, the same sort of conformational change may occur also when the protein binds to DNA.

2.3 Environment of the Fluorescent Tyrosine Residue

2.3.1 Collisional Quenching Studies

The observations described above suggested that the tyrosine residue is near a phosphate-binding site. If this is also the DNA-binding site, then one can predict that the tyrosine residue should be on the surface of the protein and near a location that is rich in basic amino acid residues. Thus, the environment of the tyrosine residue was analyzed by means of collisional quenching agents and pH titration, as described below.

Trichloroethanol, which is a nonionic collisional quenching agent, was fully effective in quenching the tyrosine fluorescence (Table 2). It is typical that quenching coefficients for surface residues in proteins are about 30% of those for the free amino acid (Eftink and Ghiron 1981). The data confirm that the fluorescent tyrosine residue is accessible and on the surface of the native HTa molecule.

Ionic quenching agents, on the other hand, gave differing results depending upon the signs of their electrical charge. For example, I^- was very effective in quenching the protein's fluorescence, but Cs^+ was almost entirely ineffective (Table 2). As a control, the quenching coefficients can be compared to the effect of each agent upon free tyrosine. The data indicate that in the native protein the tyrosine residue is "guarded" by a field of positive ionic charge (for another example see Homer and Allsopp 1976). This was not due simply to the net positive charge of the protein, since when the protein was denatured in 6 M guanidinium chloride, the quenching coefficient of Cs^+ increased to 365% of the native protein, but that of I^- decreased to 69%. Thus, the existence of the positive ionic field surrounding the tyrosine is dependent upon the native folded conformation of the protein. This is consistent with the primary structure of the protein, which shows that the tyrosine residue is not in or near a strongly basic sequence of amino acids (DeLange et al. 1981b).

It should be pointed out that the quenching measurements on the native protein were carried out in 0.75 M NaCl. In this relatively high ionic concentration, electrical

Table 2. Stern-Volmer quenching coefficients

	Preparation		Ratio:
Quenching agent	HTa	Tyrosine	HTa/tyrosine
Trichloroethanol	6.8	16	0.43
I^-	5.1	15	0.34
Cs^+	0.26	3.4	0.08
$H_{1.5}PO_4^{1.5-}$	9.2	3.9	2.36

Conditions: HTa or tyrosine (25 μM) were dissolved in 0.75 M NaCl, 1 mM EDTA, and 10 mM sodium cacodylate, pH 7.0. Excitation was at 275 nm and fluorescence at 303 nm. The Stern-Volmer coefficients were extrapolated to zero concentration of quencher, and are in units of M^{-1}. For more information on this technique see Eftink and Ghiron (1981).

charge effects should usually be screened out by the dissolved salts. Therefore, the positive ionic field on the protein surrounding the tyrosine is quite substantial.

2.3.2 pH Titration Studies

If the tyrosine residue is indeed surrounded by a fixed positive charge, then an additional prediction is that the pK_a of the tyrosine residue should be lower than normal. When the tyrosine is titrated, its ionization can be detected by either fluroescence or absorption techniques. Unexpectedly, these techniques each yielded differing results, as shown in Fig. 4.

Both the fluorescence and absorbance data agree in the initial part of the titration curve. In this part of the curve, the loss of tyrosine fluorescence correlates with the appearance of tyrosinate ion absorbance. The data indicate that the pK_a of the tyrosine is about 9.4; we believe this to be an accurate estimate of the true pK_a of the

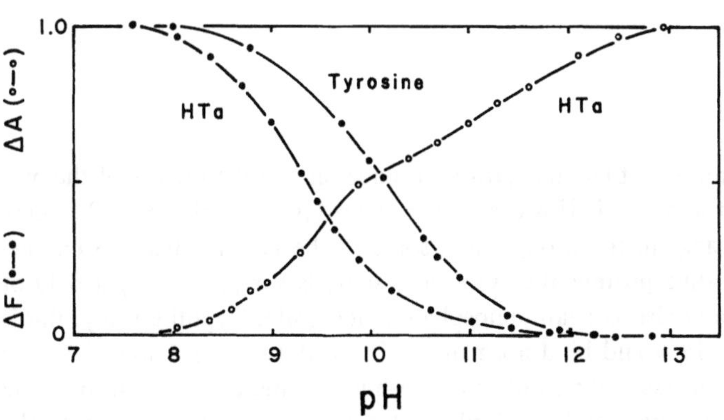

Fig. 4. Optically measured titration curves of the protein HTa and of free tyrosine. Fluorescence (•–•): ionization was measured by the decrease in fluorescence of nonionized tyrosine at 304 nm (excitation at 275 nm). Absorbance (o–o): tyrosinate ion was measured by the increase in absorbance at 300 nm. Each solution contained 25 μM tyrosine (or HTa) in 0.75 M NaCl, 10 mM Tris-chloride, pH 7, and was titrated by the addition of 5 M NaOH. The data were corrected for the "sodium error" of the pH electrode, and were normalized between pH 7 and pH 13

tyrosine in the native protein. This compares to pK_a = 10.2 for free tyrosine and to pK_a = 9.8–10.2 for surface tyrosyl residues in other typical proteins (Gurd and Wilcox 1956; Sela and Katchalski 1956). Apparently, the pK_a of the tyrosyl residue in HTa is lower than that for any other protein yet described.

At higher pH values the data become more complicated. For example, at pH 11 the absorbance measurements indicate that about one-third of the tyrosine residues are not yet ionized, but the tyrosine fluorescence is almost entirely gone. The explanation appears to be that near pH 9.5 the protein undergoes a conformational change; the nonionized tyrosine residues become buried in the hydrophobic interior of the protein, and in this environment resist ionization until quite high pH values are reached (Tanford et al. 1956). Moreover, in the hydrophobic core of the protein the tyrosine fluorescence is quenched by close proximity to peptide bonds (Cowgill 1976).

Consistent with this interpretation, at pH 10.2 the ultraviolet absorption peak of the tyrosyl residue was shifted to 284 nm. At either pH 7 or pH 13 this peak was not detectable: at pH 7 the tyrosine absorption peak was at 276 nm, and at pH 13 only the normal tyrosinate peak at 293 nm was detected. This is consistent with previous model studies of Cowgill (1976) and of Strickland et al. (1972), who showed that in a hydrophobic environment a hydrogen bond can form between a phenolic group and a nearby amide group. This results in a positive shift of the phenolic absorption spectrum, and also in the complete quenching of its fluorescence.

Nevertheless, the evidence is that at pH 7 the tyrosine residue is on the surface of the protein. The quantum yield of the tyrosine fluorescence is 0.92 relative to free tyrosine, which is so high that probably all of the tyrosine residues in the tetrameric protein complex are fluorescent. Also, the studies with collisional quenching reagents indicated that all of the fluorescent residues are surface-accessible. Finally, all of the tyrosine residues were accessible to digestion by the enzyme carboxypeptidase, and could be removed without disrupting the native core structure of the protein (D. Searcy, unpublished data).

3 Conclusion

Our present working model is that the tyrosine residues are on the surface of the protein and have a pK_a of about 9.4. However, when the pH is raised above ca. 9.5 there is a conformational change in the protein, and some of the tyrosine residues can become buried. In the native protein the tyrosine residue is surrounded by a field of fixed positive charges and also is near a phosphate-binding site. Hypothetically, this is the site where the protein would bind normally to DNA. It is not yet known whether the tyrosine residue is an essential part of the DNA-binding site, but one can speculate that the tyrosine residues might stabilize kinks in the DNA by base-stacking into the DNA helix. This possibility is currently under investigation.

The unusual fluorescence properties of the protein are partially a consequence of its amino acid composition, which in *T. acidophilum* is an adaptation to heat. In contrast to other types of protein-folding interactions, hydrophobic interactions become strongest at about 60°C (Brandts 1967). Thus, the protein HTa has evolved an exceptional hydrophobic core of 20 phenylalanine residues (in each tetrameric protein complex). The protein can be imagined as a "super oil-drop", where hydrophobic interactions provide the protein with both strength and flexibility. The flexibility may be important for conformational changes when the protein binds to DNA, as were suggested by the model studies when the protein bound to phosphate.

Finally, the point should be made that Archaebacteria, because of their distant relationship to other organisms and their adaptations to extreme environments, can provide useful materials for physical and chemical study. For example, in *T. acidophilum* adaptation to heat appears to have resulted in the evolution of cellular components that are often unusually small and stable. In addition, certain general principles of molecular structure can be developed to their greatest extent in these organisms, as illustrated in the present study by the hydrophobic core in the protein HTa. Other groups of Archaebacteria are adapted to different extreme environments, and can be expected to have different, but perhaps equally interesting adaptations.

Acknowledgments. Some of the work reported here was done while DGS was supported by the French CNRS in the laboratory of Prof. Claude Hélène, Laboratoire de Biophysique, Muséum National d'Histoire Naturelle, Paris. I thank also Thérèse Montenay-Garestier and J.-J. Toulmé for their support and criticism, Steven Searcy for help with the pH titration experiments, and David J. Laston for the collisional quenching data.

References

Brandts JF (1967) Heat effects on proteins and enzymes. In: Rose AH (ed) Thermobiology. Academic Press, New York, pp 25–72

Cowgill RW (1976) Tyrosyl fluorescence in proteins and model peptides. In: Chen RF, Edelhoch H (eds) Biochem fluorescence concepts, vol 2. Dekker, New York, pp 441–486

Darland G, Brock TD, Samsonoff W, Conti SF (1970) A thermophilic, acidophilic mycoplasma isolated from a coal refuse pile. Science 170:1416–1418

DeLange RJ, Green GR, Searcy DG (1981a) A histone-like protein (HTa) from *Thermoplasma acidophilum*. I. Purification and properties. J Biol Chem 256:900–904

DeLange RJ, Williams LC, Searcy DG (1981b) A histone-like protein (HTa) from *Thermoplasma acidophilum*. II. Complete amino acid sequence. J Biol Chem 256:905–911

Eftink MR, Ghiron CA (1981) Fluorescence quenching studies with proteins. Anal Biochem 114: 199–227

Green GR, Searcy DG, DeLange RJ (1983) Histone-like protein in the Archaebacterium *Sulfolobus acidocaldarius*. Biochim Biophys Acta 741:251–257

Gurd FRN, Wilcox PE (1956) Complex formation between metallic cations and proteins, peptides, and amino acids. Adv Protein Chem 11:311–427

Homer RB, Allsopp SR (1976) An investigation of the electronic and steric environments of tyrosyl residues in ribonuclease A and *Erwinia cartovora L*-asparaginase through fluorescence quenching by caesium, iodide, and phosphate ions. Biochim Biophys Acta 434:297–310

Jarrell KF, Sprott GD, Matheson AT (1984) Intracellular potassium concentration and relative acidity of the ribosomal proteins of methanogenic bacteria. Can J Microbiol 30:663–668

Johns EW, Forrester S (1969) Studies on nuclear proteins. The binding of extra acidic proteins to deoxyribonucleoprotein during the preparation of nuclear proteins. Eur J Biochem 8:547–551

Notbohm H (1982) Low-angle X-ray scattering analysis of the *Thermoplasma acidophilum* nucleoprotein subunit. Biochim Biophys Acta 696:223–225

Reeck GR, Swanson E, Teller DC (1978) The evolution of histones. J Mol Evol 10:309–317

Rouvière-Yaniv J (1978) Localization of the HU protein on the *Escherichia coli* nucleoid. Cold Spring Harbor Symp Quant Biol 42:439–447

Rouvière-Yaniv J, Yaniv M, Germond JE (1979) *E. coli* DNA binding protein HU forms nucleosome-like structure with circular double-stranded DNA. Cell 17:265–274

Searcy DG (1975) Histone-like protein in the prokaryote *Thermoplasma acidophilum*. Biochim Biophys Acta 395:535–547

Searcy DG (1976) *Thermoplasma acidophilum:* intracellular pH and potassium concentration. Biochim Biophys Acta 451:278–286

Searcy DG (1982) *Thermoplasma:* a primordial cell from a refuse pile. Trends Biochem Sci 7: 183–185

Searcy DG, Doyle EK (1975) Characterization of *Thermoplasma acidophilum* deoxyribonucleic acid. Int J Syst Bacteriol 25:286–289

Searcy DG, DeLange RJ (1980) *Thermoplasma acidophilum* histone-like protein. Partial amino acid sequence suggestive of homology to eukaryotic histones. Biochim Biophys Acta 609: 197–200

Searcy DG, Stein DB (1980) Nucleoprotein subunit structure in an unusual prokaryotic organism: *Thermoplasma acidophilum*. Biochim Biophys Acta 609:180–195

Searcy DG, Stein DB, Searcy KB (1981) A mycoplasma-like Archaebacterium possibly related to the nucleus and cytoplasm of eukaryotic cells. Ann NY Acad Sci 361:312–324

Sela M, Katchalski E (1956) Spectrophotometric titration of alpha-amino acid copolymers containing tyrosine. J Am Chem Soc 78:3986–3989

Stein DB, Searcy DG (1978) Physiologically important stabilization of DNA by a prokaryotic histone-like protein. Science 202:219–221

Stetter KO, Zillig W (1985) *Thermoplasma* and the thermophilic sulfur-dependent Archaebacteria. In: Woese CR, Wolfe RS (eds) The bacteria, vol 8. Academic Press, Orlando, pp 85–170

Strickland EH, Wilchek M, Horwitz J, Billups C (1972) Effects of hydrogen bonding and temperature upon the near ultraviolet circular dichroism and absorption spectra of tyrosine and *o*-methyl tyrosine derivatives. J Biol Chem 247:572–580

Tanford C, Hauenstein JD, Rands DG (1956) Phenolic hydroxyl ionization in proteins. II. Ribonuclease. J Am Chem Soc 77:6409–6413

The Structure of DNA-Binding Proteins from Eu- and Archaebacteria

J. DIJK and R. REINHARDT[1]

Introduction

The mode of packing for eukaryotic DNA is well established. A set of small basic proteins, the histones, is involved in the formation of compact DNA-protein particles in which the double-helical DNA is coiled around an octameric histone complex (Baldwin et al. 1975). A medium resolution structure of these "nucleosomes" has been obtained by crystallographic methods (Richmond et al. 1984). In bacteria, the mechanism for folding of the long circular DNA molecule into a compact form is much less clear. Although a number of proteins have been implicated for this function (Geider and Hoffmann-Berling 1981), a precise description of the composition of "bacterial chromatin" is not available.

We have searched for DNA-binding proteins with the purpose of investigating their spatial structure by X-ray analysis of protein crystals. Initially, efforts were concentrated on the model system *E. coli*, but with time our studies have progressed towards thermophilic bacteria from the genus *Bacillus* and, currently, thermoacidophilic archaebacteria. The increased stability of proteins from thermophiles generally leads to a higher rate of success in crystallisation and a better quality of the crystals.

DNA-Binding Proteins from Eubacteria

Although the structure and composition of the bacterial nucleoid are not very well defined, there is compelling evidence that bacterial DNA is compactly folded (Worcel and Burgi 1972; Delius and Worcel 1974) through the participation of at least three proteins (Pettijohn 1982).

The best characterised component is a small (M_r 10,000), basic protein that is ubiquitous in eubacteria. It has been called HU, HD or NS (reviewed by Geider and Hoffmann-Berling 1981) and supposedly binds preferentially to double-stranded DNA (Rouvière-Yaniv et al. 1979). We have adopted the name DNA-binding protein II (DNAb II), as suggested by Geider and Hoffmann-Berling (1981). The protein occurs

1 Max-Planck-Institut für molekulare Genetik, Abt. Wittmann, Ihnestraße 73, 1000 Berlin 33, FRG

Bacterial Chromatin
(ed. by C.O. Gualerzi and C.L. Pon)

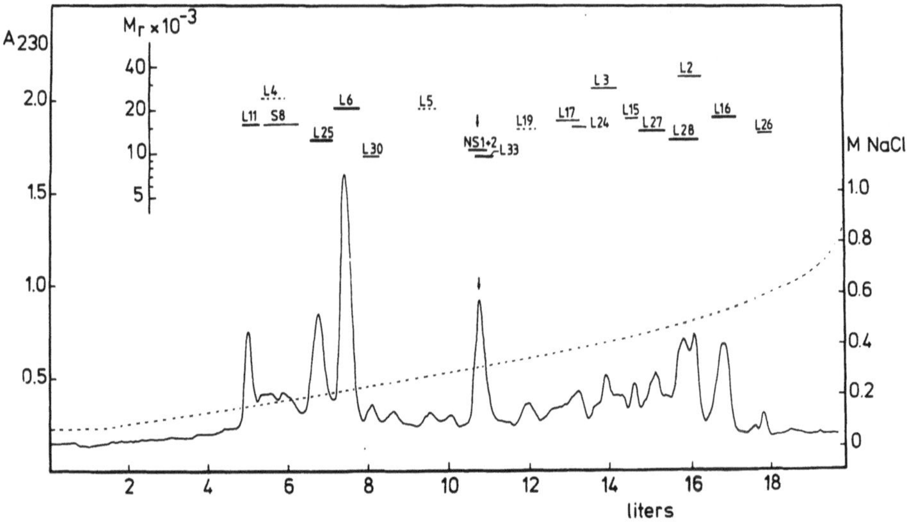

Fig. 1. The separation of ribosomal proteins from the 50S subunit of *E. coli* on CM-Sepharose Cl-6B. The *arrow* indicates the DNA-binding protein

in 50,000–100,000 copies per cell (Pettijohn 1982) and it may well have a histone-like function in the packing of DNA.

Purification

The DNA-binding protein II (named HU in these studies) from *E. coli* was identified as a heat-stable, low-molecular-weight protein (Rouvière-Yaniv and Gros 1975). The former property has been used frequently in rapid purification procedures in which a large proportion of contaminating material was removed by a brief heating step.

This procedure was obviously not suitable for our intended structural X-ray studies. The establishment of a mild purification procedure was aided by the finding that this protein can bind to ribosomes and is obtained during the isolation of ribosomal proteins. During the development of mild purification procedures for ribosomal proteins from *E. coli* (Dijk and Littlechild 1979), we observed an unidentified protein eluting from the CM-Sepharose CL-6B column just before L33 (Fig. 1). The identity of this protein was established by the comparison of tryptic fingerprints with a similar protein described by Brosius (1976) and by comparison of its N-terminal sequence with that of the proteins NS1 and NS2. The complete amino acid sequence of these two proteins, which bind to native 30S subunits (Suryanarayana and Subramanian 1978) had been determined at that time (Mende et al. 1978). It was later shown by amino acid sequence analysis of the proteins HU-α and HU-β that they are identical to NS1 and NS2 (Laine et al. 1980). The protein obtained by us during the purification of ribosomal proteins was a mixture of NS1 and NS2.

Subsequently, the association of DNA-binding protein II with the ribosome was confirmed for *B. stearothermophilus* (Dijk et al. 1983), where only one component is present. The affinity of DNAb II for ribosomes was demonstrated in reconstitution

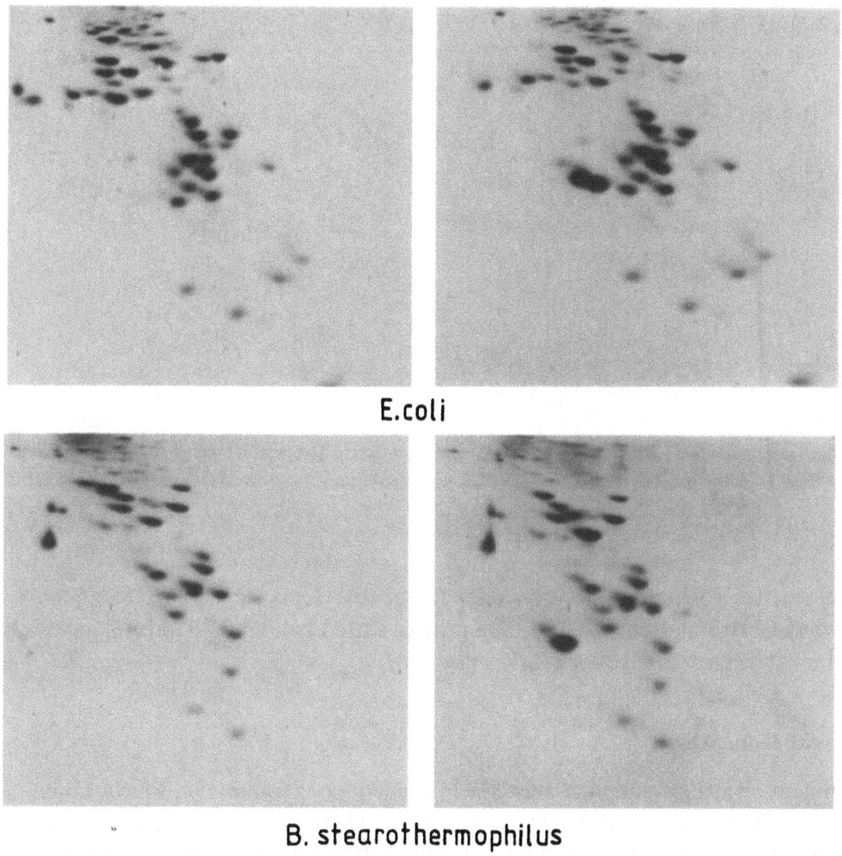

Fig. 2. Two-dimensional gel electrophoresis of 50S subunits from *E. coli* and *B. stearothermophilus* washed with 1 M NH_4Cl (*left*) and 50S subunits incubated with DNAb II followed by removal of unbound protein by gel filtration (*right*)

experiments. Ribosomes or subunits washed in the presence of 1 M salt lack DNAb II. Addition of the protein to the depleted subunits followed by separation of bound and unbound protein by gel filtration resulted in the reappearance of the DNAb II spot on 2D gels of the subunits. The intensity of the spot was comparable to that of the ribosomal proteins (Fig. 2).

Although the isolation of the DNA-binding protein II, together with ribosomal proteins, provided the opportunity of purification under mild conditions, this procedure was rather laborious and gave low yields.

The introduction of a new procedure for the preparation of ribosomes also led to a better and more efficient way of isolating DNAb II. This method, which is applicable to many bacterial species, involves the separation of ribosomes from low molecular weight cellular components by gel filtration in the presence of 1 M NH_4Cl after a preceding ammonium sulphate precipitation (Jelenc 1980). On analytical SDS gels, used to monitor these separations, a strongly dominant low molecular weight band was observed (Fig. 3). This component was identified as DNA-binding protein II. It could

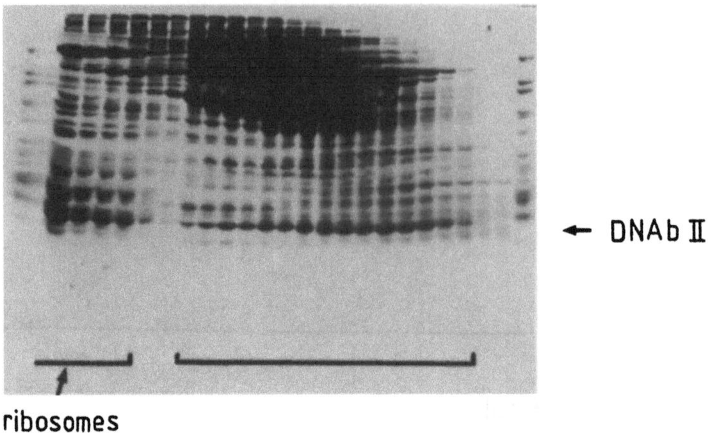

Fig. 3. The separation of ribosomes and DNA-binding protein II from *B. stearothermophilus* on Sephacryl S-300 according to Dijk et al. (1983). The proteins were detected by SDS-gel electrophoresis

easily be purified to homogeneity by two consecutive separations on CM-Sepharose CL-6B, one at pH 7 and one at pH 6. The protein eluted around 0.35 M NaCl and was obtained in a yield of 100–150 mg kg^{-1} cells.

Amino Acid Sequences

The complete or nearly complete amino acid sequences of a number of DNA-binding proteins II have by now been determined (Fig. 4). These proteins share a number of identical residues, in particular in the C-terminal half, without insertions or deletions. In addition to strict identities, many conservative replacements occur, especially with hydrophobic amino acids, in positions 7, 10, 16, 25, 28, 29, 32, 36, 42, 69, and 85. Although the N-terminal sequence of DNAb II from *Thermus thermophilus* does not show a high degree of homology with the other sequences (for example, the characteristic N-terminal sequence M-N-K is lacking), this protein shares several of the conservative hydrophobic replacements with them. Furthermore, information on the remaining part of the sequence, which is now nearing completion (R. Zierer and M. Kimura, unpublished), seems to indicate that several of the characteristic identities also occur in this protein. It, therefore, seems justified to incorporate the *T. thermophilus* protein into the family of DNAb II.

Two further DNA-binding proteins have been characterised recently. The transcription factor 1 (Greene et al. 1984) and the integration host factor (Flamm and Weisberg 1985) show a certain degree of specificity in their interaction with DNA, in contrast to DNAb II. Interestingly enough, comparison of the sequences shows conservation of the G-F and the P-X-T sequences, which are a crucial part of the DNA-binding site of DNAb II (see below), and also shows many of the conservative hydrophobic replacements. It is tempting to speculate on the existence of an extensive family of DNA-binding proteins with varying degrees of specificity based on the structural motif of DNA-binding protein II.

```
                          1       5         10        15        20        25        30
E.coli 1                  M N K S Q L I D K I A A G A D I S K A A A G R A L D A I I A
E.coli 2                  M N K T Q L I D V I A E K A E L S K T Q A K A A L E S T L A
B.stearothermophilus      M N K T E L I N A V A E T S G L S K K D A T K A V D A V F D
B.caldolyticus            M N K T E L I N T V A E I S G L S K K D A T K A V D A V F E
T.thermoph      A A K K T V T K A D L V D Q V A Q A T G L K L L E V K A M V D A L L A
C.pasteuranium            M N K A E L I T S M A E K S K L T K K D A E L A L K A L I E
R.meliloti                M N K N E L V A A V A D K A G L S K A D A S S A V D A V F E
R.leguminosarum L18       M N K N E L V S A V A E K A G L T K A D A A S A V D A V F E
R.leguminosarum L53       M N K N E L V S A V A E K A G L T K S D A A S A V D A V F D
P.aeruginosa              M N K S Q L I D A I A A S A ? ? ? K A V A G K A L D A V I E
Anabena                   M N K G E L V D A V A E K A S V T K K Q A D A V L T A A L E
Common                        K     L         A

                                  35        40        45        50        55        60
E.coli 1                  S V T E S L K E G D Q V A L V G F G T F A V K E R A A R T G
E.coli 2                  A I T E S L K E G D A V Q L V G F G T F K V N H R A E R T G
B.stearothermophilus      S I T E A L R K G D K V Q L I G F G N F E V R E R A A R K G
B.caldolyticus            S I T E A L R K G . . .
T.thermophilus            K V E E A L A N G S . . .
C.pasteuranium            S V E E A L E K G E K V Q L V G F G T F E T R E R A A R E G
R.meliloti                T I Q G E L K N G G D I R L V G F G N F S V S R R E A S K G
R.leguminosarum L18       T V Q S E L K N G G D I R L A G F G S F S V S R R E A S K G
R.leguminosarum L53       V V Q A E L K N K G D I R L A G F G S F T V S H R A A T K G
P.aeruginosa              S V T G A L K A G D ? ? ? ? V G F G T F A V K E R A A R T G
Anabena                   T I I E A V S R G D K V T L V G F G S F E S R E R K A R E G
Common                                              L   G F G   F         R         G

                                  65        70        75        80        85        90
E.coli 1                  R N P Q T G K E I T I A A A K V P S F R A G K A L K D A V N
E.coli 2                  R N P Q T G K E I K I A A A N V P A F V S G K A L K D A V K
B.stearothermophilus      R N P Q T G E E M E I P A S K V P A F K P G K A L K D A V K
C.pasteuranium            R N P R T K E V I N I P A T T V P V F K A G K E F K D K V N
R.meliloti                R N P S T G A E V D I P A R N V P K F T A G K G L K D Q V N
R.leguminosarum L18       R N P S T G A E V D I P A R N V P K F S A G K G L K D A V N
R.leguminosarum L53       R N P S T G A E V D I P A R N V P K F T P G K G L K D A V N
P.aeruginosa              R N P Q T G K ? ? ? ? ? ? ? ? ? ? ? ? ? ? ? A L K D A V N
Anabena                   R N P K T N E K M E I P A T R V P A F S A G K L F R E K V A
Common                    R N P   T           I   A     V P   F     G K           V

                                  95
C.pasteuranium            K
R.leguminosarum L18       S
R.leguminosarum L53       G
Anabena                   P P K A
```

Fig. 4. The amino acid sequences of DNA-binding proteins II. The sequences are derived from Mende et al. (1978), Laine et al. (1980), Kimura and Wilson (1983), Koch et al. (1985), Zierer (1985), Kimura et al. (1984a), Laine et al. (1983), Khanaka et al. (1985), Hawkins and Wootton (1981). The Anabena sequence is cited by Flamm and Weisberg (1985)

Solution Properties of DNA-Binding Protein II from *E. coli* and *Bacillus*

Intramolecular cross-linking studies have shown that the protein occurs as oligomers, the smallest of which is the heterodimer NS1/NS2 in *E. coli* (Rouvière-Yaniv and Kjeldgaard 1979).

In sedimentation equilibrium experiments on the *B. stearothermophilus* protein, we have observed that at low protein concentrations, the tetramer is the smallest species that can be observed: M_r 37,000–38,000. At higher concentrations, larger aggregates start to appear. The Stokes radius of the protein, determined by analytical gel filtration

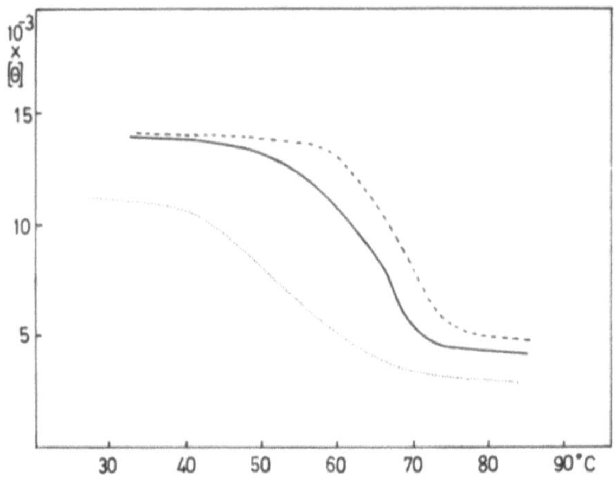

Fig. 5. Heat denaturation curves for the proteins from *E. coli* (. . . .), *B. stearothermophilus* (——), and *B. caldolyticus* (- - - -), as monitored by CD spectroscopy

on Sephadex G-50 or G-75 in 0.35 M NaCl, is 25 Å. For a tetramer this yields a frictional ratio f/fo of 1.12 and for a dimer 1.41. From the now available crystal structure we know that the molecule is somewhat asymmetric owing to the two flexible arm regions. It is, however, rather difficult to decide between the two possibilities, since both numbers are well within a reasonable range.

Circular dichroic spectroscopy in the far-UV region shows that the proteins possess a well-defined secondary structure. By analysis of the spectra using the CONTIN programme (Provencher and Glöckner 1981), 35–40% α-helix and 50–55% β-sheet were found (Dijk et al. 1983).

CD spectroscopy was used to assess the heat stability of the protein structure, using the signal intensity at 220 nm, which is mainly derived from α-helical structure. As expected, the DNA-binding protein II from *E. coli* is less stable than that from the thermophile *B. stearothermophilus* (grown at 60°C). In turn, the latter is less stable than DNAb II from the extreme thermophile *Bacillus caldolyticus* (Heinen and Heinen 1972), which is grown at 75°C (Fig. 5).

These experiments were conducted in 0.35 M KF, 5 mM phosphate, pH 7.0, in which the fluoride helps to decrease the absorbance of the sample at wavelengths below 230 nm, improving the signal-to-noise ratio in this important spectral region. Alterations in this solution have a pronounced effect on the stability of the structure, as demonstrated for the DNAb II from *B. stearothermophilus.* Changes in the concentration and the nature of the salt and in the buffer component cause a shift in the melting curve (Fig. 6). The two extremes are the buffer mentioned above and the low-salt solution 5 mM NaCl, 10 mM HEPES, pH 7.0, showing a difference of 14° in the mid-point of the melting curve.

The NMR spectra of the *E. coli* protein reveal a well-folded tertiary structure. In comparison with the spectrum taken in 8 M urea (Fig. 7), two regions in the native spectrum supply this information. The protein contains four phenylalanine residues and no other aromatic amino acids. In the denatured structure, the ring protons of phenylalanine appear as a broad resonance at 7.3 ppm. In the folded structure their resonances are strongly perturbed with an upfield shift. Also, in the 0–1 ppm region several poorly resolved, ring-current-shifted methyl resonances occur which become

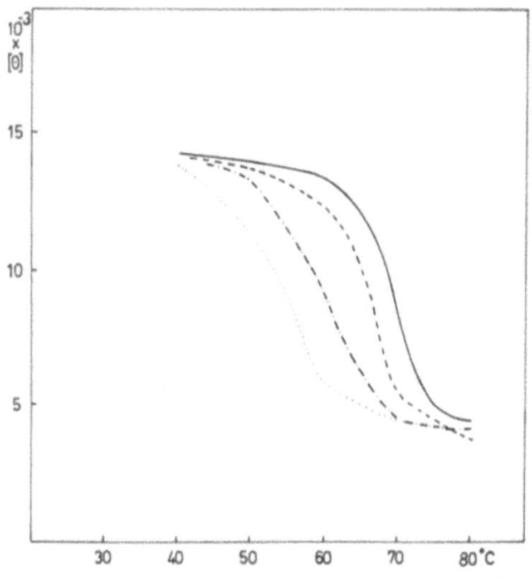

Fig. 6. The effect of different salts on the stability of the DNA-binding protein II from *B. stearothermophilus.* (. . . .) 5 mM NaCl, 10 mM HEPES pH 7.0. (· — · — ·) 100 mM NaCl, 10 mM HEPES pH 7.0. (- - - -) 350 mM NaCl, 10 mM HEPES pH 7.0. (——) 350 mM KF, 10 mM phosphate pH 7.0

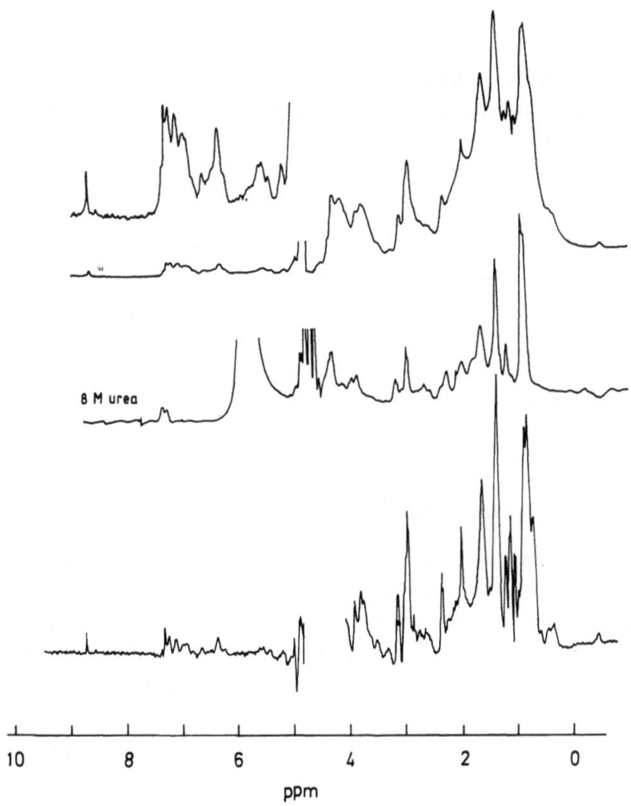

Fig. 7. 270 MHz proton NMR spectra of DNA-binding protein II from *E. coli.* The aromatic part of the native spectrum is shown with eightfold amplification. The resolution of the lower spectrum is enhanced by the convolution difference technique. The native protein was in 0.3 M NaCl, pH 7.0

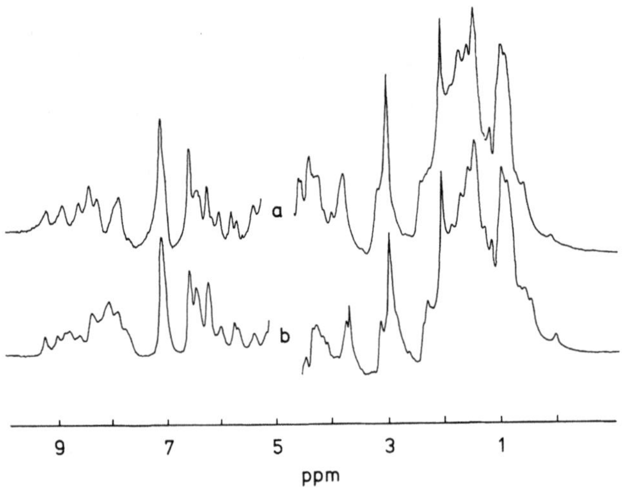

Fig. 8. 270 MHz proton NMR spectra of the DNA-binding protein II from *B. stearothermophilus* and *B. caldolyticus*, recorded in 0.1 M NaCl, pH 7.0

better visible in the resolution-enhanced spectrum (Fig. 7, bottom). These originate from methyl protons of leucine, valine and isoleucine (main resonance around 1 ppm), which are in the proximity of aromatic side chains. When in a perpendicular position to the aromatic ring, their resonances are shifted upfield and provide strong evidence for a hydrophobic core.

The NMR spectra of the proteins from *B. stearothermophilus* and *B. caldolyticus* bear a strong resemblance to that of the *E. coli* protein (Fig. 8). The main difference lies in the presence of many poorly resolved resonances in the 8–10 ppm region, which originate from non-exchanged amide protons of the polypeptide backbone. This observation is typical for thermophilic proteins. The two *Bacillus* proteins are very similar (Fig. 9), as can be expected from the high degree of sequence homology between the two.

A DNA-Binding Protein from *Thermus thermophilus*

The extreme thermophile *T. thermophilus* is more closely related to *E. coli* than *Bacillus*, on the basis of the ribosomal protein 2D gel pattern (Zierer 1985). The search for DNA-binding proteins, using the methods described above, had a somewhat unexpected outcome. A protein with an M_r of 10,000 was isolated, which was much more basic than DNAb II, i.e. it migrated in the bottom right-hand corner of the 2D gels. The partial amino acid sequence available (Fig. 4) shows that the protein has a low degree of homology to DNA-binding protein II, with a calculated α-helix content of 32% and β-sheet of 40%. These values are close to those of the *E. coli* and *B. stearothermophilus* proteins. The NMR spectrum is, however, very different (Fig. 10). In view of the low degree of homology with other DNA-binding proteins, this might reflect the large number of amino acid substitutions occurring in a similar tertiary

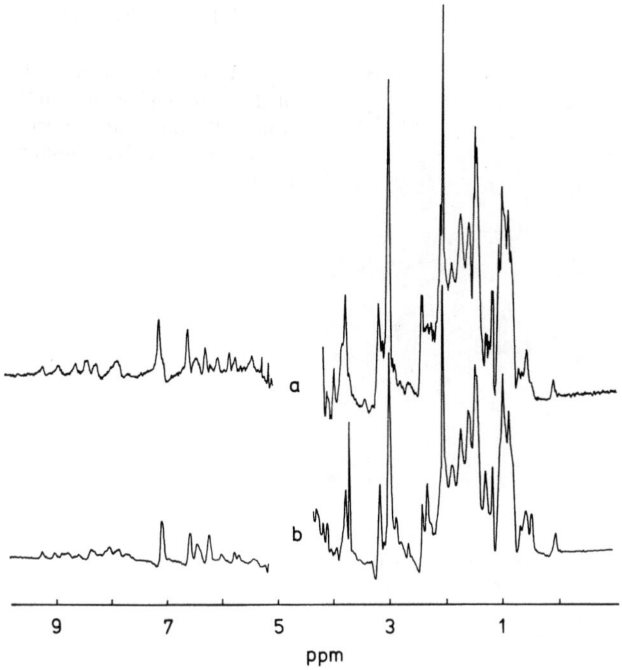

Fig. 9. NMR spectra of Fig. 8 with resolution enhancement

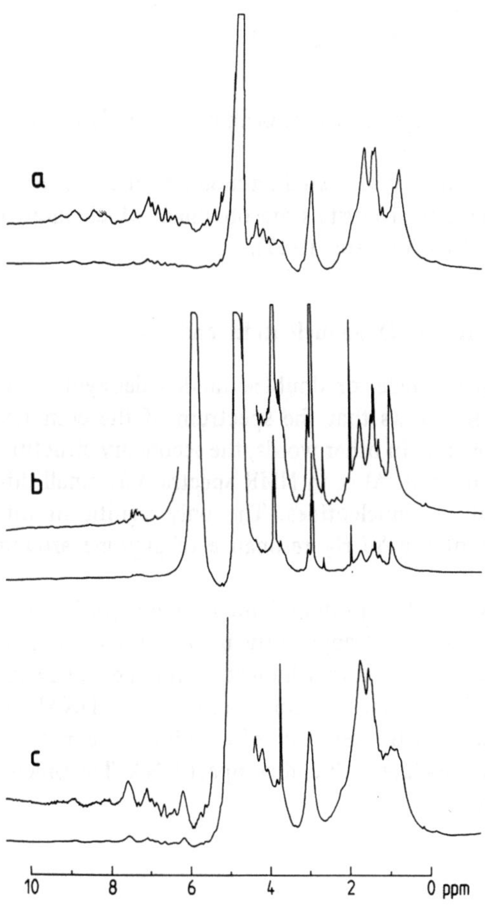

Fig. 10a–c. 270 MHz proton NMR spectra of the DNA-binding protein from *Thermus thermophilus.* The native protein (**a**) was in 0.1 M NaCl, pH 7.0. In (**b**) the spectrum in 8 M urea is shown, (**c**) shows the spectrum in the presence of $d(A)_8$-T_8

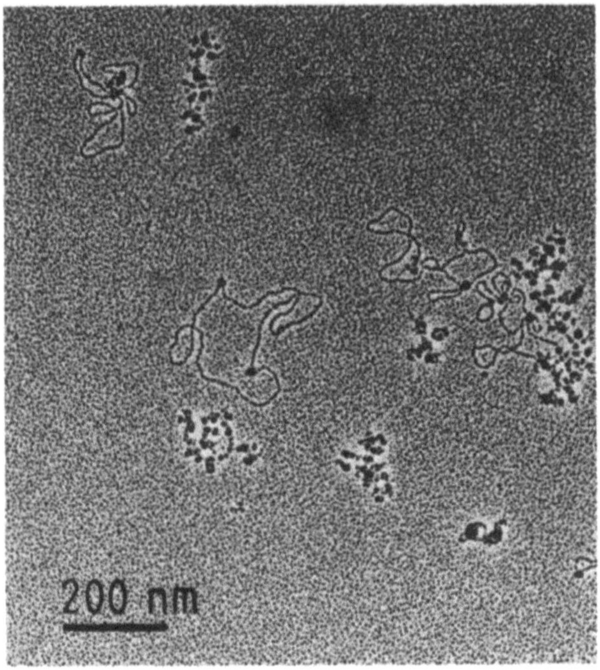

Fig. 11. Electron micrograph of the DNA-binding protein II from *Bacillus caldolyticus* bound to pBR 322 plasmid DNA

structure. The spectrum is dominated by a large resonance peak of the ϵ-CH_2 protons of lysines, which account for 15% of the residues.

The protein has been crystallised from nearly saturated ammonium sulphate using the hanging drop procedure. Large bipyramidal crystals are obtained, which show a hexagonal symmetry and diffract at least to 3 Å (Zierer 1985).

The Interaction of DNA-Binding Protein II with Oligonucleotides and DNA

CD spectroscopy of DNAb II and either single- or double-stranded deoxyoligonucleotides, such as $(dA)_8$ or $(dA)_8.(dT)_8$, shows that the spectrum of the complex is the precise sum of that of the components. In other words, the secondary structure does not change when the complex is formed. Also, in NMR spectra only small differences are observed after addition of oligonucleotides. The only significant difference appears to be the upfield shift of the δ-CH_2 resonance of arginine around 3.2 ppm.

In contrast, NMR spectra of the DNA-binding protein from *T. thermophilus* with oligonucleotides show dramatic changes (Fig. 10) apparently involving a substantial part of the protein structure. This provides yet another indication for the somewhat different behaviour and properties of this protein. Electron microscopy of DNAb II-plasmid DNA complexes (Fig. 11) yields little detailed information. The protein appears to form large aggregates, which associate with and compact DNA. The precise nature of the folding cannot be discerned.

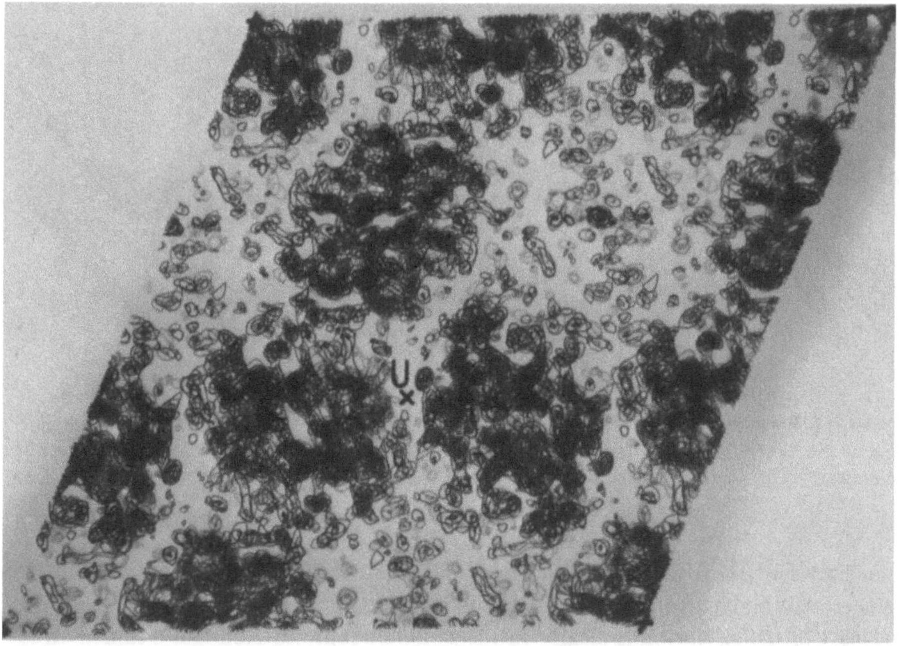

Fig. 12. Electron density map of DNA-binding protein II at 3 Å resolution. The complete cell is superimposed in the direction of the c-axis. The uranyl site (*U*) is located between two dimers

The Crystal Structure of DNA-Binding Protein II from *Bacillus stearothermophilus*

Several crystal forms of this protein have been obtained. The most useful crystals were grown from 35% 2-methyl-2,4-pentanediol in the pH range of 7–9 (Dijk et al. 1983). From the closely related organism *Bacillus caldolyticus* also several forms of well diffracting crystals have been obtained (Koch et al. 1985). All crystals have a crystallographic b-axis of approximately 37 Å.

A low-resolution 6 Å structure of the DNA-binding protein was calculated using phases from a single isomorphous uranyl derivative and anomalous scattering (Tanaka et al. 1984a). The electron density map showed that three dimers of the protein were present in the unit cell (Fig. 12). The dimers were located on parallel, independent rotation axes.

The three dimers appeared to be very similar in structure. The excellent diffraction of the crystals (well below 3 Å) and the possibility of electron density averaging to improve the image led to a more detailed structure at 3 Å resolution (Tanaka et al. 1985b). Even at this resolution, the three dimers had an essentially identical structure.

The monomer is characterised by three distinct regions. In the N-terminal half, a V-shaped structure composed of two α-helices is present (Fig. 13). Important for the contact between the two helices are the two highly conserved alanines in positions 11 and 21. The C-terminal part contains a three-stranded antiparallel β-pleated sheet which covers the top of the V-shaped helical structure. The conserved Gly-Phe-Gly sequence (46–48) is involved in the sharp turn between the first two strands and Gly-39 in the connection between the two halves of the molecule (Fig. 14).

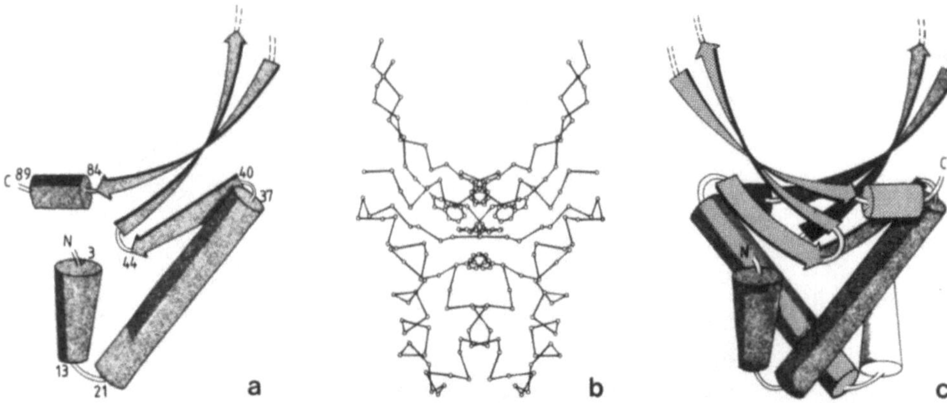

Fig. 13. a Schematic representation of the secondary structure of the monomer. b The C-α backbone of the dimer. The side chains of the phenylalanines are shown. c The secondary structure of the dimer

In between the second and third β-strands, the backbone extends away from the main protein body, and the distal region is invisible in the electron density map. The first part of the extension is clearly in the β-ribbon conformation and a similar structure for the invisible part is not unreasonable. The conserved proline residue at position 63 could be at the turning point between the out- and ingoing strands. The implications of these structural features for the function of the protein are discussed below. The molecule ends in a short α-helix.

The dimer is composed of two strongly intertwined monomers. Inspection of the structure shows that separation of the monomers is only possible after denaturation (Fig. 13). The three conserved phenylalanine residues (47, 50 and 79) of one monomer and several hydrophobic amino acids (6, 32, 36 and 44) interact strongly with the same residues of the other monomer. The stacking of the phenylalanine rings is the cause of the strong, upfield chemical shift observed in the NMR spectra (Figs. 7–9). The surface of the dimer is covered by hydrophilic side chains; the distribution of positive and negative charges is not uniform.

The extended arm regions, together with the β-strands, create a concave surface which, from its dimensions (25 Å), is complementary to double-helical B-DNA. It is tempting to speculate that this supersecondary structure is the DNA-binding site.

Elucidation of the mode of interaction with DNA using crystallographic methods has not been successful. In the crystal, the proposed site is covered by the next dimer preventing the diffusion of oligonucleotides into the potential binding site. Also, in co-crystallisation experiments the occupation of the site by nucleotides might preclude crystallisation by disturbing the important contact between two dimers in the lattice.

Therefore, the interaction with DNA was studied by model building using interactive computer graphics. The attempts were guided by several experimental observations:

```
1       5         10        15        20        25        30
M N K T E L I N A V A E T S G L S K K D A T K A V D A V F D
   <----α-helix 1-------->              <------α-helix 2----

        35        40        45        50        55        60
S I T E A L R K G D K V Q L I G F G N F E V R E R A A R K G
--α-helix 2-->    <-strand 1>      <strand2> <---arm region--

        65        70        75        80        85        90
R N P Q T G E E M E I P A S K V P A F K P G K A L K D A V K
------arm region---------------><-strand 3-><-α-helix 3->
```

Fig. 14. The distribution of secondary structure elements along the amino acid sequence of DNA-binding protein II from *B. stearothermophilus*

1. The structure of the protein is hardly changed by binding of DNA, the major difference being the shift of arginine protons in NMR spectra (see Fig. 9).
2. The arm region contains several highly-conserved arginine residues (see Fig. 4). In chemical modification studies these have been implicated in DNA binding (Lammi et al. 1984).
3. The binding between protein and DNA is largely electrostatic, in view of the inhibiting effect of increasing salt concentrations (Dijk et al. 1983).
4. The binding of the protein is not sequence-specific (Geider and Hoffmann-Berling 1981).

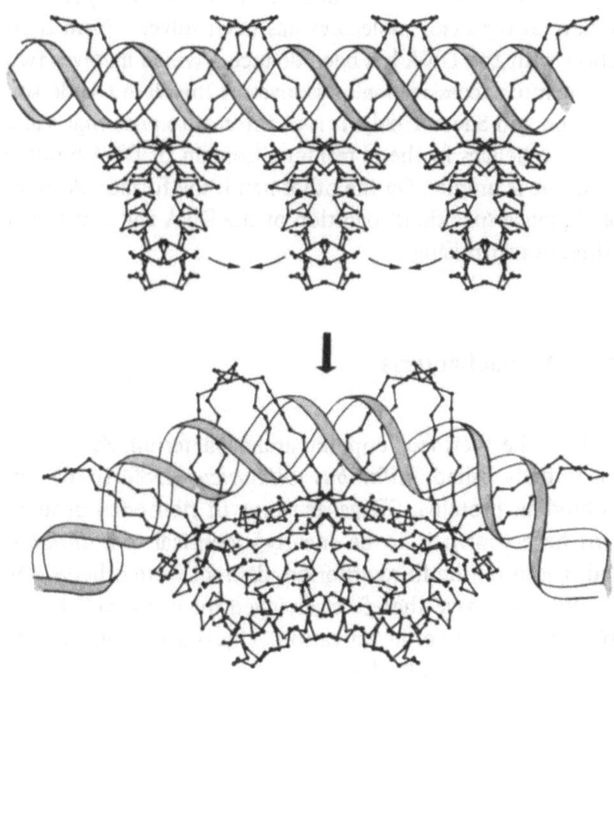

Fig. 15. The proposed mode of interaction of DNA-binding protein II with DNA

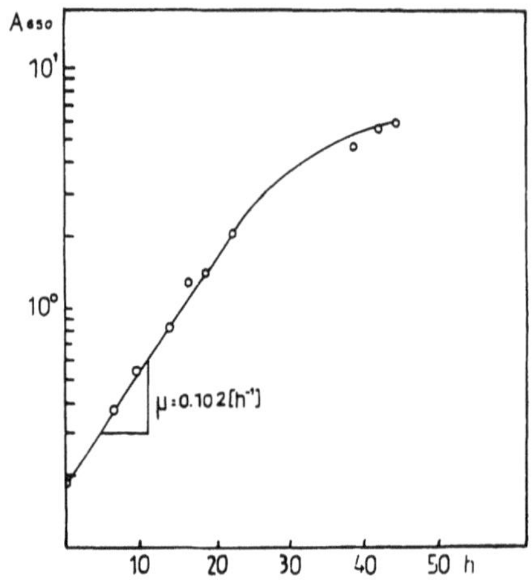

Fig. 16. The growth curve of *Sulfolubus acidocaldarius* cells at 75° and pH 4.8

These observations are incorporated into a model in which the ribbon-like arms wrap around the double helix in either the minor or the major groove of the DNA (Fig. 15). A possible mode for supercoiling is derived from the asymmetric charge distribution and the shape of the monomer. In a direction perpendicular to the two-fold rotation axis, the molecule is wedge-shaped which allows for oligomerisation into a circular assembly. The contacts between the dimers are mediated by the alternating positively and negatively charged surfaces and the N-terminal α-helix is buried in a pocket on the adjacent molecule. Such a supercoiling would yield a nucleosome-like particle with a diameter of 140 Å consisting of 8–10 dimers and around 80–100 base pairs per turn.

Recently, the structure of several repressor molecules has been solved. From their structures a mode of interaction with the DNA has been deduced, which involves two symmetry-related α-helices fitting into successive major grooves of the DNA (Anderson et al. 1982; Pabo et al. 1982; Pabo and Sauer 1984). It is of interest to note that there is a certain similarity to the two α-helices in the N-terminal domain of DNA-binding protein II and that such a structure is unique. On the other hand, the former group of proteins is characterised by a sequence-specific recognition of the DNA and a different mode of interaction is, therefore, not surprising.

DNA-Binding Proteins from Archaebacteria

Our studies have concentrated on the thermoacidophilic archaebacterium *Sulfolobus acidocaldarius.* The strain was initially called *Sulfolobus solfataricus* (DSM 1616), but the characterisation of DNA-binding proteins, 2D gel patterns of ribosomal proteins and RNA polymerase patterns have shown it to be (nearly) identical to *Sulfolobus acidocaldarius* (DSM 639) and, based on the third criterion, distinct from other *S. solfataricus* strains (Zillig et al. 1985). We have therefore, changed the name to *S. acidocaldarius.* Details of the problem are still under investigation. This organism grows well in the temperature range of 75°–85°C at pH 4.0–4.8.

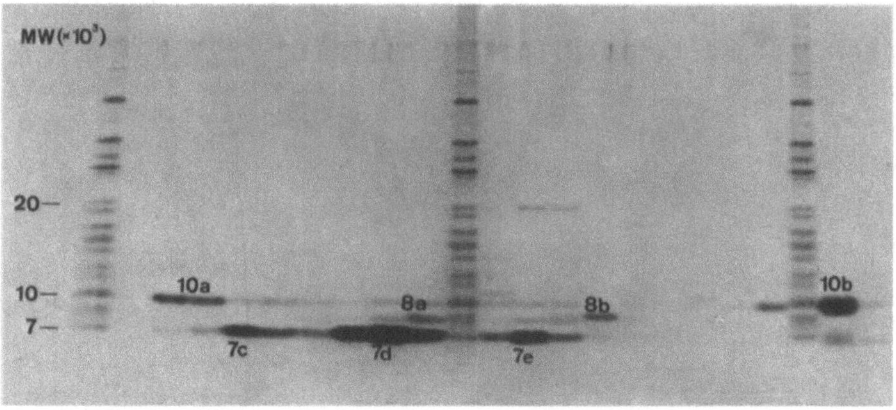

Fig. 17. The separation of small basic proteins from *Sulfolobus acidocaldarius* on CM-Sepharose CL-6B, using a NaCl gradient of 0.1–0.5 M at pH 7.0. Aliquots of the fractions were analysed by SDS-gel electrophoresis

We have been able to optimise the yield of cells by rigorous control of temperature, pH and nutrients. A feed-batch regime was established in which the culture is continuously fed with a mixture of yeast extract and glucose (ratio of 2:1 by weight), keeping the glucose level at 0.2–0.5 g l^{-1}.

By using a glass fermenter with a working volume of 130 liters, 500–700 g of rapidly growing cells (generation time 6 h) can be harvested after a period of 48 h (Fig. 16).

Purification

DNA-binding proteins have been isolated using a slightly modified procedure, which had been applied to the purification of DNA-binding proteins from eubacteria (Dijk et al. 1983). The method involves the separation of ribosomes and soluble proteins in the presence of 1.0–1.5 M salt by either gel filtration or ultracentrifugation. From the medium to low molecular weight protein fraction, basic proteins are selected by binding to CM-Sepharose CL-6B at low ionic strength and separated by elution with an NaCl gradient.

In our early experiments only two dominant proteins of low molecular weight (around 7,000) were observed, and the amino acid sequence of one of them has been determined (Kimura et al. 1984b).

Refinement of the procedure has led to the appearance of additional proteins. A representative picture is presented in Fig. 17. The proteins fall into three molecular weight classes as determined by SDS gel electrophoresis, namely 7 kD, 8 kD and 10 kD and each class can be subdivided into several species.

The yield of these proteins is not completely reproducible; this may be caused by the different growth levels of the cells and other variables. The same variation applies to the appearance of the 7 kD proteins. Recently, two additional 7 kD proteins were purified in appreciable quantities which elute very early from the CM-Sepharose

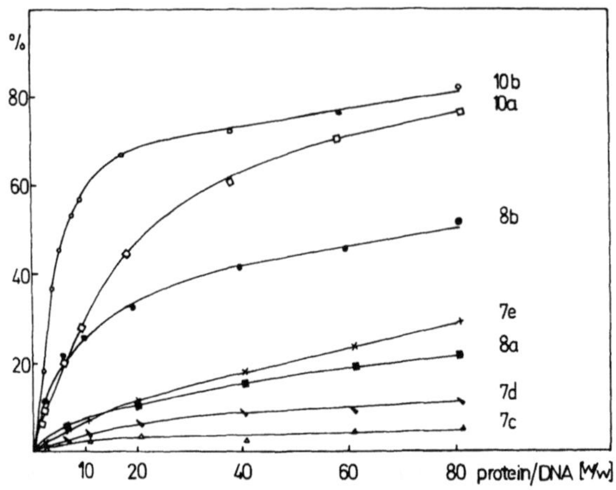

Fig. 18. Millipore filter binding assays. Increasing amounts of proteins were incubated with 0.5 μg [^{3}H]-pBR 322 DNA in the presence of 0.25 M NaCl in 0.1 × SSC

column, before component 7c. These proteins are also found in the bacterial "chromatin" fraction (see below). Although they have not been fully characterised, they obviously belong in the group of 7 kD DNA-binding proteins described here.

Nomenclature

In the absence of a meaningful nomenclature for archaebacterial DNA-binding proteins, we have introduced a numbering system, which refers to both the size and the charge of the protein. The molecular weight has been estimated from SDS gel electrophoresis, as shown in Fig. 17. These values carry a certain amount of uncertainty, as is usual for small basic proteins. On the other hand, they are corroborated by the M_r value calculated from the amino acid sequence of component 7d (Kimura et al. 1984b).

The further subdivision into several components within these classes is indicated by a, b, c, etc. in the order of their elution from CM-Sepharose CL-6B, i.e. increasing positive charge. The 7 kD proteins include 7a–e, of which proteins 7a and 7b are not shown in Fig. 17. These two proteins bind very weakly to the ion-exchange column and were discovered later. The 8 kD proteins are 8a and 8b and the 10 kD proteins 10a and 10b.

DNA-Binding Assays

In eubacteria the existence of DNA-binding protein II as the predominant DNA-binding protein is established for a number of organisms and further additions to the group can be easily identified from sequence homology.

In archaebacteria the situation is less clear and the DNA-binding character of small, abundant basic proteins has to be confirmed using various binding assays.

We have used a modification of the Millipore filter binding assay described by Lammi et al. (1984) as a first screening test. The original procedure uses a rather low ionic strength (0.1 × SSC), which encourages the binding of basic proteins to nucleic acids by electrostatic interaction. Such interactions are considered to be non-specific and do not reflect genuine DNA-binding.

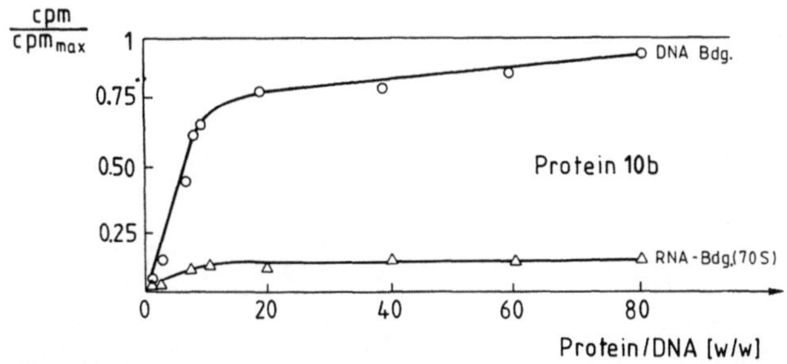

Fig. 19. Millipore filter binding assay of the binding of protein 10b to [^{3}H]-labeled 70S ribosomes

Basic control proteins, such as cytochrome c, lysozyme and the *E. coli* ribosomal proteins L6 and L16 bind very well to plasmid DNA pBR 322 (only present in the OC form) under the original assay conditions. We have, therefore, increased the NaCl concentration of the binding buffer to 0.25 M (in 0.1 × SSC), which completely suppresses the binding of these four proteins.

Under these conditions the proteins from *Sulfolobus acidocaldarius* all bind to DNA. The 7 kD proteins show the weakest binding, at the same level as the DNA-binding protein II from eubacteria (Fig. 18). Preliminary results on the newly isolated 7a and 7b proteins also indicate binding to DNA. Generally, the level of binding increases with basicity of the proteins and with increasing molecular weight. In particular, the affinity of 10b for DNA is high, it retains at least 70% of the [^{3}H]-DNA to the filter in the plateau region.

In view of the binding of DNAb II to ribosomes and subunits, filter-binding experiments with ribosomes and ribosomal RNA were carried out. Only binding of 10b was measured, an order of magnitude weaker than the binding of the same protein to DNA (Fig. 19).

Electron Microscopy

Binding experiments with circular plasmid DNA (RSF 1010 and pBR 322) were performed at relatively low ionic strength (50 mM KCl, 10 mM Tris, pH 7.2, 10 mM dithioerythritol). This condition simplifies sample preparation, but increases the danger of structural artefacts. In crucial cases control experiments at higher salt concentration were conducted in order to exclude this possibility.

The 7 and 8 kD proteins show structures very similar to those found for DNA-binding protein II (Fig. 20). The proteins bind at a single site on the DNA molecule and further condensation of DNA is obtained by association of more protein with the same site. Free DNA is observed as garland-like loops emerging from the central protein-DNA core. At higher protein levels (above 10:1 w/w to DNA) free DNA is no longer observed and the complexes have collapsed into a solid clump.

The 10 kD proteins show a different behaviour pattern. Protein 10a, at low protein levels, at first occupies a single site on each DNA molecule. Upon increase of the protein concentration the protein-DNA complexes pass through an intermediate structure, similar to that seen for the 7 and 8 kD proteins. When the DNA is saturated, a

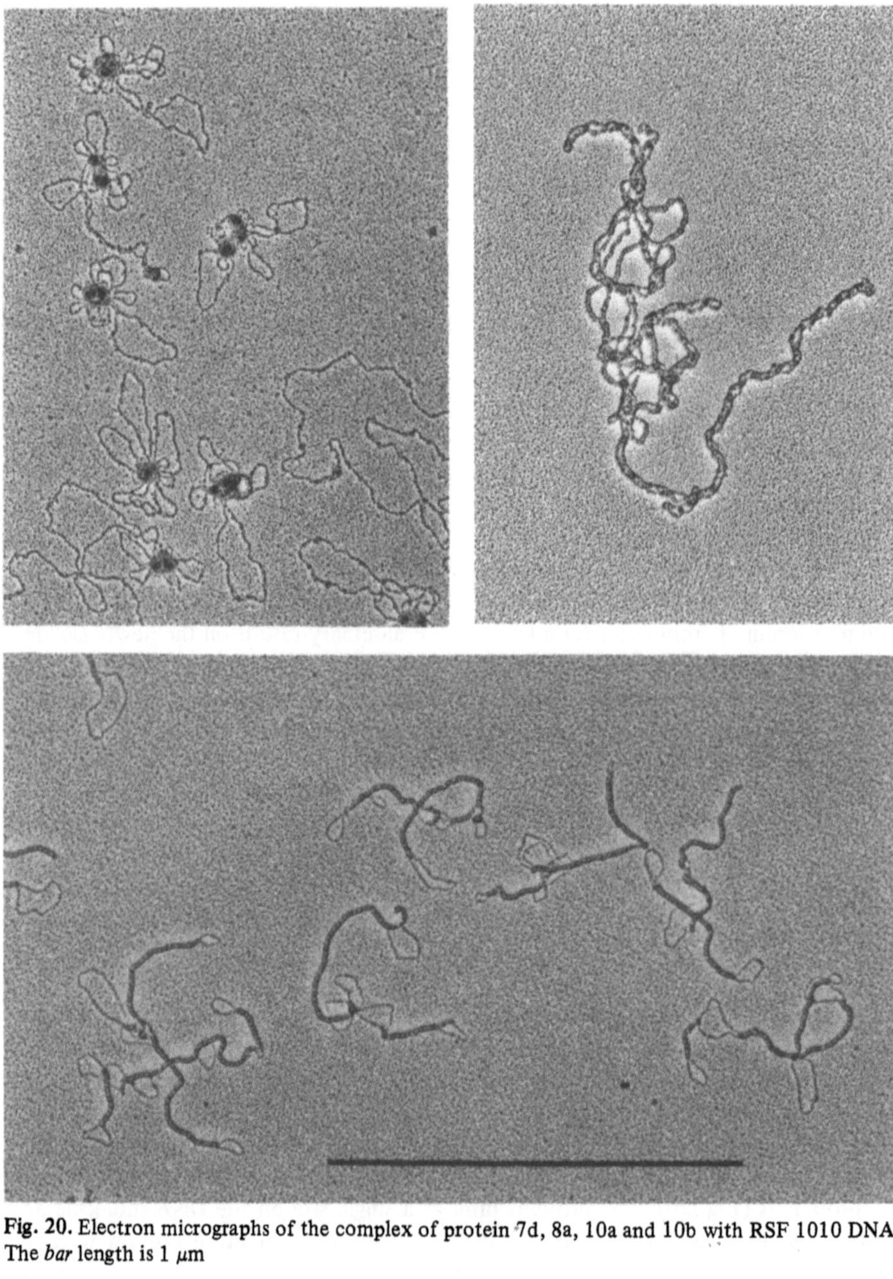

Fig. 20. Electron micrographs of the complex of protein 7d, 8a, 10a and 10b with RSF 1010 DNA. The *bar* length is 1 μm

well-defined structure emerges in which the double strand is linearly covered with protein (Fig. 20).

The interaction of protein 10b with DNA has some unusual and novel aspects. The binding to DNA shows strong cooperativity. Attachment of the first protein molecule to the DNA is followed by covering of the same DNA molecule with a continuous pro-

```
          1     5         10         15         20         25         30
7c        V K V K*F K*Y K G E E K E V D T S K I K K V W R V . . .
7d        V K V K*F K*Y K G E E K E V D T S K I K K V W R V . . .
7e        A K V R R K*Y K G E E K E V D T S K I K K V W R V . . .

------------------------------------------------------------------------

          1     5         10         15         20
8b        P A I E I G Y I G V E T R G D E A . . .

------------------------------------------------------------------------

          1       5         10         15         20         25         30
10b       G A T P S N V L Y G K K P V M N Y V L A A L T L L N Q G . .
            ↑         ↑     ↑ ↑ ↑ ↑ ↑ ↑   ↑ ↑ ↑ ↑     ↑     ↑   ↑

------------------------------------------------------------------------
```

Fig. 21. N-terminal amino acid sequences determined with a liquid-phase sequenator. K^+ indicates a modified lysine residue. The *arrows* below the sequence of 10b indicate positions of identity with a similar DNA-binding protein from *Thermococcus celer*

tein coat (Fig. 20). Closer inspection of the structures shows that the protein coat covers two double strands.

Amino Acid Sequence

7 kD Proteins: The complete amino acid sequence of component 7d was determined at an early stage of our studies (Kimura et al. 1984b). The protein is 64 amino acids long and has a rather unusual composition. Twenty percent of the positions are occupied by lysine residues with a predominance in the N-terminal region, and 40% of the amino acids have a charged side chain. Hydrophobic residues are also well represented (30%), again dominating in the N-terminal half. Although the lysine alternation in the N-terminal region is also found in other DNA-binding proteins (HMG, replicase c), this similarity does not result in significant homology to these proteins.

N-terminal amino acid sequences using liquid- and solid-phase sequencer techniques have been obtained from the other 7 kD proteins (Fig. 21).

Table 1. The Stokes radii of the DNA-binding proteins of *Sulfolobus acidocaldarius* determined by analytical gel filtration[a]

	R_8 (Å)	f/fo Monomer	f/fo Dimer
7 kD	15.3	1.20	0.95
8. kD	14.8	1.12	0.83
10 kD	19.4	1.36	1.08

[a] The frictional ratio (f/fo is calculated from the ratio of the Stokes' radius R_8 and the radius of the equivalent sphere R_{min}.

These proteins possess a surprisingly high degree of homology. The unusual lysine alternation in the N-terminal region occurs in all the proteins with the exception of the conservative replacement of Lys-4 by arginine in 7b and 7e. Other changes which could account for the difference in charge have not been found at this stage.

In the liquid-phase sequence experiments a peak, in addition to that of lysine, was observed in steps 4 and 6. In view of the clear-cut results in the neighbouring cleavages this additional component was postulated to arise from a modified lysine residue and not from overlaps. Preliminary experiments with lysine derivatives have shown that the additional peak comigrates with the Pth-derivative of N^6-monomethyl-lysine (K. Ashman, unpublished). A more complete account of the nature of the lysine modification is given in the section on NMR spectroscopy.

8 kD Proteins: Protein 8a is blocked at the N-terminus and, therefore, inaccesible to automated Edman degradation. No sequence information is currently available. For protein 8b, which is obtained in relatively small amounts, an N-terminal sequence has been obtained up to residue 17 (Fig. 21). Sequence homology to other DNA-binding proteins has not been found.

10 kD Proteins: Both proteins are blocked at the N-terminal amino group and cannot be analysed in the sequenator. By mild acid hydrolysis the N-terminal amino group of protein 10b could be freed and a partial sequence is now available (Fig. 21). Also in this case, sequence homology to other DNA-binding proteins has not been observed. However, a DNA-binding protein has recently been isolated from *Thermococcus celer.* This protein has an approximate molecular weight of 10,000 and the N-terminal region shows a convincing homology to that of protein 10b of *Sulfolobus acidocaldarius* (Fig. 21).

Several fragments of protein 10a have been sequenced and the results available exclude the possibility of a direct sequence relationship with 10b.

Stokes Radii

The degree of asymmetry and oligomerisation of proteins are easily assessed by analytical gel filtration (Ackers 1967), which allows the use of low protein concentrations, thus eliminating artefacts due to effects such as protein aggregation. The results (Table 1) show that in 0.35 M NaCl the 7 and 8 kD proteins are compact monomers and the 10 kD proteins very compact dimers or somewhat asymmetric monomers.

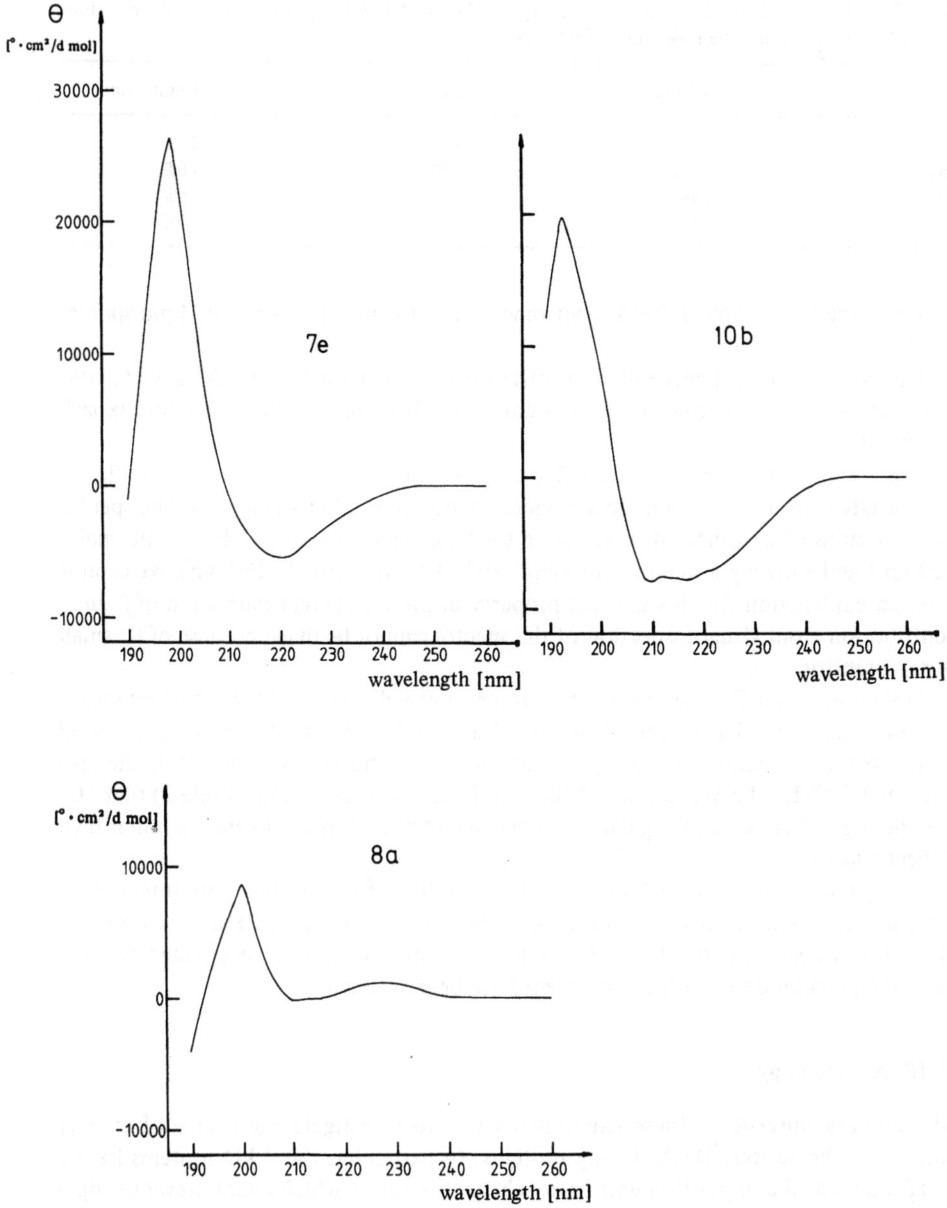

Fig. 22. Circular dichroic spectra of proteins 7e, 8a, and 10b

CD Spectroscopy

DNA-binding protein II from eubacteria has a well-defined secondary structure. The α-helix and β-sheet content estimated from CD spectra agree well with those obtained from the crystal structure. CD spectra of the DNA-binding proteins from *Sulfolobus*

Table 2. The distribution of secondary structure in the DNA-binding proteins obtained by evaluation of CD spectra using the programme CONTIN

	α-Helix	β-Sheet	Remainder
7e	15%	44%	41%
8a	12%	72%	16%
10a	19%	53%	28%
10b	33%	67%	0%

were recorded in a Jasco J-500A spectrometer at 20°C in 0.1 M KF, 5 mM phosphate, pH 7.

This solution has a lower salt concentration than that used for DNAb II of *E. coli.* It was chosen since its lower ionic strength makes it more suitable for binding experiments with DNA.

The 7 and 10 kD proteins have CD spectra which are, to a certain extent similar to that of DNAb II, i.e. they contain a moderate amount of α-helix (Fig. 22). The spectra have the unusual property of intersecting the baseline around 205–210 nm (normally 200 nm) and showing a maximum around 193–198 nm (usually 190 nm). We cannot offer an explanation for this unusual property at present. Direct estimation of β-sheet contribution from visual inspection of the spectra cannot be made because of its small signal intensity.

Protein 8a has a CD spectrum with a rather unusual shape and low signal intensity. Its resemblance to that of gene-5 protein (Day 1973) suggests that 8a may be an all β-sheet protein. Quantitative interpretation of the spectra was obtained using the programme CONTIN (Provencher and Glöckner 1981). The data show (Table 2) that 10b has the highest content of regularly ordered structure. All proteins have a rather high β-sheet content.

CD spectroscopy was used to assess the stability of the protein structure at high temperatures. A small decrease in the signal intensity at 190 and 220 nm was observed at temperatures close to 90°C. Higher temperatures could not bė pursued further, since the physical limits of the instrument have been reached.

NMR Spectroscopy

The primary purpose of these experiments was to investigate the degree of tertiary folding in the various DNA-binding proteins. In particular, the 7 kD proteins have a usual amino acid composition with many charged residues, which could prevent proper folding of the molecule.

The spectra were recorded at 270 MHz on a Bruker WH-270 spectrometer operating in the Fourier transform mode. The proteins were transferred by dialysis into 0.1 M NaCl, 5 mM phosphate, pH 7 in ^2H_2O. Protein concentration was usually around 1 mM and 1000 transients were recorded.

7 kD Proteins: These proteins have very similar NMR spectra, in accordance with the results from the amino acid sequence studies (Fig. 23). Various regions of the spectra

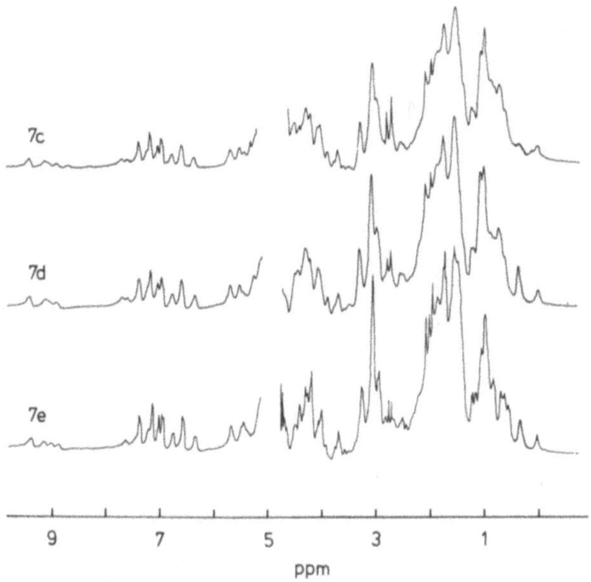

Fig. 23. 270 MHz proton NMR spectra of proteins 7c, 7d, and 7e

carry evidence for a compact tertiary fold. A comparison with the spectra of the denatured proteins in 6 M guanidinum chloride (Fig. 24) demonstrates this very clearly. In the 9 ppm region many poorly resolved amide proton resonances occur, which are exchanged after denaturation. The aromatic resonances between 6–8 ppm are strongly perturbed and several resonances of down-field shifted C-α protons are observed between 5 and 6 ppm. At the high-field side of the spectrum at 0–1 ppm, many partly resolved, ring-current-shifted methyl (RCSM) resonances are present. In protein 7e this region clearly differs from that in 7c and 7d.

Other differences between the three proteins reside in the different signal intensity of the ϵ-CH_2 resonance of lysine at 3.0 ppm and of two sharp peaks at 2.7 and 2.8 ppm. Proteins 7c and 7d appear to possess two methionine residues with resonances at 1.98 and 2.10 ppm, whereas 7e has an additional methionine with a resonance at 2.02 ppm. Preliminary sequence information on 7e suggests that the extra methionine is present in the C-terminal region of the molecule (P. Davie, unpublished).

The two sharp resonances at 2.7 and 2.8 ppm were suspected to originate from a lysine derivative. This was verified by the addition of N^6-monomethyl-lysine and N^6-dimethyl-lysine to both the native protein and protein denatured with 6 M guanidinium chloride. In the latter, the two peaks have collapsed into one. This peak has the same chemical shift as the ϵ-CH_2 protons from monomethyl-lysine, whereas the resonance from the dimethyl derivative is further downfield (Fig. 24). From NMR experiments on the *E. coli* ribosomal protein L11, we known that the trimethyl derivative gives a resonance at 3.15 ppm, excluding this possibility also. In the native structure, monomethylated lysines appear to occur in two different environments giving rise to two distinct peaks. This also indicates that not only the lysines at positions 4 and 6 are modified. In protein 7e Lys-4 is replaced by Arg-4, nevertheless the two resonances are still present.

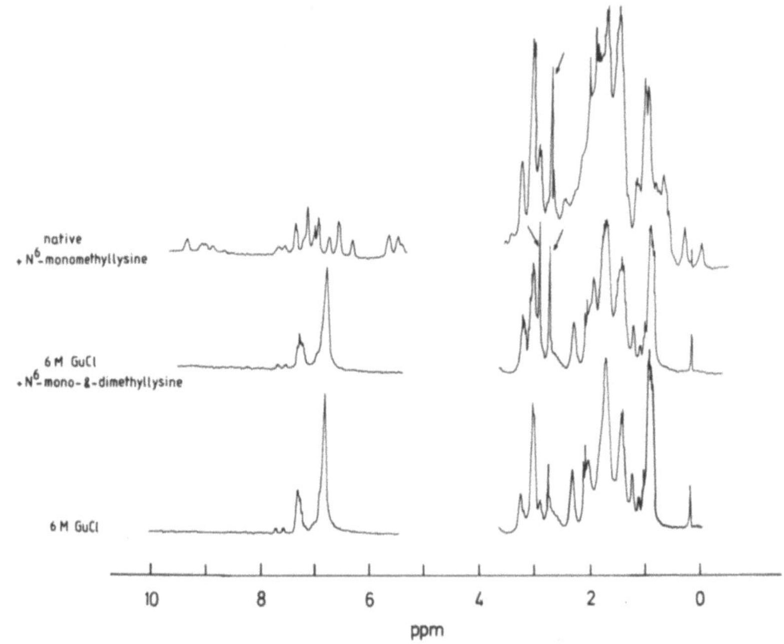

Fig. 24. NMR spectra of 7d to which N[6] -mono- and -dimethyllysine were added. The *bottom spectrum* is that of the protein denatured in 6 M guanidinium chloride

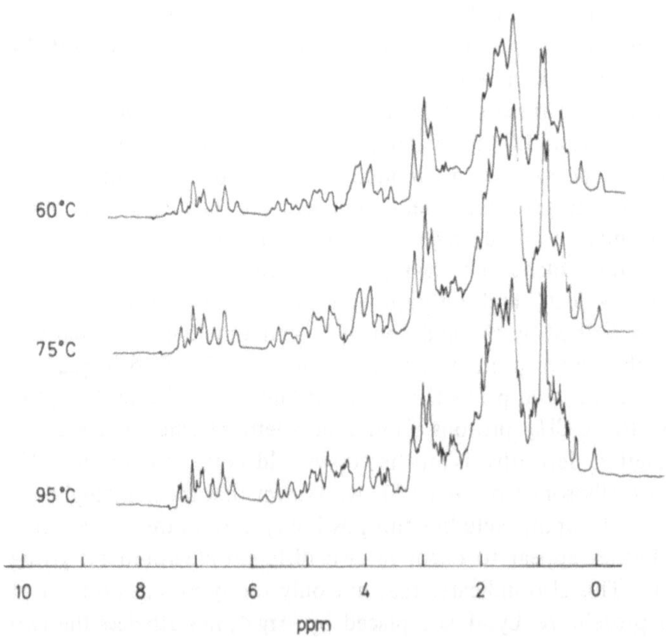

Fig. 25. The effect of increasing temperatures (at pH 7.0) on the NMR spectrum of protein 7d

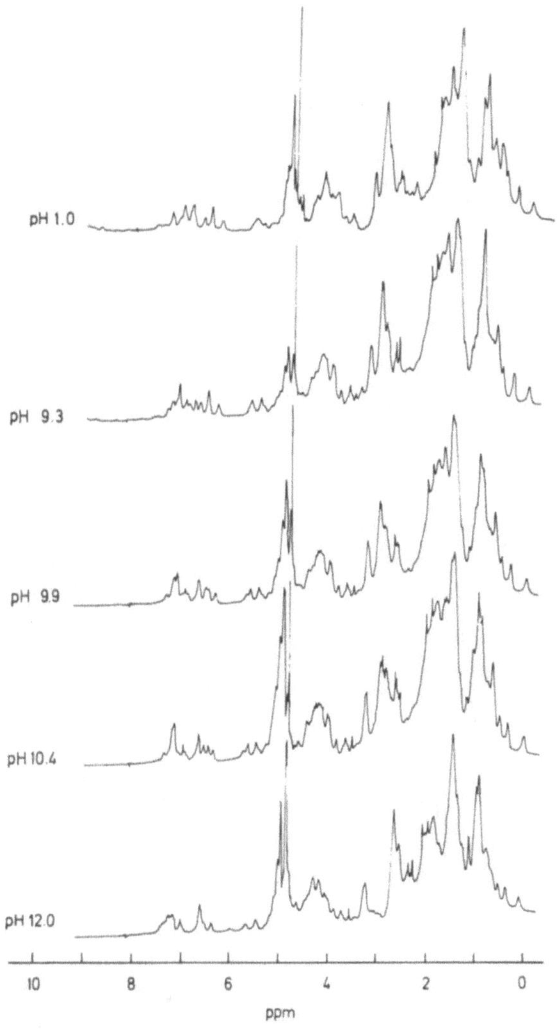

Fig. 26. 270 MHz NMR spectra of 7d recorded at extreme pH values (20°C)

The tertiary structure of the 7 kD proteins is remarkably stable. The proteins are fully unfolded in 6 M guanidinium chloride, but the introduction of 8 M urea has little effect on the folding. Also at high temperatures (up to 95°C) (Fig. 25) and extreme pH values (Fig. 26), the structure of the proteins shows only minor changes.

8 kD Proteins: As in the case of the 7 kD proteins, the NMR spectrum of 8a contains a wealth of structural information. This protein is also largely resistant to 8 M urea, but unfolds in 6 M guanidinium chloride (Fig. 27). From the denatured spectrum the presence of a histidine residue (8.8 ppm), tryptophan (7.55 and 7.70 ppm), N^6-monomethylated lysine (2.70 ppm) and one methionine (2.10 ppm) can be derived. For lack of sequence information a more detailed interpretation of the NMR spectrum cannot be given at present.

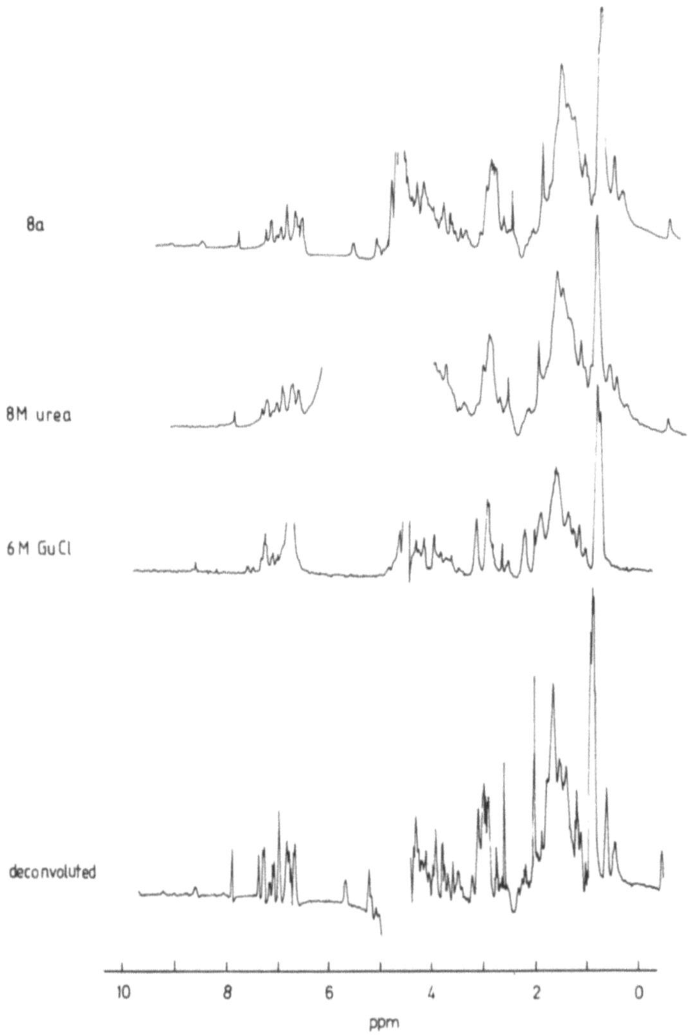

Fig. 27. 270 MHz NMR spectra of protein 8a in the folded and unfolded state

Protein 8b is obviously very different from 8a. Its spectrum shows much less evidence of folding, although the native and denatured spectrum are quite different.

10 kD Proteins: Proteins 10a and 10b have very different NMR spectra (Figs. 28 and 29). In both cases, many of the signals are strongly broadened, prohibiting a detailed interpretation. Like the smaller proteins, both are largely resistant to 8 M urea.

Protein 10a appears to have a histidine (C_2 proton at 7.75 ppm) and tryptophan residue (7.70 and 7.55 ppm). Extensively modified lysines seem to be present: N^6-monomethyl- (2.75 ppm), dimethyl- (2.90 ppm) and trimethyllysine at 3.18 ppm. Probably two methionine residues are present (2.05 and 2.10 ppm).

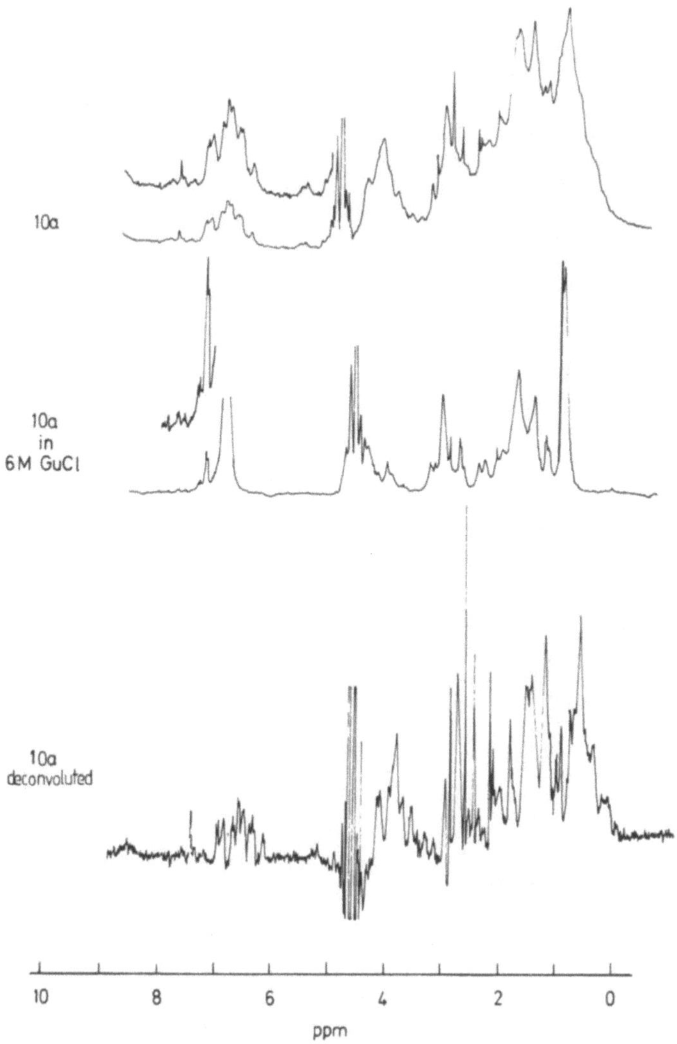

Fig. 28. 270 MHz NMR spectra of protein 10a

The spectrum of protein 10b is even more difficult to interpret, although this protein appears to have a well-folded structure. From the denatured spectrum, the absence of histidine and tryptophan is obvious.

Spectral Properties of Protein-Oligonucleotide Complexes

The interaction of DNA-binding protein II from eubacteria with nucleic acids is not easily observed. The CD spectra of complexes are the precise sum of those of the components, and changes in NMR spectra are limited to an upfield shift of the δ-CH_2 resonance of arginine.

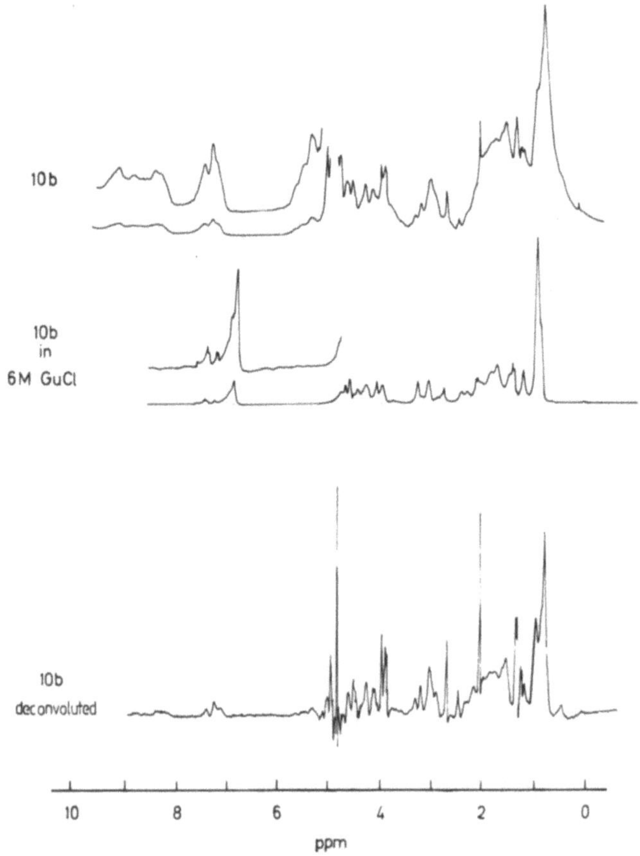

Fig. 29. 270 MHz NMR spectra of protein 10b

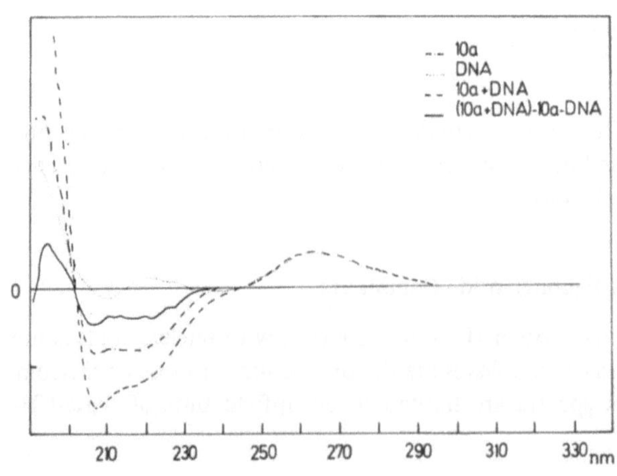

Fig. 30. Circular dichroic spectra of protein 10a in the absence and presence of DNA

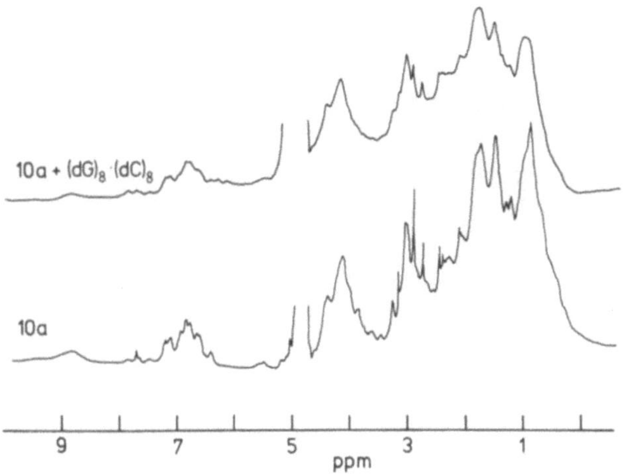

Fig. 31. 270 MHz NMR spectra of protein 10a in the presence of the double-stranded oligonucleotide $(dG)_8 \cdot (dC)_8$

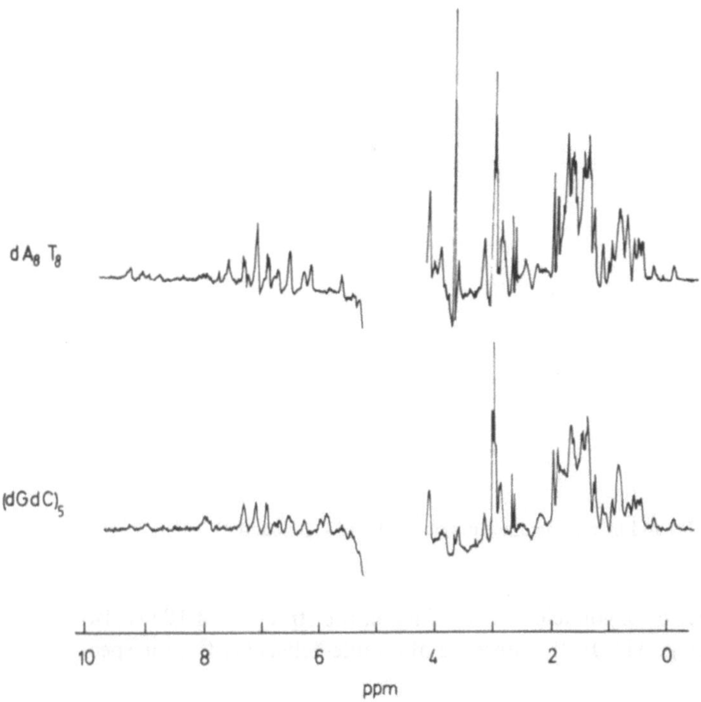

Fig. 32. NMR spectra of protein 7e in the presence of the double-stranded oligonucleotides $(dA)_8$, $(dT)_8$, and $(dGdC)_5$

It was, therefore, of interest to examine the CD and NMR spectra of complexes of archaebacterial DNA-binding proteins with DNA. For NMR spectroscopy, double-helical deoxyoligonucleotides were used in order to limit the molecular weight of the complexes and improve resolution. The only case in which CD spectroscopy shows any significant differences is that of protein 10a (Fig. 30). The difference signal suggests that complex formation is accompanied by a change in protein structure, namely an increase in α-helix.

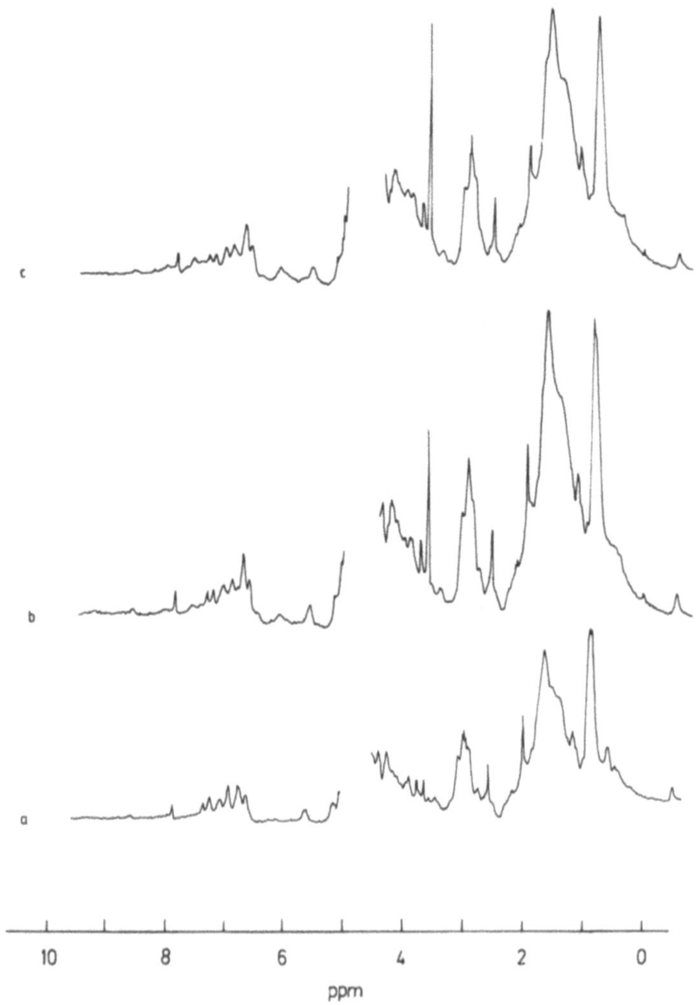

Fig. 33a–c. NMR spectra of 8a (a) and increasing amounts of $(dA)_8$, $(dT)_8$ (b and c)

A very pronounced effect of oligonucleotide binding on the structure of 10a is also observed in NMR spectra (Fig. 31). In the presence of double-helical G_8-C_8, the spectrum changes considerably.

In protein 7e the addition of A_8-T_8 and $(CG)_5$ causes pronounced changes in the ring-current-shifted methyl resonances (0–1 ppm), the methionine resonances (2 ppm) and the aromatic region (Fig. 32). These changes are reversible when the complex is dissociated in high salt. The protein structure appears to undergo a considerable change on binding of nucleic acid involving its hydrophobic core.

Similarly, the structure of protein 8a is strongly influenced by oligonucleotide binding. In this case the intensity of the ring-current-shifted methyl peaks decreases, a process resembling denaturation (Fig. 33). This too is reversible in high salt.

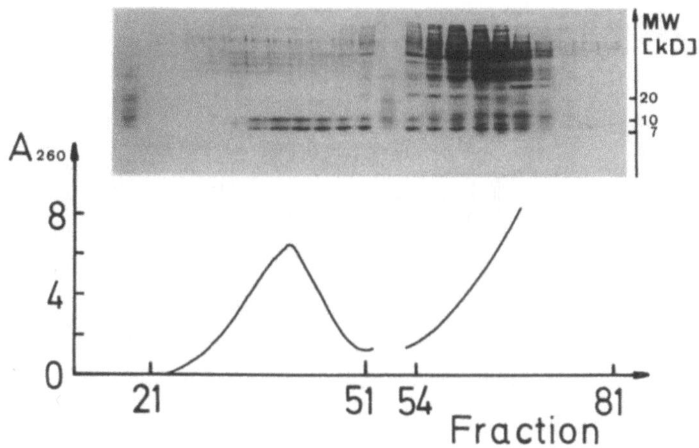

Fig. 34. The separation of DNA-protein complexes from other cellular components by gel filtration on Bio-Gel A-15 m. Analysis of aliquots from fractions of the first peak demonstrates the presence of the three protein groups

Bacterial "Chromatin"

The best evidence for the participation of these proteins in DNA folding stems from the analysis of DNA-protein complexes isolated from lysed cells. To date several conflicting reports on DNA-binding proteins in archaebacteria have been presented. From *Thermoplasma acidophilum* only one protein with a molecular weight of 10,000 has been isolated (DeLange et al. 1981), which belongs to the class of DNA-binding protein II (Kimura and Wilson 1983). In *Sulfolobus acidocaldarius* several small proteins (M_r 7,000–10,000) (Thomm et al. 1982) and two proteins with M_r values of 14,000 and 30,000 have been described (Green et al. 1983). The former result appears to agree with our findings.

We have isolated DNA-protein complexes from *Sulfolobus acidocaldarius* cells lysed in the presence of 0.2 M NaCl [buffer as described by Green et al. (1983)]. After a brief ultracentrifugation step (3 h at 40,000 rpm, 50Ti rotor) to remove ribosomes, the DNA-protein complex was eluted in the void volume of a Bio-Gel A-15 m column equilibrated with the lysis buffer. In this peak only three low-molecular-weight bands are observed (Fig. 34) as also described by Thomm et al. (1982). These three bands comigrate with representatives of the 7, 8 and 10 kD proteins.

Further identification has been obtained by HPLC analysis of the protein components. The protein was dissociated from the DNA in 1.5 M NaCl and separated from it on a Sephadex G-75 column. This step also provides a partial fractionation of the three groups of proteins. HPLC analysis on a reversed-phase column shows that the many components purified from the cytoplasm in high salt are also present in DNA-protein complexes (Fig. 35). Here, the proteins 7a and 7b are observed which, subsequently, have also been found in conventional protein purification.

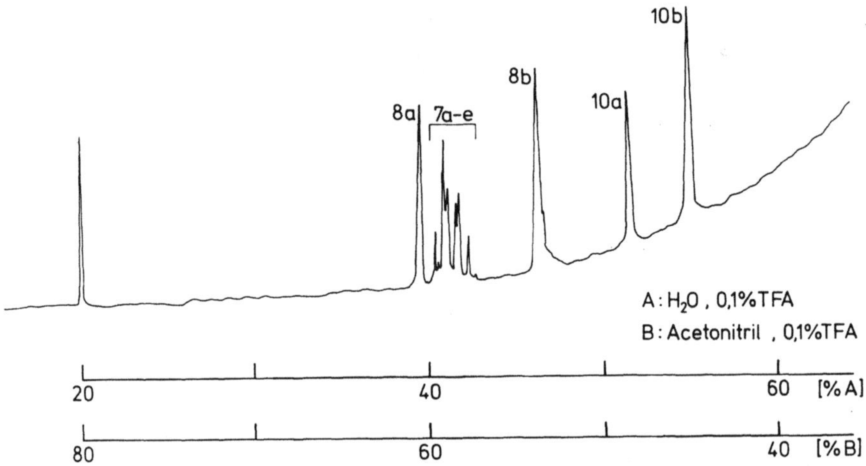

Fig. 35. Reversed-phase HPLC analysis of the proteins extracted from the DNA complex shown in Fig. 34

Conclusion

The family of DNA-binding protein II from eubacteria is a very useful system for evolutionary and structural studies. The protein (also called HU, HD and NS) is ubiquitous and easily isolated in large quantities. In the case of *E. coli,* two homologous proteins, NS1 and NS2, are present, whereas in other bacteria only one species has been detected.

The structure of several DNA-binding proteins has been characterised by CD and NMR spectroscopy. The protein from the thermophile *Bacillus stearothermophilus* has been crystallised and the structure has been solved at 3 Å resolution. The three-dimensional structure allows a more precise interpretation of the homologies in the amino acid sequences of the various proteins and suggests a possible mode of interaction with DNA. The latter involves the interaction between two flexible arms of the protein and the double-helical DNA molecule leading to supercoiling.

A related protein from the more extreme thermophile *B. caldolyticus* has also been crystallised and crystallographic studies are in progress. The results might give insight into the mechanism of stabilisation of protein structures against heat denaturation. A similar protein has been crystallised from the extreme thermophile *Thermus thermophilus.* Based on amino acid sequence studies, this protein is a more distant member of the family of DNA-binding protein II, and it will be of interest to see to which degree these differences affect its tertiary structure.

A better understanding of the packing mechanism of DNA in eubacteria will require a more detailed description of the composition of bacterial chromatin and the interaction between its components.

The contrast to eubacteria, the relationships between DNA-binding proteins in archaebacteria are less clear. The archaebacterium *Thermoplasma acidophilum* appears to have a single DNA-binding protein, which is clearly related to the family of DNA-binding protein II (DeLange et al. 1981). On the other hand, our studies on

Sulfolobus have shown the presence of at least three groups of DNA-binding proteins. The N-terminal sequences of these proteins, which have molecular weights of 7,000, 8,000 and 10,000, respectively, have been determined. The results show that the first group consists of three proteins, which have been characterised and are closely related by sequence. There is evidence for the presence of two more proteins. The two other groups can be subdivided into two proteins each, which are not related to one another. None of these proteins shows sequence homology to DNA-binding protein II or to other characterised DNA-binding proteins. On the other hand, first studies on DNA-binding proteins from the thermophile *Thermococcus celer* have revealed a DNA-binding protein, which is clearly homologous to one of the 10 kD proteins.

The proteins are partly methylated at the ϵ-amino groups of lysine which, in the case of the 7 kD proteins, contributes to the diversity. The genuine DNA-binding character of the proteins has been demonstrated by binding assays and electron microscopy of complexes with plasmid DNA. As a final proof, the same set of proteins has been identified by the analytical extraction of bacterial "chromatin". The proteins have a very stable structure as shown by CD and NMR spectroscopy. This property may well be beneficial for crystallographic studies on these proteins. Attempts along these lines are in progress.

Acknowledgements. We should like to thank our colleagues for their contributions to various parts of the work described here. In particular, we thank Philippa Davie and Mathias Grote for contributing many of their unpublished results. Free access to the Jasco J-500A CD spectrometer (Dr. W. Klabe, Biophysics Dept., Free University, Berlin) and the Bruker WH-270 NMR spectrometer (Dr. C. Crane-Robinson, Biophysics Group, Portsmouth Polytechnic) is gratefully acknowledged. These studies are supported, in part, by the Deutsche Forschungsgemeinschaft (Sfb 9).

References

Ackers G (1967) J Biol Chem 242:3237–3238
Anderson WF, Takeda Y, Ohlendorf DH, Matthews BW (1982) J Mol Biol 159:745–751
Baldwin JP, Boseley PG, Bradbury EM (1975) Nature 253:245–249
Brosius J (1976) Ph D Thesis, Free University, Berlin
Day LA (1973) Biochemistry 12:5329–5339
DeLange RJ, Williams LC, Searcy DG (1981) J Biol Chem 256:905–911
Delius H, Worcel A (1974) J Mol Biol 82:107–109
Dijk J, Littlechild JA (1979) Methods Enzymol 59:481–502
Dijk J, White SW, Wilson KS, Appelt K (1983) J Biol Chem 258:4003–4006
Flamm EL, Weisberg RA (1985) J Mol Biol 183:117–128
Geider H, Hoffmann-Berling H (1981) Annu Rev Biochem 50:233–260
Green GR, Searcy DG, DeLange RJ (1983) Biochim Biophys Acta 741:251–257
Greene JR, Brennan SM, Andrew DJ, Thompson CC, Richards SH, Heinrikson RL, Geiduschek EP (1984) Proc Natl Acad Sci USA 81:7031–7035
Hawkins AR, Wootton JC (1981) FEBS Lett 130:875–278
Heinen UJ, Heinen W (1972) Arch Mikrobiol 82:1–23
Jelenc PC (1980) Anal Biochem 105:369–374
Khanaka H, Laine B, Sautière P, Guillaume J (1985) Eur J Biochem 147:343–349
Kimura M, Wilson KS (1983) J Biol Chem 258:4007–4011
Kimura M, Kimura J, Zierer R (1984a) FEBS Lett 175:208–212

Kimura M, Kimura J, Davie P, Reinhardt R, Dijk J (1984b) FEBS Lett 176:176–178
Koch A, Reinhardt R, Dijk J (1985) (in preparation)
Laine B, Kmiecik D, Sautière P, Biserte G, Cohen-Solal M (1980) Eur J Biochem 103:447–461
Laine B, Belaiche D, Khanaka H, Sautière P (1983) Eur J Biochem 131:325–331
Lammi M, Paci M, Gualerzi CO (1984) FEBS Lett 170:99–104
Mende L, Timm B, Subramanian AR (1978) FEBS Lett 96:395–398
Pabo CO, Sauer RT (19840 Annu Rev Biochem 53:293–321
Pabo CO, Drovatin W, Jeffrey A, Sauer RT (1982) Nature 298:441–443
Pettijohn DE (1982) Cell 30: 667–669
Provencher SW, Glöckner J (1981) Biochemistry 20:33–37
Richmond TJ, Finch JT, Rushton B, Rhodes D, Klug A (1984) Nature 311:532–537
Rouvière-Yaniv J, Gros F (1975) Proc Natl Acad Sci USA 72:3428–3432
Rouvière-Yaniv J, Kjeldgaard NO (1979) FEBS Lett 106:297–300
Rouvière-Yaniv J, Yaniv M, Germond JE (1979) Cell 17:265–274
Suryanarayana T, Subramanian AR (1978) Biochim Biophys Acta 520:342–357
Tanaka I, White SW, Appelt K, Wilson KS, Dijk J (1984a) FEBS Lett 165:39–42
Tanaka I, Appelt K, Dijk J, White SW, Wilson KS (1984b) Nature 310:376–381
Thomm M, Stetter KO, Zillig W (1982) Zentralbl Bakteriol Hyg Abt I Orig C 128–139
Worcel A, Burgi E (1972) J Mol Biol 71:127–147
Zierer R (1985) Ph D Thesis, Free University, Berlin
Zillig W, Yeats S, Holz I, Böck A, Gropp F, Rettenberger M, Lutz S (1985) Nature 313:789–791

Nachwort: Converging Parallels

"What do the biologists want to know?" — A physicist, later molecular biologist.

"There must be some physics and mathematics behind all this." — A physician, later molecular biologist.

Anglicists will, I hope, excuse me for borrowing a German word to denote these concluding remarks to our conference; unfortunately, the minters of English forgot, or disdained, to coin a reverse side of "foreword". Romanists will, I trust, forgive the plagiarism in the title; it is borrowed from a recent Italian Prime Minister and presents an antinomy that is remarkable both for its insolence and for its profundity, and which, I believe, describes rather aptly the present state of research into bacterial chromatin, as epitomised by the two authentic personal communications that I have quoted above.

When many of us met three years ago and discussed the same subject, I had the impression of two lines in bacterial chromatin research that ran exactly parallel courses and showed no sign of convergence. They were, roughly speaking, the study of fundamental interactions between the chemical groups constituting proteins and nucleic acids, on the one hand, and the biochemical study of some bacterial proteins that had proved capable of binding nucleic acids, on the other. This year, a third parallel line has been added: that of the geneticists. But a bigger difference, compared with our state of knowledge in 1982, lies in the fact that these lines are clearly beginning to .... No, I cannot bring myself to write the word that would have brought down upon me the wrath of my rather strict (Euclidian) geometry master. Such is the life-long effect of British public-school discipline. Of course parallels do not converge. Railway lines appear to do so, but this lies in the eye of the beholder, who knows that in reality they are held equidistant by a never-ending row of sleepers in a single, unified, functional, three-dimensional assembly. However, it is precisely this structure that represents a dialectical synthesis of parallelism and convergence. How are the parallels beginning to build up into a coherent, stable, interconnected, many-dimensional structure?

The first line, that of the physical interaction between chemical groups, has been propagated at this conference in its purest form by Dr. Toulmé, and it was gratifying to see how it can now be extended, as more crystal structures become known, to show at least qualitatively how basic physical chemistry makes proteins and DNA stick to one another with a greater or a lesser degree of specificity. Dr. Geraci has re-introduced the almost-forgotten counter-anions as a vital structural element in highly positively-charged proteins, which tends to strengthen the suspicion, long held by physical chemists, that the amino-acid sequence of a protein is not even in principle enough to "cause" or to allow the prediction of its tertiary structure without adequate reference to the surrounding solvent.

Bacterial Chromatin
(ed. by C.O. Gualerzi and C.L. Pon)

The line of characterisation has again been well represented at this conference, and it has been both encouraging to observe the greater refinement of methods and sobering to see what complex patterns are emerging even just in the composition of the bacterial nucleoid. Dr. Le Hégarat's and co-workers' analysis of the nucleoid of *B. subtilis* provides an abundantly clear example of both of these tendencies.

Characterisation of function in vivo is well exemplified by the work of Dr. Rouvière-Yaniv and Dr. Spassky, and by the apparently incontestable, but nevertheless perplexing, results which their investigations have led to. One is remined of Piet Hein's jingle about the need to evolve problem-solvers galore, for each problem they solve unearths ten problems more. However, the paper by Dr. Gualerzi and others may provide an encouraging counter-example of a horrifyingly complex behaviour pattern that can, after all, be reduced to a small number of governing principles.

The continuing improvement in the homogeneity and structural integrity of the isolated proteins has been exploited increasingly by spectroscopists, as we have heard from Dr. Paci, Dr. Searcy and Dr. Dijk and their co-workers. The progress of increasing intricacy and subtlety of n.m.r. techniques shows no sign of abating, but fluorescence and circular dichroism, and other kinds of spectroscopy not represented at this meeting, are clearly continuing to yield unique and valuable information, especially in view of the frustrating "n.m.r. silence" of very large (e.g. DNA-protein) assemblies. None of these techniques, or X-ray crystallography, will on its own yield the key to the interactions among these complex molecules or between these and DNA. Yet there is hope that the combination of static structural definition, of the assignment of residues whose n.m.r. signal is affected by a binding interaction, of tightness of structure, and finally of denaturation and folding in solution will lead to a good mechanistic description of the workings of these proteins – in vitro, that is to say.

The third parallel line, followed by the geneticists, has been pursued by Dr. Schaechter and Dr. Drlica and their colleagues, and has taken us along on a higher level – that is, further away from physics. In spite of Dr. Drlica's dry comment "It is occasionally important to stress that physiological experiments are difficult to interpret rigorously", it is also true that many biochemical ideas can be tested, or inspired, by simple genetic results, and the apparent regulation of the degree of supercoiling strikes me as being of especial significance.

Finally, however, we have heard some contributions which may show the beginnings of some (not just chemical or light-induced) cross-links that in time could generate the multi-dimensional assembly of a theory, or at least a description, of both the global and the detailed structure, and also the detailed function, of the bacterial nucleoid. (The global functions, replication and expression, are already known.) The techniques used are not in principle new: isolation and analysis (Dr. Lossius), chemical and photochemical cross-linking (Dr. Schouten) and electron microscopy (Dr. Bjornsti and Dr. Kellenberger and others). Yet, in each case, account was taken of the fact that the energy-conscious procaryote has renounced the luxury of highly stable structures, and thus made more difficult the researcher's task of probing such structures as are there with sufficient gentleness. The development of new isolation and special fixing and staining procedures has thus brought a series of important new observations in its train, which may speedily dispose of some fundamental ontological problems, such as what goes where in the cell, and may lead to the gelling of many previously unconnected results and ideas.

Some points of possible controversy have not been given great emphasis at this meeting, thus preventing exothermic interactions, but perhaps also inhibiting some desirable discussion. For instance, it might have been valuable to assess collectively and objectively the different pictures both of the rôle and of the interaction with DNA of the protein NS alias HU alias HD alias DNAb II, even at the risk of someone's model being called wrong. And I would greatly have welcomed – though not dared to take part in – a conclave to reach agreement on what this and other proteins discussed this week should be called.

PAUL WOOLLEY

Subject Index

MIX
Papier aus verantwortungsvollen Quellen
Paper from responsible sources
FSC® C105338

If you have any concerns about our products, you can contact us on
ProductSafety@springernature.com

In case Publisher is established outside the EU, the EU authorized representative is:
Springer Nature Customer Service Center GmbH
Europaplatz 3, 69115 Heidelberg, Germany

Printed by Libri Plureos GmbH
in Hamburg, Germany